FIBER OPTICS IN ASTRONOMY III

A SERIES OF BOOKS ON RECENT DEVELOPMENTS IN ASTRONOMY AND ASTROPHYSICS

Managing Editor, D. Harold McNamara
Production Manager, Elizabeth S. Holloman

A.S.P. CONFERENCE SERIES PUBLICATIONS COMMITTEE

Sallie Baliunas, Chair
Carol Ambruster
Catharine Garmany
Mark S. Giampapa
Kenneth Janes

© Copyright 1998 Astronomical Society of the Pacific
390 Ashton Avenue, San Francisco, California 94112

All rights reserved

Printed by BookCrafters, Inc.

First published 1998

Library of Congress Catalog Card Number: 98-74355
ISBN 1-886733-72-4

Please contact proper address for information on:

PUBLISHING:
Managing Editor
PO Box 24463
211 KMB
Brigham Young University
Provo, UT 84602-4463 USA
801-378-2298

pasp@astro.byu.edu
Fax: 801-378-2265

ORDERING BOOKS:
Astronomical Society of the Pacific
CONFERENCE SERIES
390 Ashton Avenue
San Francisco, CA 94112 - 1722 USA
415-337-2624

catalog@aspsky.org
Fax: 415-337-5205

A SERIES OF BOOKS ON RECENT DEVELOPMENTS IN ASTRONOMY AND ASTROPHYSICS

Vol. 1-Progress and Opportunities in Southern Hemisphere Optical Astronomy: CTIO 25th Anniversary Symposium
ed. V. M. Blanco and M. M. Phillips　　　　　　　　　　　ISBN 0-937707-18-X

Vol. 2-Proceedings of a Workshop on Optical Surveys for Quasars
ed. P. S. Osmer, A. C. Porter, R. F. Green, and C. B. Foltz　　ISBN 0-937707-19-8

Vol. 3-Fiber Optics in Astronomy
ed. S. C. Barden　　　　　　　　　　　　　　　　　　　　ISBN 0-937707-20-1

Vol. 4-The Extragalactic Distance Scale: Proceedings of the ASP 100th Anniversary Symposium
ed. S. van den Bergh and C. J. Pritchet　　　　　　　　　　ISBN 0-937707-21-X

Vol. 5-The Minnesota Lectures on Clusters of Galaxies and Large-Scale Structure
ed. J. M. Dickey　　　　　　　　　　　　　　　　　　　　ISBN 0-937707-22-8

Vol. 6-Synthesis Imaging in Radio Astronomy: A Collection of Lectures from the Third NRAO Synthesis Imaging Summer School
ed. R. A. Perley, F. R. Schwab, and A. H. Bridle　　　　　　ISBN 0-937707-23-6

Vol. 7-Properties of Hot Luminous Stars: Boulder-Munich Workshop
ed. C. D. Garmany　　　　　　　　　　　　　　　　　　ISBN 0-937707-24-4

Vol. 8-CCDs in Astronomy
ed. G. H. Jacoby　　　　　　　　　　　　　　　　　　　ISBN 0-937707-25-2

Vol. 9-Cool Stars, Stellar Systems, and the Sun. Sixth Cambridge Workshop
ed. G. Wallerstein　　　　　　　　　　　　　　　　　　　ISBN 0-937707-27-9

Vol. 10-Evolution of the Universe of Galaxies: Edwin Hubble Centennial Symposium
ed. R. G. Kron　　　　　　　　　　　　　　　　　　　　ISBN 0-937707-28-7

Vol. 11-Confrontation Between Stellar Pulsation and Evolution
ed. C. Cacciari and G. Clementini　　　　　　　　　　　　ISBN 0-937707-30-9

Vol. 12-The Evolution of the Interstellar Medium
ed. L. Blitz　　　　　　　　　　　　　　　　　　　　　ISBN 0-937707-31-7

Vol. 13-The Formation and Evolution of Star Clusters
ed. K. Janes　　　　　　　　　　　　　　　　　　　　　ISBN 0-937707-32-5

Vol. 14-Astrophysics with Infrared Arrays
ed. R. Elston　　　　　　　　　　　　　　　　　　　　　ISBN 0-937707-33-3

Vol. 15-Large-Scale Structures and Peculiar Motions in the Universe
ed. D. W. Latham and L. A. N. da Costa　　　　　　　　　ISBN 0-937707-34-1

Vol. 16-Proceedings of the 3rd Haystack Observatory Conference on Atoms, Ions and Molecules: New Results in Spectral Line Astrophysics
ed. A. D. Haschick and P. T. P. Ho　　　　　　　　　　　ISBN 0-937707-35-X

Vol. 17-Light Pollution, Radio Interference, and Space Debris
ed. D. L. Crawford　　　　　　　　　　　　　　　　　　ISBN 0-937707-36-8

Vol. 18-The Interpretation of Modern Synthesis Observations of Spiral Galaxies
ed. N. Duric and P. C. Crane　　　　　　　　　　　　　　ISBN 0-937707-37-6

Vol. 19-Radio Interferometry: Theory, Techniques, and Applications, IAU Colloquium 131
ed. T. J. Cornwell and R. A. Perley　　　　　　　　　　　ISBN 0-937707-38-4

Vol. 20-Frontiers of Stellar Evolution: 50th Anniversary McDonald Observatory (1939-1989)
ed. D. L. Lambert　　　　　　　　　　　　　　　　　　　ISBN 0-937707-39-2

Vol. 21-The Space Distribution of Quasars
ed. D. Crampton
ISBN 0-937707-40-6

Vol. 22-Nonisotropic and Variable Outflows from Stars
ed. L. Drissen, C. Leitherer, and A. Nota
ISBN 0-937707-41-4

Vol. 23-Astronomical CCD Observing and Reduction Techniques
ed. S. B. Howell
ISBN 0-937707-42-4

Vol. 24-Cosmology and Large-Scale Structure in the Universe
ed. R. R. de Carvalho
ISBN 0-937707-43-0

Vol. 25-Astronomical Data Analysis Software and Systems I
ed. D. M. Worrall, C. Biemesderfer, and J. Barnes
ISBN 0-937707-44-9

Vol. 26-Cool Stars, Stellar Systems, and the Sun, Seventh Cambridge Workshop
ed. M. S. Giampapa and J. A. Bookbinder
ISBN 0-937707-45-7

Vol. 27-The Solar Cycle: Proceedings of the National Solar Observatory/Sacramento Peak 12th Summer Workshop
ed. K. L. Harvey
ISBN 0-937707-46-5

Vol. 28-Automated Telescopes for Photometry and Imaging
ed. S. J. Adelman, R. J. Dukes, Jr., and C. J. Adelman
ISBN 0-937707-47-3

Vol. 29-Viña Del Mar Workshop on Catacysmic Variable Stars
ed. N. Vogt
ISBN 0-937707-48-1

Vol. 30-Variable Stars and Galaxies
ed. B. Warner
ISBN 0-937707-49-X

Vol. 31-Relationships Between Active Galactic Nuclei and Starburst Galaxies
ed. A. V. Filippenko
ISBN 0-937707-50-3

Vol. 32-Complementary Approaches to Double and Multiple Star Research, IAU Collouquium 135
ed. H. A. McAlister and W. I. Hartkopf
ISBN 0-937707-51-1

Vol. 33-Research Amateur Astronomy
ed. S. J. Edberg
ISBN 0-937707-52-X

Vol. 34-Robotic Telescopes in the 1990s
ed. A. V. Filippenko
ISBN 0-937707-53-8

Vol. 35-Massive Stars: Their Lives in the Interstellar Medium
ed. J. P. Cassinelli and E. B. Churchwell
ISBN 0-937707-54-6

Vol. 36-Planets Around Pulsars
ed. J. A. Phillips, S. E. Thorsett, and S. R. Kulkarni
ISBN 0-937707-55-4

Vol. 37-Fiber Optics in Astronomy II
ed. P. M. Gray
ISBN 0-937707-56-2

Vol. 38-New Frontiers in Binary Star Research: Pacific Rim Colloquium
ed. K. C. Leung and I.-S. Nha
ISBN 0-937707-57-0

Vol. 39-The Minnesota Lectures on the Structure and Dynamics of the Milky Way
ed. Roberta M. Humphreys
ISBN 0-937707-58-9

Vol. 40-Inside the Stars, IAU Colloquium 137
ed. Werner W. Weiss and Annie Baglin
ISBN 0-937707-59-7

Vol. 41-Astronomical Infrared Spectroscopy: Future Observational Directions
ed. Sun Kwok
ISBN 0-937707-60-0

Vol. 42-GONG 1992: Seismic Investigation of the Sun and Stars
ed. Timothy M. Brown
ISBN 0-937707-61-9

Vol. 43-Sky Surveys: Protostars to Protogalaxies
ed. B. T. Soifer   ISBN 0-937707-62-7

Vol. 44-Peculiar Versus Normal Phenomena in A-Type and Related Stars, IAU Colloquium 138
ed. M. M. Dworetsky, F. Castelli, and R. Faraggiana   ISBN 0-937707-63-5

Vol. 45-Luminous High-Latitude Stars
ed. D. D. Sasselov   ISBN 0-937707-64-3

Vol. 46-The Magnetic and Velocity Fields of Solar Active Regions, IAU Colloquium 141
ed. H. Zirin, G. Ai, and H. Wang   ISBN 0-937707-65-1

Vol. 47-Third Decennial US-USSR Conference on SETI
ed. G. Seth Shostak   ISBN 0-937707-66-X

Vol. 48-The Globular Cluster-Galaxy Connection
ed. Graeme H. Smith and Jean P. Brodie   ISBN 0-937707-67-8

Vol. 49-Galaxy Evolution: The Milky Way Perspective
ed. Steven R. Majewski   ISBN 0-937707-68-6

Vol. 50-Structure and Dynamics of Globular Clusters
ed. S. G. Djorgovski and G. Meylan   ISBN 0-937707-69-4

Vol. 51-Observational Cosmology
ed. G. Chincarini, A. Iovino, T. Maccacaro, and D. Maccagni   ISBN 0-937707-70-8

Vol. 52-Astronomical Data Analysis Software and Systems II
ed. R. J. Hanisch, R. J. V. Brissenden, and Jeannette Barnes   ISBN 0-937707-71-6

Vol. 53-Blue Stragglers
ed. Rex A. Saffer   ISBN 0-937707-72-4

Vol. 54-The First Stromlo Symposium: The Physics of Active Galaxies
ed. Geoffrey V. Bicknell, Michael A. Dopita, and Peter J. Quinn   ISBN 0-937707-73-2

Vol. 55-Optical Astronomy from the Earth and Moon
ed. Diane M. Pyper and Ronald J. Angione   ISBN 0-937707-74-0

Vol. 56-Interacting Binary Stars
ed. Allen W. Shafter   ISBN 0-937707-75-9

Vol. 57-Stellar and Circumstellar Astrophysics
ed. George Wallerstein and Alberto Noriega-Crespo   ISBN 0-937707-76-7

Vol. 58-The First Symposium on the Infrared Cirrus and Diffuse Interstellar Clouds
ed. Roc M. Cutri and William B. Latter   ISBN 0-937707-77-5

Vol. 59-Astronomy with Millimeter and Submillimeter Wave Interferometry, IAU Colloquium 140
ed. M. Ishiguro and Wm. J. Welch   ISBN 0-937707-78-3

Vol. 60-The MK Process at 50 Years: A Powerful Tool for Astrophysical Insight: A Workshop of the Vatican Observatory
ed. C. J. Corbally, R. O. Gray, and R. F. Garrison   ISBN 0-937707-79-1

Vol. 61-Astronomical Data Analysis Software and Systems III
ed. Dennis R. Crabtree, R. J. Hanisch, and Jeannette Barnes   ISBN 0-937707-80-5

Vol. 62-The Nature and Evolutionary Status of Herbig Ae / Be Stars
ed. P. S. Thé, M. R. Pérez, and E. P. J. van den Heuvel   ISBN 0-937707-81-3

Vol. 63-Seventy-Five Years of Hirayama Asteroid Families: The role of Collisions in the Solar System History
ed. R. P. Binzel, Y. Kozai, and T. Hirayama   ISBN 0-937707-82-1

Vol. 64-Cool Stars, Stellar Systems, and the Sun, Eighth Cambridge Workshop
ed. Jean-Pierre Caillault   ISBN 0-937707-83-X

Vol. 65-Clouds, Cores, and Low Mass Stars
ed. Dan P. Clemens and Richard Barvainis ISBN 0-937707-84-8

Vol. 66- Physics of the Gaseous and Stellar Disks of the Galaxy
ed. Ivan R. King ISBN 0-937707-85-6

Vol. 67-Unveiling Large-Scale Structures Behind the Milky Way
ed. C. Balkowski and R. C. Kraan-Korteweg ISBN 0-937707-86-4

Vol. 68-Solar Active Region Evolution: Comparing Models with Observations
ed. K. S. Balasubramaniam and George W. Simon ISBN 0-937707-87-2

Vol. 69-Reverberation Mapping of the Broad-Line Region in Active Galactic Nuclei
ed. P. M. Gondhalekar, K. Horne, and B. M. Peterson ISBN 0-937707-88-0

Vol. 70-Groups of Galaxies
ed. Otto G. Richter and Kirk Borne ISBN 0-937707-89-9

Vol. 71-Tridimensional Optical Spectroscopic Methods in Astrophysics, IAU Colloquium 149
ed. G. Comte and M. Marcelin ISBN 0-937707-90-2

Vol. 72-Millisecond Pulsars: A Decade of Surprise.
ed. A. A. Fruchter, M. Tavani, and D. C. Backer ISBN 0-937707-91-0

Vol. 73-Airborne Astronomy Symposium on the Galactic Ecosystem: From Gas to Stars to Dust
ed. M. R. Haas, J. A. Davidson, and E. F. Erickson ISBN 0-937707-92-9

Vol. 74-Progress in the Search for Extraterrestrial Life: 1993 Bioastronomy Symposium
ed. G. Seth Shostak ISBN 0-937707-93-7

Vol. 75-Multi-Feed Systems for Radio Telescopes
ed. D. T. Emerson and J. M. Payne ISBN 0-937707-94-5

Vol. 76-GONG '94: Helio- and Astero-Seismology from the Earth and Space
ed. Roger K. Ulrich, Edward J. Rhodes, Jr., and Werner Däppen ISBN 0-937707-95-3

Vol. 77-Astronomical Data Analysis Software and System IV
ed. R. A. Shaw, H. E. Payne, and J. J. E. Hayes ISBN 0-937707-96-1

Vol. 78-Astrophysical Applications of Powerful New Databases: Joint Discussion No. 16 of the 22nd General Assembly of the IAU
ed. S. J. Adelman and W. L. Wiese ISBN 0-937707-97-X

Vol. 79-Robotic Telescopes: Current Capabilities, Present Developments, and Future Prospects for Automated Astronomy
ed. Gregory W. Henry and Joel A. Eaton ISBN 0-937707-98-8

Vol. 80-The Physics of the Interstellar Medium and Intergalactic Medium
ed. A. Ferrara, C. F. McKee, C. Heiles, and P. R. Shapiro ISBN 0-937707-99-6

Vol. 81-Laboratory and Astronomical High Resolution Spectra
ed. A. J. Sauval, R. Blomme, and N. Grevesse ISBN 1-886733-01-5

Vol. 82-Very Long Baseline Interferometry and the VLBA
ed. J. A. Zensus, P. J. Diamond, and P. J. Napier ISBN 1-886733-02-3

Vol. 83-Astrophysical Applications of Stellar Pulsation. IAU Colloquium 155
ed. R. S. Stobie and P. A. Whitelock ISBN 1-886733-03-1

Vol. 84-The Future Utilisation of Schmidt Telescopes, IAU Colloquium 148
ed. Jessica Chapman, Russell Cannon, Sandra Harrison, and Bambang Hidayat ISBN 1-886733-05-8

Vol. 85-Cape Workshop on Magnetic Cataclysmic Variables
ed. D. A. H. Buckley and B. Warner ISBN 1-886733-06-6

Vol. 86-Fresh Views of Elliptical Galaxies
ed. Alberto Buzzoni, Alvio Renzini, and Alfonso Serrano ISBN 1-886733-07-4

Vol. 87-New Observing Modes for the Next Century
ed. Todd Boroson, John Davies, and Ian Robson   ISBN 1-886733-08-2

Vol. 88- Clusters, Lensing, and the Future of the Universe
ed. Virginia Trimble and Andreas Reisenegger   ISBN 1-886733-09-0

Vol. 89-Astronomy Education: Current Developments, Future Coordination
ed.John R. Percy   ISBN 1-886733-10-4

Vol. 90-The Origins, Evolution, and Destinies of Binary Stars in Clusters
ed. E. F. Milone and J. C. Mermilliod   ISBN 1-886733-11-2

Vol. 91-Barred Galaxies, IAU Colloquium 157
ed. R. Buta, D. A. Crocker, and B. G. Elmegreen   ISBN 1-886733-12-0

Vol. 92-Formation of the Galactic Halo–Inside and Out
ed. H. L. Morrison and A. Sarajedini   ISBN 1-886733-13-9

Vol. 93-Radio Emission from the Stars and the Sun
ed. A. R. Taylor and J. M. Paredes   ISBN 1-886733-14-7

Vol. 94-Mapping, Measuring, and Modelling the Universe
ed. Peter Coles, Vincent Martinez, and Maria-Jesus Pons-Borderia   ISBN 1-886733-15-5

Vol. 95-Solar Drivers of Interplanetary and Terrestrial Disturbances: Proceedings of 16th
International Workshop, National Solar Observatory/Sacramento Peak
ed. K. S. Balasubramaniam, S. L. Keil, and R. N. Smartt   ISBN 1-886733-16-3

Vol. 96- Hydrogen-Deficient Stars
ed. C. S. Jeffery and U. Heber   ISBN 1-886733-17-1

Vol. 97-Polarimetry of the Interstellar Medium
ed.W. G. Roberge and D. C. B. Whittet   ISBN 1-886733-18-X

Vol. 98-From Stars to Galaxies: The Impact of Stellar Physics on Galaxy Evolution
ed. Claus Leitherer, Uta Fritze-von Alvensleben, and John Huchra   ISBN 1-886733-19-8

Vol. 99-Cosmic Abundances: Proceedings of the 6th Annual October Astrophysics Conference
ed. Stephen S. Holt and Geroge Sonneborn   ISBN 1-886733-20-1

Vol. 100-Energy Transport in Radio Galaxies and Quasars
ed. P. E. Hardee, A. H. Bridle, and J. A. Zensus   ISBN 1-886733-21-X

Vol. 101-Astronomical Data Analysis Software and Systems V
ed. George H. Jacoby and Jeannette Barnes   ISSN   1080-7926

Vol. 102-The Galactic Center, 4th ESO/CTIO Workshop
ed. Roland Gredel   ISBN 1-886733-22-8

Vol. 103-The Physics of Liners in View of Recent Observations
ed. M. Eracleous, A. Koratkar, C. Leitherer, and L. Ho   ISBN 1-886733-23-6

Vol. 104-Physics, Chemistry, and Dynamics of Interplanetary Dust, IAU Colloquium 150
ed. Bo A. S. Gustafson and Martha S. Hanner   ISBN 1-886733-24-4

Vol. 105-Pulsars: Problems and Progress, IAU Colloquium 160
ed. M. Bailes, S. Johnston, and M. A. Walker   ISBN 1-886733-25-2

Vol. 106-Minnesota Lectures on Extragalactic Neutral Hydrogen
ed. Evan D. Skillman   ISBN 1-886733-26-0

Vol. 107-Completing the Inventory of the Solar System: A Symposium held in conjuunction with
the 106th Annual Meeting of the ASP
ed. Terrence W. Rettig and Joseph M. Hahn   ISBN 1-886733-27-9

Vol. 108-M. A. S. S. Model Atmospheres and Spectrum Synthesis: 5th Vienna Workshop
ed. S. J. Adelman, F. Kupka, and W. W. Weiss   ISBN 1-886733-28-7

Vol. 109-Cool Stars, Stellar Systems, and the Sun, Ninth Cambridge Workshop
ed. Roberto Pallavicini and Andrea K. Dupree   ISBN 1-886733-29-5

Vol. 110-Blazar Continuum Variability
ed. H. R. Miller, J. R. Webb, and J. C. Noble   ISBN 1-886733-30-9

Vol. 111-Magnetic Reconnection in the Solar Atmosphere: Proceedings of a Yohkoh Conference
ed. R. D. Bentley and J. T. Mariska   ISBN 1-886733-31-7

Vol. 112-The History of the Milky Way and Its Satellite System
ed. A. Burkert, D. H. Hartmann, and S. R. Majewski   ISBN 1-886733-32-5

Vol. 113-Emission Lines in Active Galaxies: New Methods and Techniques, IAU Colloquium 159
ed. B. M. Peterson, F. Z. Cheng, and A. S. Wilson   ISBN 1-886733-33-3

Vol. 114-Young Galaxies and QSO Absorption-Line Systems
ed. Sueli M. Viegas, Ruth Gruenwald, and Reinaldo R. de Carvalho ISBN 1-886733-34-1

Vol. 115-Galactic and Cluster Cooling Flows
ed. Noam Soker   ISBN 1-886733-35-X

Vol. 116-The Second Stromlo Symposium: The Nature of Elliptical Galaxies
ed. M. Arnaboldi, G. S. Da Costa, and P. Saha   ISBN 1-886733-36-8

Vol. 117- Dark and Visible Matter in Galaxies
ed. Massimo Persic and Paolo Salucci   ISBN 1-886733-37-6

Vol. 118-First Advances in Solar Physics Euroconference: Advances in the Physics of Sunspots
ed. B. Schmieder, J. C. del Toro Iniesta, and M. Vázquez   ISBN 1-886733-38-4

Vol. 119-Planets Beyond the Solar System and the Next Generation of Space Missions
ed. David R. Soderblom   ISBN 1-886733-39-2

Vol. 120-Luminous Blue Variables: Massive Stars in Transition
ed. Antonella Nota and Henny J. G. L. M. Lamers   ISBN 1-886733-40-6

Vol. 121-Accretion Phenomena and Related Outflows, IAU Colloquium 163
ed. D. T. Wickramasinghe, G. V. Bicknell and L. Ferrario   ISBN 1-886733-41-4

Vol. 122-From Stardust to Planetesimals: Symposium held as part of the 108th Annual Meeting of the ASP
ed. Yvonne J. Pendleton and A. G. G. M. Tielens   ISBN 1-886733-42-2

Vol. 123-The 12th 'Kingston Meeting': Computational Astrophysics
ed. David A. Clarke and Michael J. West   ISBN 1-886733-43-0

Vol. 124-Diffuse Infrared Radiation and the IRTS
ed. Haruyuki Okuda, Toshio Matsumoto, and Thomas L. Roellig   ISBN 1-886733-44-9

Vol. 125- Astronomical Data Analysis Software and Systems VI
ed. Gareth Hunt and H. E. Payne   ISBN 1-886733-45-7

Vol. 126-From Quantum Fluctuations to Cosmological Structures
ed. D. Vallis-Gabaud, M. A. Hendry, P. Molaro, and K. Chamcham   ISBN 1-886733-46-5

Vol. 127-Proper Motions and Galactic Astronomy
ed. Roberta M. Humphreys   ISBN 1-886733-47-3

Vol. 128- Mass Ejection from AGN (Active Galactic Nuclei)
ed. N. Arav, I. Shlosman, and R. J. Weymann   ISBN 1-886733-48-1

Vol. 129-The George Gamow Symposium
ed. E. Harper, W. C. Parke, and G. D. Anderson   ISBN 1-886733-49-X

Vol. 130-The Third Pacfic Rim Conference on Recent Development on Binary Star Research
ed. Kam-Ching Leung   ISBN 1-886733-50-3

Vol. 131-Boulder-Munich II: Properties of Hot, Luminous Stars
ed. Ian D. Howarth   ISBN 1-886733-51-1

Vol. 132-Star Formation with the Infrared Space Observatory (ISO)
ed. João L. Yun and René Liseau   ISBN 1-886733-53-X

Vol. 133-Science with the NGST
ed. Eric P. Smith and Anuradha Koratkar   ISBN 1-886733-53-8

Vol. 134-Brown Dwarfs and Extrasolar Planets
ed. Rafael Rebolo, Eduardo L. Martin,
and Maria Rosa Zapatero Osorio   ISBN 1-886733-54-6

Vol. 135-A Half Century of Stellar Pulsation Interpretations: A Tribute to Arthur N. Cox
ed. P. A Bradley and J. A. Guzik   ISBN 1-886733-55-4

Vol. 136- Galactic Halos: A UC Santa Cruz Workshop
ed. Dennis Zaritdky   ISBN 1-886733-56-2

Vol. 137-Wild Stars in the Old West: Proceedings of the 13th North American Workshop
on Cataclysmic Variables and Related Objects
ed. S. Howell, E.Kuulkers, and C. Woodward   ISBN 1-886733-57-0

Vol. 138-1997 Pacific Rim Conference on Stellar Astrophysics
ed. Kwing L. Chan, K. S. Cheng, and Harinder P. Singh   ISBN 1-886733-58-9

Vol. 139-Preserving the Astronomical Windows: Proceedings of Joint Discussion
No. 5 of the 23rd General Assembly of the IAU
ed. Syuzo Isobe and Tomohiro Hirayama   ISBN 1-886733-59-7

Vol. 140-Synoptic Solar Physics
ed. K. S. Balasubramaniam, J. W. Harvey, and D. M. Rabin   ISBN 1-886733-60-0

Vol. 141-Astrophysics from Antarctica
ed. Giles Novak and Randall H. Landsberg   ISBN 1-886733-61-9

Vol. 142-The Stellar Initial Mass Function, 38th Herstmonceux Conference
ed. Gerry Gilmore and Debbie Howell   ISBN 1-886733-62-7

Vol. 143-The Scientific Impact of the Goddard High Resolution Spectrograph
ed. John C. Brandt, Thomas B. Ake III, and Carolyn Collins Petersen   ISBN 1-886733-63-5

Vol. 144- Radio Emission from Galactic and Extragalactic Compact Sources
ed. J. Anton Zensus, G. B. Taylor, and J. M. Wrobel   ISBN 1-886733-64-3

Vol. 145-Astronomical Data Analysis Software and Systems VII
ed. Rudolf Albrecht, Richard N. Hook, and Howard A. Bushouse ISBN 1-886733-65-1

Vol. 146-The Young Universe: Galaxy Formation and Evolution at Intermediate and High Redshift
ed. S. D'Odorico, A. Fontana, and E. Giallongo   ISBN 1-886733-66-X

Vol. 147-Abundance Profiles: Diagnostic Tools for Galaxy History
ed. Daniel Friedli, Mike Edmunds, Carmelle Robert, and Laurent Drissen   ISBN 1-886733-67-8

Vol. 148-Origins
ed. Charles E. Woodward, J. Michael Shull, and Harley A. Thronson, Jr.   ISBN 1-886733-68-6

Vol. 149-Solar System Formation and Evolution
ed. D. Lazzaro, R. Vieira Martins, S. Ferraz-Mello, J. Fernández, and C. Beaugé   ISBN 1-886733-69-4

Vol. 150-New Perspectives on Solar Prominences, IAU Colloquium 167
ed. David Webb, David Rust, and Brigitte Schmieder   ISBN 1-886733-70-8

Vol. 151-Cosmic Microwave Background and Large Scale Structure of the Universe
ed. Yong-Ik Byun and Kin-Wang Ng   ISBN 1-886733-71-6

Vol. 152-Fiber Optics in Astronomy III
ed. S. Arribas, E. Mediavilla, and F. Watson   ISBN 1-886733-72-4

Inquiries concerning these volumes should be directed to the:
 Astronomical Society of the Pacific
 CONFERENCE SERIES
 390 Ashton Avenue
 San Francisco, CA 94112-1722 USA
 415-337-2126
 catalog@aspsky.org
 Fax: 415-337-5205

ASTRONOMICAL SOCIETY OF THE PACIFIC
CONFERENCE SERIES

Volume 152

FIBER OPTICS IN ASTRONOMY III

Proceedings of a meeting held in Puerto de la Cruz,
Canary Islands, Spain
2-4 December 1997

Edited by
S. Arribas, E. Mediavilla, and F. Watson

Table of Contents

Preface xviii

Acknowledgements xx

Conference Participants xxi

Conference Photograph xxv

Part 1: New Advances in Fiber Technology and Testing

The Astronomical Uses of Optical Fibers................................. 3
   I. Parry

Review of Fiber-Optic Properties for Astronomical Spectroscopy 14
   S. C. Barden

New Silica Fiber for Broad-Band Spectroscopy 20
   G. F. Schötz, J. Vydra, G. Lu and D. Fabricant

Use and Development of Fiber Optics on the VLT 32
   J. Baudrand, I. Guinouard, L. Jocou and M. Casse

Results on Fiber Characterization at ESO 44
   G. Avila

Fiber Sky Subtraction Revisited 50
   F. Watson, A. R. Offer, I. J. Lewis, J. A. Bailey and K. Glazebrook

Review of Fiber-Optic Instrumentation at NOAO 60
   S. C. Barden and T. E. Ingerson

FRD Optimization for PMAS 64
   J. Schmoll, E. Popow and M. M. Roth

Part 2: Multi-Object Spectroscopy

The Anglo–Australian Observatory 2dF Project: Current Status and the
  First Year of Science 71
  I. J. Lewis, K. Glazebrook and K. Taylor

6dF: an Automated Multi-Object Fiber Spectroscopy System for the
  UKST 80
  Q. A. Parker, F. G. Watson and S. Miziarski

Performance of the Fiber-Positioning System for the Sloan Digital Sky
  Survey 92
  *W. A. Siegmund, R. E. Owen, J. Granderson, R. French Leger,
  E. J. Mannery, P. Waddell and C. L. Hull*

Status of the Fiber Feed for the Sloan Digital Sky Survey 98
  *R. E. Owen, M. J. Buffaloe, R. French Leger, E. J. Mannery,
  W. A. Siegmund, P. Waddell and C. L. Hull*

The WYFFOS/AUTOFIB-2 Multi-Fiber Spectrograph on the WHT:
  Description and Science Results 104
  T. Bridges

Fiber-to-Object Allocation Algorithms for Fiber Positioners 111
  F. Sourd

Multi-Fiber Spectroscopy at the Observatorio "Guillermo Haro" 117
  *B. E. Carrasco, S. Vázquez, D. Ren, R. M. Sharples, R. Langarica,
  I. J. Lewis and I. R. Parry*

Part 3: Two-Dimensional Fiber Spectroscopy

Integral-Field Spectroscopy with Optical Fibers on Medium-Size
  (1.5–4-m) Telescopes 123
  C. Vanderriest

The ARGUS Mode of the ALBIREO Spectrograph: Evaluation of Its
  Performance and First Results 135
  G. Herpe, J. Sánchez del Río, C. Vanerriest and A. del Olmo

Mapping the Structure and Dynamics of Late-Stage Mergers 141
  E. T. Chatzichristou

INTEGRAL: an Optical-Fiber System for 2-D Spectroscopy on the 4.2-m
  William Herschel Telescope 149
  *S. Arribas, C. del Burgo, D. Carter, L. Cavaller, R. Edwards, J. Fuentes,
  B. García-Lorenzo, A. García-Marín, B. Gentles, J. M. Herreros, L. Jones,
  E. Mediavilla, M. Pi, D. Pollacco, J. L. Rasilla, P. Rees and N. Sosa*

2-D Spectroscopy of the Gravitational Lens System Q2237+0305 with
INTEGRAL 155
    M. Serra-Ricart, E. Mediavilla, S. Arribas, C. del Burgo, A. Oscoz,
    D. Alcalde, L. J. Goicoechea, B. García-Lorenzo and J. Buitrago

Dynamics and Star Formation Properties of Giant Luminous Arcs 161
    G. Soucail

The PMAS Fiber Spectrograph 168
    M. M. Roth and U. Laux

Two-Dimensional Stellar Kinematics and Population Analysis of
Galaxies 174
    R. Peletier, S. Arribas, C. del Burgo, B. García-Lorenzo, E. Mediavilla,
    C. Gutiérrez, F. Prada and A. Vazdekis

Two-Dimensional Spectroscopy of Active Galaxies 180
    E. Mediavilla, S. Arribas, B. García-Lorenzo,
    and C. del Burgo

Two-Dimensional Spectroscopy with Optical Fibers: the Kinematics
of NGC 1068 185
    B. García-Lorenzo, E. Mediavilla, S. Arribas and C. del Burgo

Two-Dimensional Spectroscopy of M 31 with INTEGRAL 189
    C. del Burgo, S. Arribas, E. Mediavilla and B. García-Lorenzo

TEIFU: an Integral-Field Unit Optimized for Use with the ELECTRA
Adaptive-Optics System at the WHT 193
    R. Haynes, R. Content, J. Allington-Smith and P. Doel

Two-Dimensional Fiber-Bundle Manufacture and FRD Characterization 197
    J. L. Rasilla Ana Belén-Fragoso-López and A. García-Marín

Studies on Bundle Design 203
    A. García-Marín, J. L. Rasilla, S. Arribas and E. Mediavilla

ESPRIT D'ARGUS: an IRAF-based Software Package
for the Treatment of ARGUS Data 206
    P. Teyssandier and C. Vanderriest

Part 4: Projects for Large Telescopes

Integral-Field Spectroscopy with the GEMINI Multi-Object
Spectrographs 213
    J. Allington-Smith, R. Haynes and R. Content

The Original FUEGOS Project on the VLT 220
    P. Felenbok

A Wide-Field Integral-Spectroscopy Unit for the VLT-VIRMOS 229
    E. Prieto, O. Le Fevre, M. Saisse, L. Hill, C. Voet, D. Mancini,
    D. Maccagni, J. P. Picat and G. Vettolani

Fiber-Optic Instrumentation and the Hobby–Eberly Telescope 235
L. W. Ramsey

The Hobby–Eberly Telescope Fiber-Instrument Feed 247
S. D. Horner, L. G. Engel and L. W. Ramsey

Galaxy Kinematics with Integral-Field Spectroscopy and the
Hobby–Eberly Telescope 253
M. A. Bershady, D. Andersen, L. Ramsey and S. Horner

Part 5: Multi-Object and 2-D Infrared Fiber Spectroscopy

AUSTRALIS: a Multi-Fiber Near-IR Spectrograph for the VLT 261
K. Taylor

SINFONI: a High-Resolution Near-Infrared Imaging Spectrometer for the
VLT 271
M. Tecza and N. Thatte

AOIFU: AOB OSIS Infrared Fiber Unit 282
J. Guerin

SMIRFS-II: Multi-Object and Integral-Field Unit Spectroscopy at the
UKIRT 289
R. Haynes, J. Allington-Smith and D. Lee

COHSI: a Lens Array and Fiber Feed for the Near Infrared 300
M. A. Kenworthy, I. R. Parry and K. A. Ennico

Part 6: Other Applications

Optical Fibers in Astronomical Interferometry 309
V. Coudé du Foresto

The Impact of Fiber Optics on Photometry: the Design of Two High-Speed
Multichannel Instruments 320
H. Barwig, K. H. Mantel and S. Kiesewetter

Overview of Fiber Instruments at ESO 329
G. Avila

A Two-Beam Two-Slice Image Slicer for Fiber-Linked Spectrographs ... 337
A. Kaufer

A Fiber-Linked Four Stokes-Parameter Polarimeter for the SOFIN
Spectrometer on the Nordic Optical Telescope 343
B. Pettersson, E. Stempels and N. Piskunov

Part 7: Overview

Conference Overview 349
  Paul Felenbok

Preface

In December, 1979, a group of astronomers led by Roger Angel and John Hill made the first-ever observations with a multi-fibre spectroscopy system. Their target was the galaxy cluster Abell 754, and the novel instrument they used was called *Medusa*. Although single fibres had been used previously for astronomy, the 20-object capability of *Medusa* caught the imagination of instrument-builders everywhere, highlighting the possibilities for optical fibres in astronomy. Today many of the world's observatories use fibres in multi-object or other applications: integral-field spectroscopy, interferometry or photometry.

In its short history, the topic of astronomical fibre optics has been punctuated by a number of milestones. Among the most significant—and certainly among the most agreeable—have been the conferences in the series *Fiber Optics in Astronomy*, the third of which is encapsulated in these proceedings.

It is to Sam Barden and his colleagues at NOAO that thanks are due for initiating the series: the first *Fiber Optics in Astronomy* took place in Tucson during a glorious week in April, 1988. Despite the relative maturity which the technique had by then assumed, the conference had about it a memorable pioneering spirit. By the time Peter Gray chaired *Fiber Optics in Astronomy II* in Sydney in November, 1991, real consolidation was under way. Progress reports on existing instruments jostled for space with descriptions of major new facilities.

Now, after a break of six years, *Fiber Optics in Astronomy III* has been held. Hosted by the Instituto de Astrofísica de Canarias in Tenerife from 2nd to 4th December, 1997, the conference attracted 62 participants to the pleasant winter sunshine of the Canary Islands.

Once again, the spirit of the conference was one of consolidation rather than revolution. This time, however, the major new facilities centred around 8-m class telescopes rather than those of 4-m, a reflection of the speed with which the larger instruments have become the benchmark. Nonetheless, with fibre optics still offering one of the best ways of turning small telescopes into giants, many other projects were also represented.

As with the previous meetings, the emphasis was very much on instrumentation and instrumental techniques, though recent scientific results also had their place. *Fiber Optics in Astronomy III* was divided into six sessions representing the functional areas in which fibres are used in astronomy with an introductory session on the technology of fibres themselves. New advances were presented at all the sessions, and it is clear that the topic still has no shortage of novel and innovative ideas.

Thanks to the efforts of all its contributors, this book represents the current state-of-the-art in astronomical fibre optics. It is a worthy successor to the other two white volumes that adorn the bookshelves of everyone with a serious interest in the subject. But, unlike those venerable publications, it makes a firm stand on the one major issue that divides the international community. Are those marvellous components at the heart of our instruments 'fibres' or 'fibers'...? The sharp-eyed will have noticed that in this preface, both British and American spellings are used, highlighting the inconsistency. Throughout the rest of the book, though, American spelling has been adopted to accord with the book's publication in the USA. To those authors who are particularly ardently attached to their 'fibres', the editors apologise.

Santiago Arribas and Evencio Mediavilla
Instituto de Astrofísica de Canarias

Fred Watson
Anglo–Australian Observatory

July 1998

Cover illustration: Two-dimensional distribution of the spectra and reconstructed image of the gravitational-lens system Q2237+0305 (the Einstein Cross).

Acknowledgements

Many thanks are due to Gerardo Avila, Sam Barden, Paul Felenbok and Roger Haynes, who helped the editors in the scientific organization of this conference.

We are deeply grateful to Terry Mahoney for his invaluable work in the edition of these proceedings. He read, corrected, and organized the material of this volume and standardized the presentation of the different authors.

We owe very special thanks to Monica Murphy, Judith de Araoz and Beatriz Mederos for their hard work in the organization of the conference. The contribution of Begoña García-Lorenzo and Carlos del Burgo is also very much appreciated.

This meeting was made possible through support from the *Dirección General de Educación*, the *Cabildo* of Tenerife, and the Consejería de Cultura of the Town Hall of Puerto de la Cruz, Tenerife.

Santiago Arribas and Evencio Mediavilla
Instituto de Astrofísica de Canarias

Fred Watson
Anglo–Australian Observatory

Conference Participants

ALLINGTON-SMITH, Jeremy (J.R.Allington-Smith@durham.ac.uk), University of Durham, South Road, Durham DH1 3LE, United Kingdom

ARRIBAS, Santiago (sam@ll.iac.es), Instituto de Astrofísica de Canarias, E38200 La Laguna, Tenerife, Spain

AVILA, Gerardo (gavila@web3.hq.eso.org), European Southern Observatory, Karl-Schwarzschild-Strasse 2, 85748 Garching be München, Germany

BARDEN, Samuel (barden@noao.edu), National Optical Astronomy Observatory, PO Box 26732, Tucson AZ 85726-6732, USA

BARWIG, Heinz (hbarwig@usmu01.usm.uni-muenchen.de), Institute for Astronomy & Astrophysics, University of Munich, Scheinerstr. 1, 81679 München, Germany

BAUDRAND, Jacques (baudrand@obspm.fr), Observatoire de Paris, 5 Place Jules Janssen, 92195 Meudon, France

BERSHADY, Matthew A. (mbershady@astro.psu.edu), University of Wisconsin, Department of Astronomy, 475 N. Charter Street, Madison WI 53706, USA

BRIDGES, Terry (tjb@ast.cam.ac.uk), Royal Greenwich Observatory, Madingley Road, CB3 0EZ Cambridge, United Kingdom

DEL BURGO, Carlos (cburgo@ll.iac.es), Instituto de Astrofísica de Canarias, E-38200 La Laguna, Tenerife, Spain

CAMLIBEL, Irfan (AliCam@compuserve.com), Fiberguide Industries, Inc., 07980 Stirling, New Jersey, USA

CARRASCO, Esperanza (bec@inaoep.mx), Instituto Nacional Astrofisica, Optica y Electrónica, Luis Enrique Erro 1, Tonantzintla, Mexico

CHATZICHRISTOU, Eleni (chatzich@strw.LeidenUniv.nl), Leiden Observatory, 2300 RA Leiden, The Netherlands

COUDE DU FORESTO, V. (foresto@hplyot.obspm.fr), Observatoire de Paris, 5 Place Jules Janssen, 92195 Meudon, France

FELENBOK, Paul (felenbok@DAEC.ObsPM.Fr), Observatoire de Paris, 5 Place Jules Janssen, 92195 Meudon, France

FRAGOSO-LOPEZ, Ana (afragoso@ll.iac.es), Instituto de Astrofísica de Canarias, E-38200 La Laguna, Tenerife, Spain

GARCIA LOPEZ, Ramón (rgl@ll.iac.es), Instituto de Astrofísica de Canarias, E-38200 La Laguna, Tenerife, Spain

GARCIA-LORENZO, Begoña (bgarcia@ll.iac.es), Instituto de Astrofísica de Canarias, E-38200 La Laguna, Tenerife, Spain

GARCIA MARIN, Adolfo (agm@ll.iac.es), Instituto de Astrofísica de Canarias, E-38200 La Laguna, Tenerife, Spain

GARCIA VARGAS, María Luisa (mgarcia@ll.iac.es), Instituto de Astrofísica de Canarias, E-38200 La Laguna, Tenerife, Spain

GNEIDING, Clemens (clemens@lna.br), CNP Laboratorio Nacional de Astrofísica, 37500-000 Itajuba–MG, Brazil

GUERIN, Jean (Jean.Guerin@obspm.fr), Observatoire de Paris, 5 Place Jules Janssen, 92195 Meudon, France

GUTIERREZ, Carlos (cgc@ll.iac.es), Instituto de Astrofísica de Canarias, E-38200 La Laguna, Tenerife, Spain

HAYNES, Roger (Roger.Haynes@durham.ac.uk), University of Durham, South Road, Durham DH1 3LE, United Kingdom

HERPE, Georges (HERPE5 Place Jules Janssen, 92195 Meudon, France

HORNER, Scott (horner@astro.psu.edu), Pennsylvania State University, University Park, 16803 Pennsylvania, USA

JIANQIAO, Xue, Beijing Astronomical Observatory, China

JOCOU, Laurent (jocou@gin.obspm.fr), Observatoire de Paris, 5 Place Jules Janssen, 92195 Meudon, France

KAROJI, Hiroshi (karoji@optik.mtk.nao.ac.jp), National Astronomical Observatory, 96720 Hilo, Hawaii, USA

KAUFER, Andreas (A.Kaufer@lsw.uni-heidelber.de), Landessternwarte Heidelberg, Königstuhl 12, 69117 Heidelberg, Germany

KENWORTHY, Matthew (mak@ast.cam.ac.uk), Institute of Astronomy, Madingley Road, CB3 0HA Cambridge, United Kingdom

LEWIS, Ian (ijl@aaocbn.aao.GOV.AU), Anglo–Australian Observatory, Siding Spring Mountain, NSW 2357, Coonabarabran, Australia

LU, Grant (Grantlu@aol.com), Heraeus Amersil, 3473 Satellite Blvd., Duluth GA 30096, USA

MARTIN FLEITAS, Juan Manuel (jmartin@ll.iac.es), Instituto de Astrofísica de Canarias, E-38200 La Laguna, Tenerife, Spain

MEDIAVILLA, GRADOLPH, Evencio (emg@ll.iac.es), Instituto de Astrofísica de Canarias, E-38200 La Laguna, Tenerife, Spain

MIGNOLI, Marco (MIGNOLI@astbo3.bo.astro.it), Osservatorio Astronomico di Bologna, Via Zamboni 33, 40126 Bologna, Italy

MONAI Sergio (monai@sigfrido.oat.ts.astro.it), Osservatorio Astronomico di Trieste, Via Tiepolo 11, 34131 Trieste, Italy

NOUMARU, Junichi (noumaru@subaru.naoj.org), Subaru Telescope, National Astronomical Observatory, 96720 Hilo, Hawaii, USA

OSCOZ ABAD, Alejandro (aoscoz@ll.iac.es), Instituto de Astrofísica de Canarias, E-38200 La Laguna, Tenerife, Spain

PARKER, Quentin andrew (qap@aaocbn1.aao.GOV.AU), Anglo–Australian Observatory, Siding Spring Mountain, NSW 2357 Coonabarabran, Australia

PARRY, Ian (irp@ast.cam.ac.uk), Institute of Astronomy, Madingley Road, CB3 0HA Cambridge, United Kingdom

PELETIER, Reynier (R.F.Peletier@dur.ac.uk), Department of Physics, University of Durham, South Road, DH1 3LE Durham, United Kingdom

PETTERSSON, Bertil (bertil@vela.astro.uu.se), Uppsala Astronomiska Observatorium, 751 20 Uppsala, Sweden

PFEIFFER, Michael (pfeiffer@usmu01.usm.uni-muenchen.de), Institute of Astronomy & Astrophysics, University of Munich, Scheinerstr. 1, Munich, Germany

POLLACCO, Don (dlp@ing.iac.es), Isaac Newton Group, Apdo. Correos 368, 38700 S/C de La Palma, Spain

PRIETO, Eric (prieto@astrsp-mrs.fr), Laboratoire d'Astronomie Spatiale, 13012 Marseille, France

RAMSEY, Lawrence (lwr@astro.psu.edu), Pennsylvania State University, University Park, PA 16830, USA

RASILLA PIÑEIRO, José Luis (jlr@ll.iac.es), Instituto de Astrofísica de Canarias, E-38200 La Laguna, Tenerife, Spain

ROTH, Martin (mmr@tina.aip.de), Astrophysical Institute, Potsdam an der Sternwarte 16, D-14482, Potsdam, Germany

RUSSELL, Owen (owen@astro.washington.edu), Dept. of Astronomy, University of Washington, Seattle WA98195-1580, USA

RUTTEN, Rene (rgmr@ing.iac.es), Isaac Newton Group, ORM, Edificio Mayantigo, Alvarez de Abreu 68, 38780 S/C de La Palma, Spain

SANCHEZ DEL RIO, Justo (justo@iaa.es), Instituto de Astrofísica de Andalucía, Camino Bajo de Huetor 24, 18080 Granada, Spain

SCHMOLL, Juergen (jschmoll@aip.de), Astrophysical Institute, Potsdam an der Sternwarte 16, 14482 Potsdam, Germany

SCHÖTZ, Gerhard (gerhard.schoetz@europe.heraeus.com), Heraeus Quarzglas GmbH, Quarzstrasse, 63450 Hanau, Germany

SERRA RICART, Miquel (mserra@ll.iac.es), Instituto de Astrofísica de Andalucía, Camino Bajo de Huetor 24, 18080 Granada, Spain

SIEGMUND, Walter (siegmund@dirac.phys.washington.edu), Dept. of Astronomy, University of Washington, PO Box 351580, Seattle WA 98195-1580, USA

SOUCAIL, Genevieve (soucail@obs-mip.fr), Observatoire Midi-Pyrenees, 14 Avenue Edouard Belin, 31400 Toulouse, France

SOURD, Francis (Francis.Sourd@lip6.fr), Observatoire de Paris, 5 Place Jules Janssen, 92195 Meudon, France

STEMPELS, Eric (stempels@astro.uu.se), Uppsala Astronomiska Observatorium, 751 20 Uppsala, Sweden

TAYLOR, Keith (kt@aaoeep2.aao.GOV.AU), Anglo–Australian Observatory, PO Box 296, NEW 2121 Epping, Australia

TECZA, Matthias (tecza@mpeis1.mep-garching.mpg.de), MPI f"ur Extraterrestrische Physik, Giessenbachstrasse, 85748 Garching, Germany

VANDERRIEST, Christian (Christian.Vanderriest@DAEC.ObsPM.Fr), Observatoire de Paris, 5 Place Jules Janssen, 92195 Meudon, France

WATSON, Fred (fgw@aaocbnu1.aao.gov.au), Anglo–Australian Observatory, Siding Spring Mountain, NWS 2357 Coonabarabran, Australia

WOCHE, Manfred (woche@physics.uch.gr), Dept. of Physics, University of Crete, PO Box 2208, GR 71003 Heraklion, Greece

Part 1: New Advances in Fiber Technology and Testing

The Astronomical Uses of Optical Fibers

Ian R. Parry

Institute of Astronomy, University of Cambridge, Madingley Road, Cambridge CB3 0HA, United Kingdom

Abstract. The various ways in which optical fibers are used in astronomy are reviewed and the relevant properties of fibers are discussed. Applications include multi-object spectroscopy, integral-field spectroscopy, interferometry, high-precision radial-velocity determination, and photometry. Future developments and trends in fiber instrumentation are also briefly discussed.

1. Introduction

Optical fibers were first used to carry out astronomical observations in the late 1970s and have been used very successfully ever since. In this paper we review the various ways in which fibers have been used to benefit astronomical instrumentation. Only applications where the light propagating down the fiber has come from the sky are considered—applications where fibers are used to provide a data link are not discussed.

Several dozens of instruments have been built for astronomy that use optical fibers, and the range of scientific applications is very broad. The conference for which this paper is a contribution is the third of a series specifically for the use of fibers for astronomical instrumentation, and over the last 20 years the field has grown considerably. Many papers from the proceedings for the first two conferences are cited in this paper. A review of optical fibers in astronomical instruments is given by Heacox & Connes (1992). Fibers are used for multi-object spectroscopy, bi-dimensional spectroscopy, high-precision radial-velocity spectroscopy, interferometry, and photometry. The wavelengths covered extend from the UV ($\sim$350 nm) to the IR ($\sim$2.3 μm).

In §2 we discuss the important properties of fibers that govern how they are used in their various applications. Section 3 summarizes the important area of multi-object spectroscopy. Integral-field spectroscopy is an application which has seen a lot of recent activity and is reviewed in §4. Other applications of fibers are discussed in §5. Finally, the way in which this field could develop in the future is considered in §6.

2. Properties of Optical Fibers

Invariably, the fibers used in astronomy are of the step-index type—gradient-index fibers are seldom if ever used. Figure 1 shows how such a fiber is constructed. The refractive index of the cladding is lower than that of the core. The buffer is for mechanical protection and many options are available. The highest quality fibers for optical wavelengths have a silica core and a doped silica cladding. A simple way to think about how fibers work is to imagine rays traveling in the core within a small angular range with respect to the fiber axis. These rays will be totally reflected at the core/cladding boundary and no light will escape. However, to fully understand how fibers work it is necessary to treat them as waveguides and consider how the waves propagate rather than simply consider the propagation of rays. An allowed wave form is called a mode and for most of the fibers used in astronomy the light is carried using several hundred modes. The fibers are often referred to as multi-mode fibers.

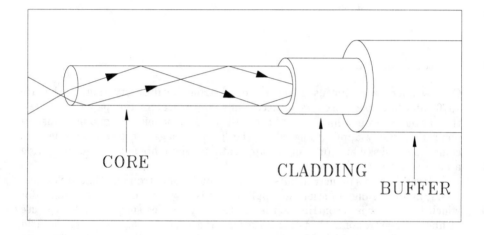

Figure 1.   The construction of a step-index optical fiber.

The main features of optical fibers that make them useful for astronomical observations are their:

- Spatial reformatting capability
- Ability to scramble the light in an aperture
- Convenience
- Fixed path length (for single-mode fibers)

The flexibility of fibers is of course their most unique and powerful feature. In particular, this allows fibers to be used to spatially reformat the light at some image plane. A spectrograph requires a long thin entrance aperture but often the parts of the telescope focal plane that one wishes to observe are arranged in some other geometry. In the case of multi-object spectroscopy the areas of interest in the focal plane are randomly scattered whereas for integral-field spectroscopy the area of interest is a two-dimensional region.

The light distribution at the output face of an individual fiber has radial symmetry i.e. the intensity in any annular zone centered on the center of the fiber face is the same at every point along the annulus. In many cases the intensity across the output face is completely uniform. The detailed distribution of light at the output face is largely independent of the distribution of light at the input end, and there is always radial symmetry. This property is particularly useful for spectroscopy, where a very high precision wavelength calibration is required. This is because the centroid of the spectral features obtained does not shift when the image at the input to the fiber moves due to seeing or telescope shake. Fibers can be made to scramble very well by stressing them but this introduces focal-ratio degradation (FRD) which is an undesirable property of fibers (see below). The scrambling versus FRD balance can be adjusted for the particular application concerned.

Fibers are tremendously convenient. Often the simplest way to get light from one place to another is to use an optical fiber. For example a fiber may be used to feed light from the telescope to a large bulky echelle spectrograph. On a telescope that does not have a coudé or Nasmyth feed, using a fiber is a very convenient way forward. Fibers can inject or collect light in the most inaccessible places and offer more possibilities when designing instruments. For example, a spectrograph collimator can have an internal slit. Fibers are also particularly useful when a spectrograph is to be shared on several different telescopes because they make it very simple to interface it to any telescope.

When the fiber core diameter is of the order of the wavelength of the light only one mode of propagation is allowed and the fiber is called a single-mode fiber. In this case the optical path length for the fiber is a fixed quantity and it is possible to use the fiber for interferometry. The fiber replaces the path-length compensation optics normally used in interferometers.

However, fibers are not perfect, and there are several features which significantly compromise their performance. These include:

- Focal-ratio degradation

- Variable transmission properties

- Internal absorption losses

- Performance at cryogenic temperatures

In a perfect world light would emerge from a fiber at exactly the same f-ratio as it entered. However, in practice this is not the case, and the emergent light is more spread out in angle (lower f-ratio) than the input beam. This effect is known as FRD and is a problem because it means that a fiber does not preserve the A-omega product (the product of the beam area and its solid angle—normally conserved in an optical system). Although the amount of energy in the beam is essentially preserved, FRD is undesirable because the energy density (surface brightness) is reduced. Because the total A-omega is increased (without the amount of information being increased) a fiber-fed spectrograph will suffer a loss in performance: at least one of throughput, spectral resolution, multiplex gain, and wavelength coverage will be reduced. For a spectrograph where the collimator is adjusted to collect all the light but the camera f-ratio is kept constant, the effect of FRD is that the measurement of a single spectral resolution element requires more detector pixels. For detectors with significant detector noise (dark current, read-noise, cosmic-ray events, etc.) this compromises the signal-to-noise ratio in the spectral-resolution element, as well as reducing the number of simultaneous measurements that can be made.

Many workers have evaluated the FRD properties of fibers (e.g. Ramsey 1988; Avila 1988; Carrasco & Parry 1994). To minimize FRD it is important to purchase a high-quality fiber and take great care in constructing the fiber feed. The construction process must ensure that the stresses on the fiber from termination ferrules and bends, etc., are kept to an absolute minimum, and the quality of the input and output end faces must be very high. It is also important to feed the light into the fiber at a fast f-ratio ($\sim f/2$–$f/6$) because FRD is less of a problem for fast input beams.

Because FRD depends on the stresses on the fiber the transmission can vary as the fiber geometry changes. When the telescope moves the fibers also move, and this can lead to unwanted variations, which can lead to systematic errors in the data, such as poor sky subtraction.

The transparency of the core material in optical fibers is, generally speaking, extremely good, which is why they can be used in great lengths for communication applications. However, for some wavelengths the transmission is not ideal, and this depends on the type of fiber materials used. For the wavelength regime 0.3–2.0 μm there are generally two types of all-silica fibers. Wet fibers (high OH content) have good transmission in the 0.3–0.55-μm region but suffer from absorption bands in the red and IR. Dry fibers (low OH), in contrast, work well in the 0.55–2.0-μm region but suffer scattering losses in the UV and blue part of the spectrum. Fibers which combine the best of both of these types are now becoming available (see Schötz & Vydra, these proceedings).

Optical fibers that have good performance in the IR are also commercially available. For the near-IR (1.0–2.0 μm) low-OH silica fibers work very well and it is possible to purchase fibers with an ultra-low OH content. Zirconium fluoride fibers can be used for longer wavelengths (1–5 μm). These fibers are very expensive, extremely fragile and are best used in short lengths to maintain efficiency. For even longer wavelengths (3–8 μm) it is possible that chalcogenide fibers could be used, but very little work on the exploitation of these for astronomy has been done.

With the recent improvements in IR detector arrays there has been much more demand for fiber systems to operate in the IR. This in turn demands that for wavelengths greater than $\sim$1.6 μm the fibers themselves should be cooled to minimize the thermal background. Although there is no fundamental reason why fibers should not work at cryogenic temperatures, the current experience of astronomers is very limited, and it is unclear how well the fibers will work. Clearly, the stresses induced by cooling, which are potentially quite large, must be minimized to reduce FRD.

3. Multi-Object Spectroscopy

Optical fibers have had their biggest impact on observational astronomy through the technique of multi-object spectroscopy. Ever since they were first used in 1979 at Steward Observatory (Hill et al. 1980) many observatories have built and used multi-fiber spectrographs because they are vital for many branches of astronomy that require large spectroscopic samples. For a review of multi-object spectroscopy using fibers which lists 32 multi-fiber instruments see Parry (1997).

Often it is necessary to obtain spectra for many (>50) objects in a sample to answer an important question. And because observers are working at the faintest limits, long (> 1 hr) exposure times are required to give a sufficient signal-to-noise ratio, even on the largest telescopes. Clearly, by observing the objects simultaneously rather than one at a time much larger samples can be observed. To be able to observe many objects simultaneously requires sufficient objects in the telescope field of view (FOV), and enough detector pixels to record all the spectra. For very many important observational programs these two criteria are met, and multi-object spectroscopy is of great value.

The range of scientific discoveries made with multi-fiber spectrographs is enormous. Examples include the discovery of the Sagittarius Dwarf Galaxy (Ibata et al. 1995), the evolution of the QSO luminosity function (Boyle et al. 1988), the Local Galaxy Luminosity Function (Lin et al. 1996), evolution of the Galaxy Luminosity Function (Ellis et al. 1996), and the globular cluster binary fraction (Yan & Cohen 1996). The large-scale structure of the universe will be probed with unprecedented accuracy by the 2dF and SLOAN surveys in the near future.

In a fiber-fed multi-object spectrograph the fibers at the telescope focal plane are placed on the objects of interest, and the other ends of the fibers are arranged at the input to the spectrograph in a single line which is often referred to as the fiber "slit" (see Fig. 2). This decoupling of the telescope and the spectrograph makes fibers particularly good for wide-field and/or high spectral resolution multi-object spectroscopy, and they are better than multi-slit devices in this respect. However, sky subtraction with fibers is more problematic than for slit spectrographs, and up to now fiber systems have not been used for very faint spectroscopy (Watson, this conference; Parry 1997; Parry & Carrasco 1990; Cuby & Mignoli 1994; Wyse & Gilmore 1992; Elston & Barden 1989).

The first fiber-fed multi-object spectrograph ever used was Medusa, which was developed by John Hill and his collaborators on the Steward Observatory 2.3 meter (Hill et al. 1980). It had 37 fibers and was first used to get galaxy spectra in 1979 December. The fibers were held in position at the focal plane by a plug-plate which was made specifically for the field being observed. The plug-plate consisted of a plate of material with holes drilled at the positions of the target objects, and the observer had to plug the fibers into the plate at the telescope, carefully noting down which fiber was used on which object. All the plug-plates had to be manufactured in advance of the observing run using target astrometry obtained from photographic plates. Detailed descriptions of Medusa are given by Hill et al. (1982, 1983). The plug-plate concept has been used for many multi-fiber spectrographs since then and is still being used today. The biggest and most ambitious example is the plug-plate system for the SLOAN survey. This will have 640 fibers in its 3-deg field (Seigmund, these proceedings;

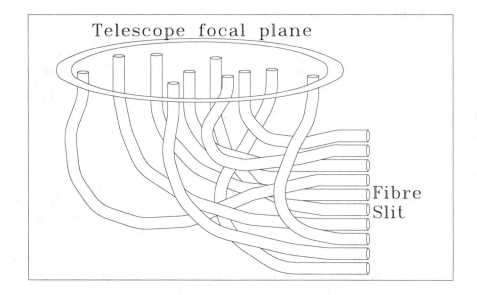

Figure 2. Fibers being used for multi-object spectroscopy.

Owen, these proceedings; Kron 1997), and one million redshifts will be obtained with this facility.

The amount of work required to prepare all the plug-plates and the lack of versatility at the telescope drove several groups to develop automated fiber positioners. The MX spectrometer was built as a replacement for Medusa and was the first robotic fiber positioner (Hill et al. 1982; Hill 1984; Hill & Lesser 1986). It used a "fishermen round the pond" arrangement to position the fibers with each fiber being held at the tip of an arm (fishing rod) which was moved by a 2-axis actuator. MX had 32 independent arms.

An alternative robot positioner which could have a higher multiplex gain than MX was first proposed by Parry (1986). AUTOFIB was built for the Anglo-Australian Telescope (AAT) and used a set of fibers which were designed to be manipulated by a single pick-and-place robot system (Parry & Gray 1986; Parry & Sharples 1988). Each fiber was held magnetically onto a steel plate (the field plate) at the telescope focal plane. The fibers lay on the illuminated side of the field plate and each had a microprism on the end to fold the light from the telescope into the fiber. The tip of each fiber also acted as a handle for the X-Y robot. Because the fibers were physically smaller than the arms required for an MX-type system, it was possible to have more of them. AUTOFIB had 64 spectroscopic fibers and eight guide fibers, and was first used on the AAT in 1987. The most ambitious and powerful AUTOFIB-type system built so far is the 2dF which has 400 fibers on the sky and a 2-deg FOV (Lewis et al., these proceedings)

SMIRFS (Spectroscopic Multi-object Infra-Red Fiber System) was the first IR fiber system for multi-object spectroscopy and was first used on UKIRT in

1995 (Haynes & Parry 1994; Haynes 1995; Haynes et al. 1995). It is a plug-plate system designed as an add-on for CGS4 and can be used in the J, H, and K (1–2.5 μm) windows. The 4-arcmin FOV and the 256×256 detector allow 14 fibers to be used. Two fiber bundles of ultra-low OH silica and zirconium fluoride were used. The system was not cooled (although the spectrograph CGS4 is) but has specific features to help reduce the thermal background.

4. Integral-Field Spectroscopy

Integral field spectroscopy is the technique of obtaining the spectra for an array of spatial samples in a two-dimensional field of view. Each spatial point gives a spectrum, so the resulting data can be thought of as a data-cube with two spatial dimensions, and the other dimension corresponding to wavelength. Bi-dimensional spectroscopy can be used to study the internal dynamics of extended objects (galaxies, star-forming regions, planetary nebulae, etc.), search for emission-line sources, and obtain the spectra of several objects in very crowded fields. Sometimes the way in which the spectral information varies from one pixel to the next is not of interest, and the integral-field technique is simply used as an image slicer to provide higher spectral resolution.

The reformatting capability of fibers make them ideal for this purpose. All that is needed is a fiber bundle with the input ends arranged in a two-dimensional array, and the output ends arranged as a single line of fibers at the spectrograph entrance slit. To minimize cross-talk the bundle is usually arranged so that fibers that are adjacent at the slit are also adjacent on the sky. Because the individual fiber spectra are related, and each fiber is small and samples a relatively small patch of sky (∼0.1–1.0 arcsec across) more fibers can be accommodated by the spectrograph than for multi-object spectroscopy. The first integral-field fiber device was ARGUS, which has 397 fibers. It was built between 1982 and 1984 and first used on 2-m telescopes in France (Vanderriest & Lemonnier 1988). From 1986 it was used with SILFID at the CFHT where it provided a 16-arcsec diameter FOV with 0.7-arcsec sampling. Another pioneering fiber-based integral-field unit (IFU) was Densepak for the KPNO 4 meter (Barden & Scott 1986), which had 49 fibers in a 16×19-arcsec FOV. A further example is the HEXAFLEX system (Arribas et al. 1991).

Recently, fiber-based IFUs have incorporated lens arrays. An array of bare fibers suffers packing-fraction losses because the fibers are circular and the cladding prevents the fiber cores from touching. If a lens array is used to sample the image at the focal plane it creates an array of well separated micro-pupil images. The light in each of these micro-pupils can be collected by a fiber of matching diameter and so the packing fraction losses are eliminated. The lenses in the lens array can also feed the light into the fibers at a suitably fast f-ratio (independent of the f-ratio at the image plane) thus reducing FRD losses. If the lens array utilizes relatively large lenses it is necessary to use some fore-optics to appropriately convert the telescope image scale. This approach has several advantages, including maximizing lens-array quality, allowing optimal AR coatings, providing telecentric micro-pupils, enabling high-precision fiber alignment, and offering extra operational versatility (e.g. image scale options, and portability between telescopes).

The first fiber IFU to use a lens array was the Multi-Pupil Fiber Spectrograph for the Russian 6-m telescope and was built by Afanasiev and his co-workers. SPIRAL (Parry et al. 1996) was first used on the AAT in 1997 and uses a macro-lens IFU. The SMIRFS IFU for CGS4 on UKIRT was the first fiber-based integral-field unit to be used in the IR (Haynes et al., these proceedings). Another macro-lens IFU based on SPIRAL has been built for COHSI (the Cambridge OH Suppression Instrument) which was first used at UKIRT in 1998 March (Piché et al. 1997; Kenworthy et al., these proceedings; Ennico et al. 1998).

5. Other Applications

Most of the astronomical instruments built using fibers have been for multi-object or integral-field spectroscopy. However, there are other applications; namely, high-precision radial-velocity work, interferometry, and photometry, and these are mentioned briefly in this section. Fibers are also used for support systems such as autoguider probes.

High-precision radial-velocity measurements rely on the scrambling properties of optical fibers (see §2 above). Heacox (1988) reports a velocity accuracy of 30 ms^{-1}, obtained by providing the UoH 88-inch telescope's coudé spectrograph with a fiber feed. Brown et al. (1994) achieve a velocity accuracy of 3 m s^{-1} with AFOE (Advanced Fiber Optic Echelle) on the 1.5-m telescope of the Whipple Observatory.

By coherently recombining the light from several independent telescopes it is possible to achieve spatial resolutions limited not by the aperture of the individual telescopes but by the spacing between them. Such interferometers need to maintain a fixed path length as the telescopes track their target, and this can be provided by single-mode fibers (see Coude du Forresto et al., these proceedings). Compared to mirror-based path-length compensators fibers are simple and convenient. They also have the advantage of being unaffected by atmospheric turbulence.

Virtually all photometry is done without fibers but there are a few photometric systems that use fibers for good reasons. Barwig et al. (1987) and Barwig et al. (these proceedings) describe fiber-based instrumentation for multi-channel, multi-color, high time-resolution photometry. The advantage of using fibers in this case is that they allow the light from two well separated stars to be split up into $UBVRI$ colors and monitored simultaneously by several high-speed photon-counting detectors. By observing a reference star at the same time, atmospheric variations can be very accurately calibrated out.

Multi-fiber photometry has also been proposed for projects where a very large number of objects spread over a large FOV (much larger than can be seen with a conventional camera) need to be monitored for variability. A particular example is the search for extra-solar planets by looking for planetary transits. Initially it was thought that fibers would not be accurate enough to do this work (Borucki et al. 1988) but later, similar proposals, showed that the photometric capability of fibers was in fact adequate (Parry et al. 1994). This concept has not yet been put into practice.

6. Future Prospects

First of all, it worth mentioning the enormous spectroscopic surveys that will be carried out by the 2dF and SLOAN facilities. At the time of writing, the 2dF survey is already under way, and the SLOAN survey is close to starting. These surveys will be of an unprecedented scale involving large teams of astronomers and instrument scientists.

More generally though, we are now entering an era where large telescopes (apertures >6 m) will dominate ground-based astronomy, and undoubtedly fibers will play an important role in exploiting the full potential of these telescopes. Curiously, the initial complement of instruments for these telescopes does not include very many fiber instruments—there are no facility fiber instruments for Keck or Subaru, and the only one for Gemini is the IFU add-on for GMOS. First-generation fiber instruments for large telescopes include FUEGOS (Felenbok et al. 1994) for the VLT and Hectospec (Fabricant et al. 1994) for the MMT. The Hobby Eberly Telescope (Ramsey et al., these proceedings) is specifically designed to exploit fibers, and most of the instruments are fiber fed. The second generation of instruments currently being planned is more ambitious in its use of fibers and includes AUSTRALIS (Taylor et al., these proceedings) for the VLT, FMOS (Karoji et al., this conference) for SUBARU, and possibly new instruments for GEMINI.

An important trend for fiber spectroscopy on large telescopes is the push to higher spectral resolution. For many programs the light-grasp of a 4-m telescope is simply insufficient to get to the required faint limits at high spectral resolution. Typically, the signal-to-noise is photon limited rather than background limited, and the gain from a larger aperture is considerable. FUEGOS and Hectospec will both offer spectral resolutions that are much higher than has typically been available with existing multi-object spectrographs. Programs that require high ($R > 10000$) spectral resolution include the determination of the velocity dispersions of dwarf spheroidals, QSO studies, and detailed studies of globular clusters.

New multi-object spectrographs with very large multiplex gain will become available. For example, AUSTRALIS and FMOS will have a multiplex gain of 400, and FMOS will have a 30-arcmin FOV—a very large FOV for an 8-m telescope. The large budgets associated with the new large telescopes, and the availability of larger-format detectors are important factors in this trend. The importance of solving the problems associated with fiber sky subtraction becomes even more acute given the potential rewards, and the unprecedented level of investment in these new instruments.

Integral-field spectroscopy with fibers is not a new technique but it has recently seen a strong resurgence. There are many reasons for this, including the problems of matching spectrographs to large telescopes, the availability of large-array detectors, and scientific demand (such as the desire to study the internal dynamics of very distant galaxies). Fiber-fed spectrographs are no longer being seen as add-ons to slit spectrographs, and several dedicated fiber spectrographs are being built. The versatility of this approach leads to designs which can do both integral-field spectroscopy and multi-object spectroscopy.

Some of the instruments currently being planned (e.g. FMOS and AUSTRALIS) will combine integral-field spectroscopy and multi-object spectroscopy simultaneously and offer a multi-IFU mode. Many of the galaxies already known to be at high redshift from existing surveys have images which show evidence for mergers. The fastest way to map a large sample of these objects spectroscopically to confirm the merger hypothesis is with a multi-IFU. Optical fibers are probably the only way forward in this area.

We are only just beginning to use fibers for spectroscopy in the IR, and this is very much a technique for the future. It is important that we can successfully use fibers at cryogenic temperatures (below 220 K) to get the maximum benefit in the IR. COHSI, GOHSS, and CIRPASS are examples of multi-fiber systems currently being built or planned for the J and H bands which include cryogenically cooled fibers. SINFONI (Tecza et al., these proceedings) is an integral-field spectrograph for the J, H and K bands with fully cooled ($\sim$77 K) optical fibers and is the most ambitious K-band fiber spectrograph to date.

Fibers do not figure in the current or future plans of the *HST*, and yet it is very clear that a space-based telescope would benefit greatly from some of the fiber techniques pioneered in ground-based astronomy. Some of the instruments in the early planning stages for *NGST* are now beginning to consider optical fibers. Maybe one day we will decide to put a fiber-based integral-field unit or even a robotic fiber positioner on a telescope in space.

References

Arribas, S., Mediavilla, E., & Rasilla, J. L. 1991, ApJ, 369, 260

Avila, G. 1988, in ASP Conf. Ser. Vol. 3, Fiber Optics in Astronomy, ed. S. C. Barden (San Francisco: ASP), 63

Barden, S. C., & Scott, K. 1986, B.A.A.S., 18, 951

Borucki, W. J., Torbet, E. B., & Pham, P. C. 1988, in ASP Conf. Ser. Vol. 3, Fiber Optics in Astronomy, ed. S. C. Barden (San Francisco: ASP), 247

Boyle, B. J., Shanks, T., & Peterson, B. A. 1988, MNRAS, 235, 935

Brown T. M., Noyes R. W., Nisenson, P., Korkzennik, S. G., & Horner, S. 1994, PASP, 106, 1285

Carrasco, B. E., & Parry, I. R. 1994, MNRAS, 271, 1

Cuby J., & Mignoli, M. 1994, Proc. SPIE, 2198, 99

Ellis, R. S., Colless, M., Broadhurst, T., Heyl, J., & Glazebrook, K. 1996, MNRAS, 280, 235

Elston, R., & Barden, S. C. 1989, NOAO newsletter 19

Ennico, K. A., Parry, I. R., Kenworthy, M. A., Ellis, R. S., Mackay, C. D., Beckett, M. G.,Aragon-Salamanca, A., Glazebrook, K., Brinchmann, J., Pritchard, J. M., Medlen, S. R., Piche, F., McMahon, R. G., & Cortecchia, F. 1998, Proc. SPIE 3354, in press

Fabricant, D. G., Hertz, E. H., & Szentgyorgyi, A. H. 1994, Proc. SPIE, 2198, 251

Felenbok, P., Cuby, J-G., Lemonnier, J., Baudrand, M., Casse, M., Andre, M., Czarny, J., Daban, J-B., Marteaud, M., & Vola, P. 1994 Proc. SPIE, 2198, 115

Haynes, R. 1995, PhD Thesis, University of Durham

Haynes, R., & Parry, I. R. 1994, Proc. SPIE, 2198, 572

Haynes, R., Sharples, R. M., Ennico, K. A., & Parry, I. R. 1995, Spectrum (newsletter of Royal Observatories), 7, 4

Heacox, W. D. 1988, in ASP Conf. Ser. Vol. 3, Fiber Optics in Astronomy, ed. S. C. Barden (San Francisco: ASP), 204

Heacox, W. D., & Connes, P. 1992, A&ARev, 3, 169

Hill, J. M. 1984, PhD Thesis, University of Arizona

Hill, J. M., Angel, J. R. P., & Scott, J. S. 1983, Proc. SPIE, 380, 354

Hill, J. M., Angel, J. R. P., Scott, J. S., Lindley, D., & Hintzen, P. 1980, ApJ 242, L69

Hill, J. M., Angel, J. R. P., Scott, J. S., Lindley, D., & Hintzen, P. 1982, Proc. SPIE, 331, 279

Hill, J. M., & Lesser, M. P. 1986, Proc. SPIE, 627, 303

Ibata, R. A., Gilmore, G., & Irwin, M. J. 1995, 277, 781

Kron, R. 1997, in Astrophysics and Space Science Lib., Vol. 212, Wide Field Spectroscopy, ed. E. Kontizas, M. Kontizas, D. H. Morgan, & G. P. Vettonani (Dordrecht: Kluwer), 41

Lin, H., Kirshner, R. P., Shectman, S. A., Landy, S. D., Oemler, A., Tucker, D. L., & Schechter, P. L. 1996, ApJ, 464, 60

Parry, I. R. 1986, PhD Thesis, University of Durham

Parry, I. R. 1997, in Astrophysics and Space Science Lib., Vol. 212, Wide Field Spectroscopy, ed. E. Kontizas, M. Kontizas, D. H. Morgan, & G. P. Vettonani (Dordrecht: Kluwer), 3

Parry, I. R., & Carrasco, B. E. 1990, Proc. SPIE, 1235, 702

Parry, I. R., & Gray, P. M. 1986, Proc. SPIE, 627, 118

Parry, I. R., Kenworthy, M. A., & Taylor, K. 1996, Proc. SPIE, 2871, 1325

Parry, I. R., Lewis, I. J., Sharples, R. M., Dodsworth, G. N., & Webster, J. 1994, Proc. SPIE, 2198, 125

Parry, I. R., & Sharples, P. M. 1988, in ASP Conf. Ser. Vol. 3, Fiber Optics in Astronomy, ed. S. C. Barden (San Francisco: ASP), 93

Piché, F., Parry I. R., Ennico K. A., Ellis R. S., Pritchard J., Mackay C. D., & McMahon R. G. 1997, Proc. SPIE, 2871, 1332

Ramsey, L. W. 1988, in ASP Conf. Ser. Vol. 3, Fiber Optics in Astronomy, ed. S. C. Barden (San Francisco: ASP), 26

Vanderreist, C., & Lemonnier, J. P. 1988, in Instrumentation for Ground Based Optical Telescopes: Present and Future, ed. L. B. Robinson (New York: Springer), 304

Wyse, R. F. G., & Gilmore, G. 1992, MNRAS, 257, 1

Yan, L. & Cohen, J. G. 1996, AJ, 112, 1489

Fiber Optics in Astronomy III
ASP Conference Series, Vol. 152, 1998
S. Arribas, E. Mediavilla, and F. Watson, eds.

Review of Fiber-Optic Properties for Astronomical Spectroscopy

Samuel C. Barden

National Optical Astronomy Observatories, PO Box 26732, Tucson, AZ 85726-6732, USA

Abstract. I give a brief review of the basic fiber-optic properties for step-index, multimode fibers that are important for their use in astronomical spectrographs as a prelude for the papers which follow. I also give results from an on-sky performance evaluation for some typical fibers.

1. Fiber Transmission

The spectral-transmission efficiency of a fiber is the primary property of concern for its implementation in astronomical spectrographs. Figure 1 shows the spectral performance for two types of fibers. The top figure gives the transmission of a 25- long high-OH fiber. High-OH fibers typically give the best blue or ultraviolet transmission at the expense of OH absorption bands redward of 700 nm. The lower plot shows the transmission for fibers in which the OH has been removed resulting in good transmission out to 1.8 μm but with a loss in spectral transmission blueward of 400 nm. For the past decade, these were the only available choices.

Hydrogen doping of a low-OH fiber can enhance the blue transmission. However, over time the hydrogen can escape from the fiber resulting in possible degradation of the blue performance. The paper by Avila et al. (these proceedings) gives some evaluation of hydrogen-doped fibers.

A new fiber has recently become available which maintains the good red and near-infrared performance of a low-OH fiber while preserving good blue transmission. Although this new fiber does not transmit quite as well as a high-OH fiber shortward of 350 nm, it does allow instruments to span a larger spectral window (350 nm to 1.8 μm) with only a single fiber. Please see the paper following this for further information about this new and exciting fiber (Lu et al., these proceedings).

2. Focal-Ratio Degradation

Another critical behavior of fibers that is of concern is the property called focal-ratio degradation or FRD. Fibers tend to scatter the angle of propagation of a ray of light to larger angles as a result of micro-bends, scattering centers, and stresses within the core of the fiber. This scattering has a direct impact on the etendue, or resolution-throughput product of the spectrograph fed by the fibers. Ideally, the light exiting the fiber should be in a cone with angular

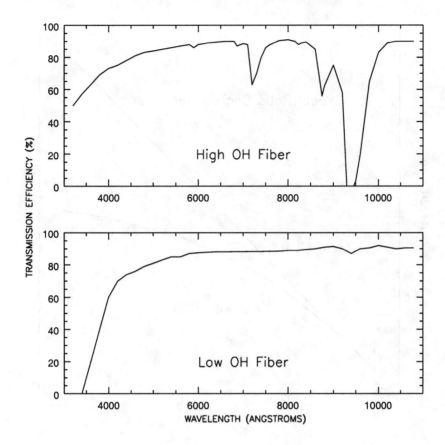

Figure 1. Spectral transmission for two types of fibers in the optical bandpass. The top figure is for a blue transmissive fiber, the bottom for a red transmissive fiber. The fiber lengths are 25 m.

extent equal to the focal ratio of the light illuminating the fiber. However, as can be seen by the example shown in Fig. 2, that condition is met only for very fast focal ratios. At slow input focal ratios, the fiber will either decrease the resolving power and/or the throughput efficiency of the spectrograph. The amount of FRD produced by a fiber depends on factors due to the fiber-drawing process (how well the diameter was controlled), fiber cabling (stress introduced by an outer jacket), end termination (how good the polished end is), and fiber mounting (epoxy-induced stress levels).

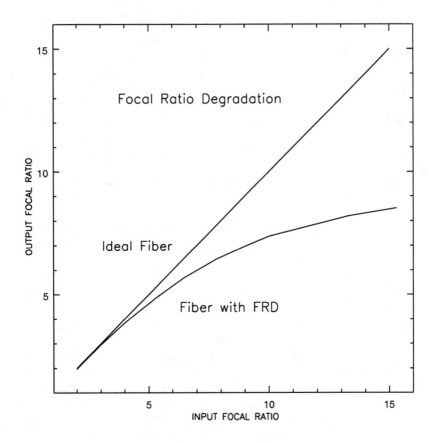

Figure 2. A plot representing the focal-ratio degradation in a fiber.

3. Image Scrambling

Fiber optics tend to scramble the spatial information of the image produced on the input end of the fiber. Spectrographs illuminated by fiber optics show significant stability over spectrographs illuminated only by an aperture or slit since seeing and guiding variations are smoothed out by the fiber optic. Azimuthal image scrambling of the near-field output image is almost always complete. Radial scrambling, however, may not be complete and the near-field image may retain some memory of the radial position of the input image on the fiber. Figure 3 shows that the fraction of light in the output image that retains the radial position of the input increases as the fiber is fed with faster-input focal ratios. As a consequence, for spectrographs in which image stability is of importance, a trade-off must be made between the level of allowable FRD versus the level of image scrambling desired.

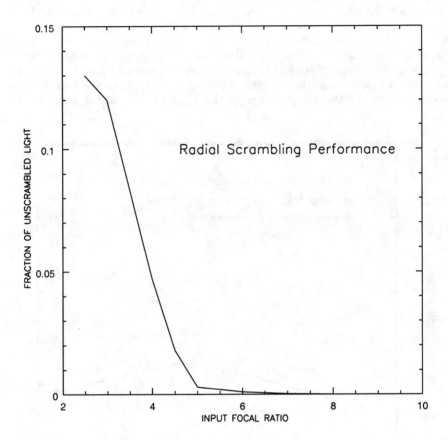

Figure 3. A plot representing the fraction of light on the output end of a fiber which still retains information regarding the radial location of the image on the input end of the fiber.

There are also effects on the far-field image of the fiber which are influenced by the position of the image and fluctuations in the input focal ratio due to variable seeing conditions. The double-fiber scramblers mentioned in some of the articles in these proceedings (e.g. Avila et al.) are designed to increase the level of scrambling and decrease the sensitivity of the spectrograph to such variable conditions.

4. Measured On-Sky Performance of Two Fibers

As part of a feasibility study for fiber coupling the Kitt Peak Coudé Spectrograph, two fiber cables were recently evaluated for their on-sky performance of throughput and focal-ratio degradation. The fiber cables were originally fabricated for the Penn State Fiber Optic Echelle which was in use at Kitt Peak for several years in the late 1980s. Each cable is 22 m in length and contains a high-OH fiber with a polyimide buffer from Polymicro, one with a 100-μm core, the other with a 200-μm core. The cables were fabricated in the mid-1980s.

Only the 200-μm fiber had transmission measurements made by observing a standard star through broad-band filters. Figure 4 shows the transmission measurements for the 200-μm fiber in the U, B, V, R, and I bands. The

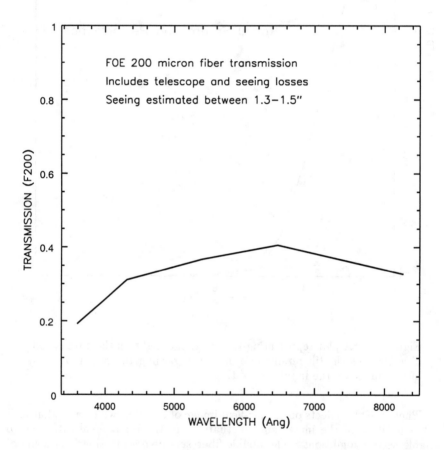

Figure 4. Plot of measured telescope, seeing, aperture, and fiber efficiency for the 200-μm fiber cable in U, B, V, R, and I.

figure includes telescope and seeing/aperture losses. The seeing was estimated to be about 1.3 to 1.5 arcsec and the fiber subtended a diameter of 2.5 arcsec. Simulations of a circular aperture on top of a Moffat profile suggest that the aperture loss was about 40%. Assuming about 15% loss on the telescope mirrors yields a fiber transmission of about 75%.

The FRD for an $f/8$ input for both cables were evaluated by illumination of starlight at the $f/8$ Cassegrain focus of the KPNO 2.1-m telescope. The 200-

μm fiber delivered about 70% of its light into an output focal ratio equivalent to that of the telescope and 95% into an output focal ratio of $f/6.3$. The 100-μm fiber showed considerably higher FRD in which 95% was contained in an $f/5$ output cone. A faint $f/2$ halo was also seen with the 100-μm fiber (not visible in the figure, however). Images of the far-field pattern for both fiber cables are shown in Figure 5. Note that there is a hint of the secondary central shadow in

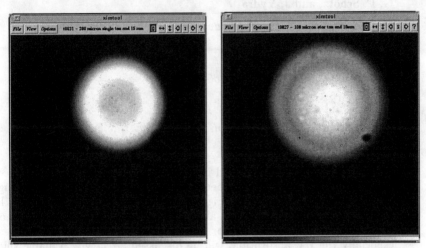

Figure 5. Far-field pattern for a 200- (left) and 100- (right) μm fiber cable illuminated by an $f/8$ input beam.

the 200-μm far-field pattern.

Acknowledgments. I would like to express my thanks to Di Harmer for her assistance in collecting the data to evaluate the FOE fibers.

New Silica Fiber for Broad-Band Spectroscopy

G. F. Schötz, and J. Vydra

Heraeus Quarzglas, 63450 Hanau, Germany

G. Lu

Heraeus Amersil, Duluth GA 30096, USA

D. Fabricant

Harvard-Smithsonian Center for Astrophysics, Cambridge MA 02138, USA

Abstract. Optical fibers used in spectroscopic applications have been limited by the ability of silica fibers to transmit well from the UV through the IR. High-OH fibers have high transmission in the UV but the IR transmission is limited by OH absorption bands. Low-OH fibers show UV absorption. The causes for transmission losses in all silica fibers, especially the UV absorption, will be discussed. A newly available silica glass with good transmission from the UV through the IR for astronomical broad-band spectroscopy applications will be presented.

1. Introduction

Multi-object spectroscopy with optical fibers at large field-of-view telescopes requires high fiber transmission typically in the spectral range between 350 and 1000 nm, and sometimes up to 2000 nm. The most important factor for transmission is the fiber material. Among the different types of fibers, all-silica fibers provide the highest transmission in this range; nevertheless, the transmission is restricted by the OH absorption bands with a fundamental absorption at 2722 nm and the corresponding higher-order bands at 2212, 1383, 1246, 925, and 724 nm (Humbach 1996). In order to avoid the negative influence of these OH bands, fibers with an OH content of only several ppm have to be used. Unfortunately, the UV transmission of such a fiber is lower than in fibers with high OH content.

All optical fibers consist of a core and an outer cladding, where the cladding has a lower index of refraction than the core. This geometry is essential for waveguiding, since the propagating wave is totally reflected at the interface between core and cladding.

All silica fibers are typically produced in a two-step process. In the first step, a preform with a diameter of several centimeters is produced; this is then drawn into a fiber in the second step. Different preform production methods are reviewed by Koel (1983). The plasma outside deposition (POD) process used at Heraeus Quarzglas for the production of preforms for multimode fibers will be explained in the next section. In §3 the factors that affect the optical properties of a multimode fiber including the cladding-to-core diameter ratio (CCDR), the refractive-index difference and the core material are discussed. In §4 a newly available fiber with good transmission from the UV through the IR is discussed, and in §5 recommendations for choosing the appropriate fiber for different spectral regions of interest are given.

2. Production of All-Silica Multimode Fibers

The production of all-silica fibers starts with the manufacturing of preforms which have the same CCDR value and refractive-index step as the final fiber, but an outer diameter of several centimeters.

In principle there are several ways of producing a refractive-index step in an all-silica fiber: 1) doping the core material (e.g. with Ge or P) to increase refractive index relative to undoped silica; 2) doping the cladding (e.g. with F or B) to decrease the refractive index relative to undoped silica; and 3) a combination of 1) and 2).

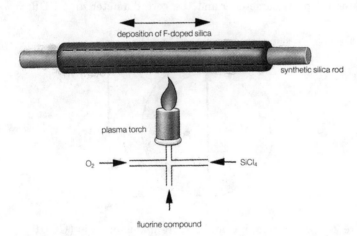

Figure 1. Deposition of a fluorine-doped cladding layer.

Preforms can be fabricated by a variety of techniques (Koel 1983). The most common methods for preform production are modified chemical vapor deposition (MCVD), vapor-phase axial deposition (VAD), and outside vapor deposition (OVD). The method which is used for the production of the Heraeus Fluosil preforms is a modification of the OVD method. The preforms are manufactured by the plasma outside deposition process. Rods of extremely pure synthetic fused silica are coated with fluorine doped silica layers (see Fig. 1) to obtain preforms with step-like refractive-index profiles. Strong thermal gradients combined with the high temperature plasma lead to chemical-deposition conditions, which allow very high fluorine concentrations to be incorporated in the fused silica network. Refractive-index differences of 0.027, corresponding to numerical apertures in excess of 0.28, have been realised.

A Fluosil preform is manufactured as a large ingot which is pulled and cut to yield rods suitable for fiber drawing. A sample fiber is drawn from every manufactured ingot to determine the attenuation by the cut-back method.

3. Optical Properties of Fibers

The fiber characteristics which determine their optical properties, and in particular, their transmission losses will be discussed in this section. For an introduction to fiber optics see also (Nelson 1988).

3.1. Cladding/Core Diameter Ratio

The guided light in an optical fiber is totally reflected at the core/cladding interface. At every reflection a small amount of the light intensity penetrates into the cladding (evanescent field). In the case of a thin cladding, the evanescent field can reach the outer diameter of the fiber, thus leading to outcoupling and hence to attenuation of the guided wave. The cladding thickness relative to the core is given by the cladding-to-core diameter ratio, which is defined as the ratio of the outer preform diameter b and the core diameter a: CCDR = b/a (see Fig. 2.

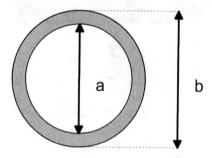

Figure 2. Definition of cladding/core diameter ratio (CCDR).

Figure 3 shows the attenuation spectra of two low-OH fibers (Fluosil SWU), both with a core diameter of 200 µm but with different cladding thicknesses. The CCDR values of the two SWU fibers are 1.1 and 1.2. The attenuation K, in dB is linearly proportional to the fiber length and is commonly used to describe telecommunication fiber. It is defined as:

$$K[\text{dB}] = -10 \cdot \log T,$$

where T is the transmission of the fiber. The transmission of a fiber with an attenuation of 10 dB km^{-1} is 10% after 1 km of fiber length.

The additional losses at higher wavelengths in the SWU1.1 fiber compared to the SWU1.2 fiber are due to the higher guiding losses with the smaller cladding thickness. As a rule of thumb, the cladding thickness should be at least ten times the wavelength of the guided light to avoid significant losses. As an example, for

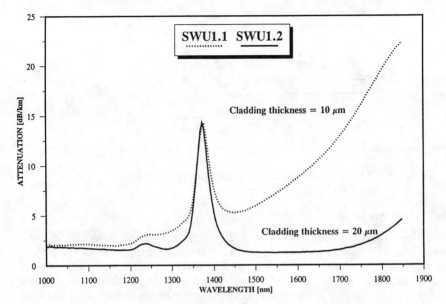

Figure 3. Transmission of fibers with different cladding thicknesses.

the transmission of light with the wavelength of 1 μm in a fiber with 200 μm core diameter the CCDR value should not be less than 1.1. The validity of this rule can be seen in Figure 3. For wavelengths up to around 1 μm the attenuation of both fibers is the same, but for longer wavelengths additional losses are observed in the SWU1.1 fiber, which has a cladding thickness of 10 μm.

3.2. Refractive-Index Difference

The refractive index difference of a step-index fiber defines the numerical aperture, NA, of the fiber. The acceptance-cone half angle, Θ, is defined as:

$$\sin \Theta = NA = \sqrt{n_1^2 - n_2^2},$$

where n_1 and n_2 denote the refractive indices of the core, and the cladding material respectively. A typical value for the refractive-index difference of a Fluosil all-silica fiber is $\Delta n = 0.017$, which corresponds to an NA of 0.22. The value is defined by the fluorine concentration in the cladding. With the POD process, numerical apertures in excess of 0.28 have been realized. In choosing the optimum NA for a certain application, considerations include the geometry of coupling light into and out of the fiber, as well as sensitivity to bend losses. If fibers undergo strong bending, the angle of total reflection can be exceeded by the light traveling in the fiber at the point of the strongest bending, see beam B in Figure 4. In order to minimize these losses a large NA (and/or a large CCDR) is required (Thyagarajan 1987).

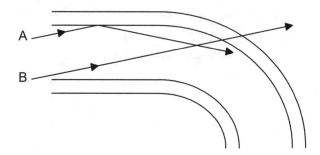

Figure 4. Mechanism of transmission loss in bent fiber.

3.3. Core Material

While the CCDR and refractive-index difference can be adjusted to fit most applications, the most important factor for the optical properties of the fiber is the core, and to some extent also the cladding material, since a small fraction of the light travels in the cladding. The best transmissivity is achieved in all-silica fibers due to the advantage of silica compared to other materials used for fiber production, such as plastic or liquids. In plastic-clad silica (PCS) the influence of the plastic cladding leads to very significant additional losses compared to all-silica fibers. But even in all-silica fibers manufactured with extremely pure synthetic fused-silica core material, the transmission is restricted due to the physical properties of silica, which will be discussed below.

Pure Silica The glass matrix of fused silica consists of an irregular network of SiO_4 tetrahedrons. The attenuation spectrum of pure silica is V-shaped with a minimum at 1550 nm (see dashed line in Fig. 5). The increase in attenuation at small wavelengths is caused by Rayleigh scattering due to density fluctuations in the glass matrix. The dependence of Rayleigh scattering on wavelength, λ, is proportional to λ^{-4}. At wavelengths below 160 nm very strong absorption takes place, which is due to the Urbach tail of the excitonic resonance of the electronic band transition (Griscom 1991). The increasing attenuation with wavelength in the IR is due to the resonance absorption of the SiO_4 tetrahedron vibration.

Intrinsic Defects In addition to the absorption in "perfect" silica, intrinsic defect centers in the glass matrix of "real" silica cause further absorption bands mainly in the UV (Griscom 1991). The structure of the most important defects is listed in Table 1, together with the absorption bands. Depending on the stoichiometry of the silica, oxygen-rich or -deficient material, different types of defects have been observed in silica.

New Silica Fiber for Broad-Band Spectroscopy

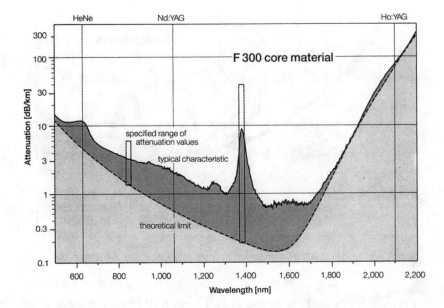

Figure 5. Typical attenuation of a Fluosil SWU fiber (OH content in the core well below 1 ppm). The dashed line gives the theoretical limit of the attenuation in silica due to Rayleigh scattering in the UV and VIS, and the SiO_4 tetrahedron vibration resonance in the IR.

Table 1. Defect absorption in fused silica

Name	Defect center	Absorption band
E'	$O_3 \equiv Si^\bullet$	210/215 nm
Oxygen-rich silica:		
NBOH	$O_3 \equiv Si\text{-}O^\bullet$	630 nm
	$O_3 \equiv Si\text{-}O^-$	265 nm
Peroxy radical	$O_3 \equiv Si\text{-}O\text{-}O^\bullet$	163 nm
Peroxy linkage	$O_3 \equiv Si\text{-}O\text{-}O\text{-}Si \equiv O_3$	330 nm
Oxygen-deficient silica:		
Twofold coordinated silicon	$O\text{-}_\bullet Si^\bullet\text{-}O$	247 nm
Oxygen deficiency	$O_3 \equiv Si\text{-}Si \equiv O_3$	243 nm, 163 nm

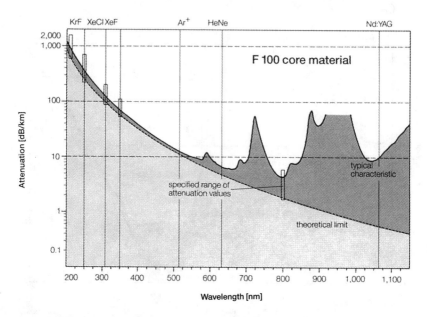

Figure 6. Typical attenuation of a Fluosil SSU fiber (OH content in the core around 700 ppm). The dashed line gives the theoretical limit of the attenuation in silica due to Rayleigh scattering.

Impurities Another reason for absorption bands in silica is the occurrence of impurities.

Metal impurities cause absorption bands, mainly in the UV and visible. The concentration of metal impurities is very low in state-of-the-art synthetic fused silica, so they have only a minor influence on fiber transmission.

The presence of bound hydroxyl (SiOH) in silica produces optical absorption bands due to the OH vibration (Humbach 1996). The fundamental resonance at 2722 nm has a very strong attenuation of 10 dB m^{-1} at an OH concentration of 1 ppm. Overtones of the fundamental mode and combined modes with the SiO$_4$ tetrahedron vibration show smaller but still significant absorption. The next bands in order of decreasing attenuation are located at 2212, 1383, 1246, 943, and 724 nm. These bands can be seen in the attenuation spectrum of a Fluosil SSU fiber (Fig. 6) having a silica core with an OH content of around 700 ppm.

Due to the production process, synthetic silica has a certain chlorine content, which is especially high in low-OH material. Chlorine can be found covalently bound to silicon (Si–Cl) as well as Cl$_2$ molecules in interstices in the glass matrix. According to Khalilov et al. (1994) there is no absorption band corresponding to the Si–Cl defect, but the interstitial Cl$_2$ causes a broad absorption band with a maximum around 320 nm.

4. Low-OH Fiber with Improved UV Transmission

As discussed in the last section, a fiber with high transmission in a broad spectral range (e.g. 350–1000 nm [2000 nm]) must have a low OH content, to avoid the strong vibrational absorption in the OH bands.

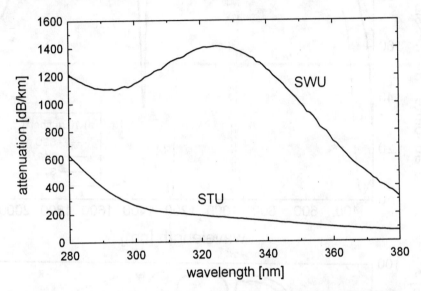

Figure 7. Attenuation of the standard low-OH Fluosil fiber SWU, and the recently developed STU in the region of the Cl_2 absorption band.

Unfortunately, low-OH materials have a much higher concentration of optically active intrinsic-defect centers than high-OH materials; this causes, depending on the type of silica, several absorption bands in the UV (see Table-1). In addition, low-OH materials usually have a higher chlorine content than high-OH materials.

In this paper we present the recently developed low-OH fiber Fluosil STU, which has an extended UV transmission range. This is achieved by using a core material with a reduced interstitial chlorine content avoiding the absorption band at 330 nm (see Fig. 7). The UV optical properties of this new core material do not show temporal changes, and hydrogen doping was not used for preform or fiber production. In Fig. 8 a) and b) the transmission after 10 m and 100 m of the STU fiber is compared with the transmission of standard low-OH and high-OH Fluosil fibers SWU and SSU.

The OH content in the STU preform is between 5 and 20 ppm, which causes higher OH absorption in the fiber compared to SWU, but the attenuation around 350 nm is much lower than in SWU fibers (Fig. 7). If low OH absorption is an important factor for the application, an STU-D preform is available with a similar core material as the STU but with an OH content less than 1 ppm (similar to SWU preforms). For this reason the IR transmission of STU-D is nearly the same as in SWU fibers, but the UV transmission is much better and

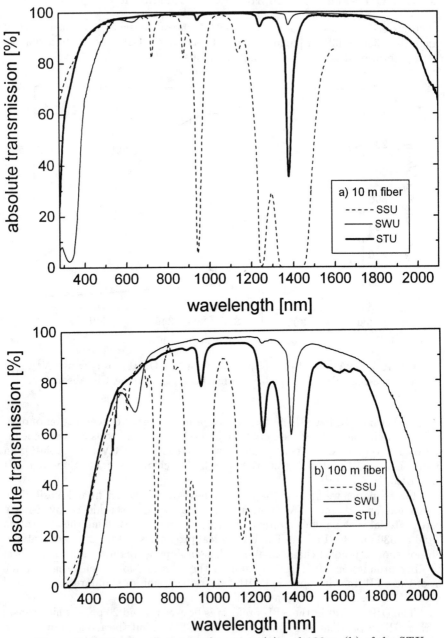

Figure 8. The transmission after 10 m (a) and 100 m (b) of the STU fiber compared with the transmission of standard low-OH fiber (SWU) and high-OH fiber (SSU).

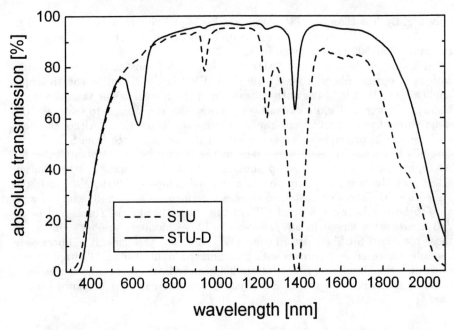

Figure 9. Transmission after 100 m of the STU fiber (5–20 ppm OH) compared with the transmission of STU-D fiber (1 ppm OH).

Table 2. Typical transmission and attenuation values of different fiber types

Wavelength (nm)	STU	STU-D	SWU	SSU
Transmission [%] after 10 m fiber				
300	54	29	6	71
350	79	66	6	85
400	91	87	71	91
450	94	93	81	94
500	97	97	91	97
Transmission [%] after 100 m fiber				
300	0.2	0.0	0.0	3.2
350	10.0	1.6	0.0	20.0
400	39.8	25.1	3.2	39.8
450	56.2	50.1	12.6	56.2
500	70.8	70.8	39.8	70.8
Attenuation in dB km^{-1}				
300	270	540	1200	150
350	100	180	1200	70
400	40	60	150	40
450	25	30	90	25
500	15	15	40	15

only slightly less than in STU fibers (see Fig. 9). The transmission after 10 m and 100 m, and the attenuation of the different fibers at certain wavelengths in the range 300–500 nm are given in Table 2.

Focal-ratio degradation (FRD) is a concern for most astronomical applications of optical fibers (Ramsey 1988). If FRD is significant in spectroscopic applications (that is, if the output beam from the fiber is faster than the input beam), a larger and more expensive spectrograph is required to maintain the same throughput, spectral coverage, and spectral resolution. Although there is no reason in principle to expect inferior FRD performance from STU fibers, careful FRD measurements were made at the Harvard–Smithsonian Center for Astrophysics, using a test setup similar to that shown in fig. 4 of Ramsey (1988). Production 25-m long polyimide-coated samples of 200-, 250-, and 300-μm core STU fibers demonstrated excellent performance when illuminated with an $f/6$ beam. Between 95 and 97% of the transmitted light (measured in an $f/3$ aperture) emerged in an $f/6$ beam. These numbers are very similar to those measured for fibers drawn from SWU and SSU preforms. The fibers were carefully mounted in v-grooves with a minimal amount of silica-filled epoxy and coiled in large diamter loops ($\sim$ 100- to 200-mm radius) for these measurements. Careless mounting and handling can easily degrade FRD preformance for any fiber.

5. Choosing the Optimal Fiber Type

SSU fibers offer excellent transmission, essentially without absorption bands in the wavelength range 190 to 550 nm. At longer wavelengths OH absorption bands (OH content is 600–800 ppm) limit the performance (Figs. 7–8).
SWU fibers contain core material with an OH content significantly below 1 ppm and have excellent transmission properties in the visible, near- and mid-IR spectral ranges, but at wavelengths shorter than 500 nm the transmission is lower than in SSU fibers.
STU fibers have a similar transmission spectrum to SWU fibers in the visible and IR with higher OH absorption bands due to the slightly higher OH content (between 5 and 20 ppm), but the transmission between 280 and 500 nm is much better than in SWU fibers. These fibers cover a very broad spectral range between 300 and 2100 nm, or even higher wavelengths if short fibers are sufficient for the application.
STU-D fibers should be chosen instead of STU if low OH absorption is important for the application. The OH content is less than 1 ppm and is therefore similar to SWU preforms. For this reason the IR transmission of STU-D is very similar to SWU, except the UV transmission which is much better than SWU and only slightly less than in STU.

6. Conclusions

Several factors affecting the optical properties of all-silica fibers have been discussed. Guiding losses in fibers can be minimized by the proper choice of CCDR and numerical aperture for each application. Transmission losses due to absorption in the core material have to be minimized by changing the core material. A new fiber with low OH content has been presented which has an extended transmission region in the UV combined with the excellent transmission of low-OH silica in the visible and IR. The improvement in the wavelength range 280–500 nm has been achieved by using a core material with reduced interstitial Cl_2 content. The UV (down to 280 nm) optical properties of this new core material are absolutely stable over time because hydrogen doping was not used in preform or fiber production.

Acknowledgments. The authors would like to thank K. H. Wörner, and G. Reinel (Heraeus Quarzglas GmbH) for performing the fiber-attenuation measurements.

References

Griscom, D. L. 1991, J. Ceram. Soc. Jap., 99, 923

Humbach, O. 1996, J. Non-Cryst. Solids, 203, 19

Khalilov, V. Kh., Dorfman, G. A., Danilov, E. B., Guskov, M. I. & Ermankov, V. E. 1994, J. Non.-Cryst. Solids, 169, 15

Koel, G. J. 1983, Ann. Télécommun., 38, 36

Nelson, G. 1988, in ASP Conf. Ser. Vol. 3, Fiber Optics in Astronomy, ed. S. C. Barden (San Francisco: ASP), 2

Ramsey, L.W. 1988, in ASP Conf. Ser. Vol. 3, Fiber Optics in Astronomy, ed. S. C. Barden (San Francisco: ASP), 26

Thyagarayan, K. 1987, Opt. Lett., 12, 296

Fiber Optics in Astronomy III
ASP Conference Series, Vol. 152, 1998
S. Arribas, E. Mediavilla, and F. Watson, eds.

Use and Development of Fiber Optics on the VLT

J. Baudrand, I. Guinouard, and L. Jocou

Observatoire de Meudon, 5 Place J. Janssen, 92195 Meudon, France

M. Casse

European Southern Observatory, 85748 Garching, Germany

Abstract. The European Very large Telescope (VLT) is steadily progressing towards completion still foreseen around the turn of the century. The many instruments proposed by the VLT instrumentation plan are also in progress and among them a high-resolution multi-aperture fiber spectrograph (FUEGOS) has entered its final design stage. We report here, more specifically, on the results of the thorough studies that have been engaged in regarding the fiber links for this instrument. The technical solutions that have been ultimately chosen are discussed and justified, and the resulting photometric performances foreseen for FUEGOS are also presented.

1. Introduction

We present here the main results carried out so far in relation to the design and development of the fiber link for FUEGOS, the multi-aperture fiber-fed spectrograph destined to equip one of the Nasmyth foci of the unit telescope 2 of the VLT, the European large telescope currently under construction on Mount Paranal in Chile and still expected for first light around the turn of the century (Felenbok et al. 1994). The FUEGOS instrument is required to work in the 370–900-nm domain under two modes of operation, the integral-field mode (ARGUS mode) for single, large, extended objects and the multi-object mode (MEDUSA mode) for point-like objects distributed across the field of view. In both modes of operation the spectral resolution was demanded at the start of the project to reach at least the 30000 level. These requirements, in conjunction with the fact that we are involved here with an 8-m telescope aperture, were the main drivers for the design of the fiber link for FUEGOS that will be presented in detail here.

2. Presentation and Justification of the Fiber-Link Design

2.1. Individual Fiber Link for the MEDUSA Mode

The individual fiber link for one object in the Medusa mode (IFL) is the basic element of the whole architecture of the fiber network for FUEGOS. It consists of a kevlar armored plastic cable, 7 m long and 1.5 mm in diameter, sheathing a bundle of 7 step-index silica fibers with a 100-μm core diameter. Both extremities are flat polished and equipped with silica microlens arrays bonded on top of the fiber ferrules. (Baudrand et al. 1994). This fiber bundle is terminated at the input extremity by a small magnetic button adapted for operation inside the instrument fiber-positioner system and at the other end by a square slitlet designed to be fitted inside the slit mechanism of the spectrograph. Eighty-six such output slitlets can be stacked side by side to form a pseudo-slit for the instrument (Fig. 1a). This is the multiplex capacity of FUEGOS, a rather moderate figure that comes as the price to pay for the sky image slicing performed to satisfy the spectral- resolution requirement, as explained below.

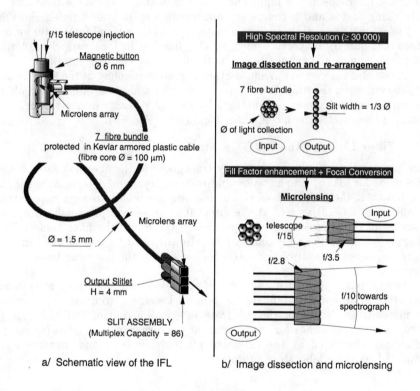

Figure 1. Presentation and justification of the IFL design.

2.2. Justification for the IFL Design

This architecture was essentially justified by the original project requirement to reach a spectral resolution of at least 30000. As is readily expressed by elementary considerations, it is a difficult task to achieve such performance on a very large aperture telescope, and one is usually compelled to resort to some

2.2. Justification for the IFL Design

This architecture was essentially justified by the original project requirement to reach a spectral resolution of at least 30000. As is readily expressed by elementary considerations, it is a difficult task to achieve such performance on a very large aperture telescope, and one is usually compelled to resort to some kind of image dissection and re-arrangement between the sky image and the spectrograph entrance to safeguard the instrument's photometric throughput.

On FUEGOS this was done by way of the fiber image dissector scheme, in this instance a coherent seven-fiber bundle with a round compacted input aperture reshaped in a line at the other extremity and providing a slit width one third of the aperture diameter on the sky. This simple solution was somehow improved by the complement of a honeycomb-type microlens array cemented on top of the fiber tips at the input end, which provides at the same time a near-unity filling factor and a convenient and cheap way of transforming the slow $f/15$ telescope Nasmyth beam into a faster $f/3.5$ (an optimal focal ratio for our chosen 0.22 numerical aperture fiber, which minimizes its focal ratio degradation and avoids any significant scattering at the fiber core-cladding interface).

It was also decided to provide another microlens coupling at the fiber bundle output extremity, this time to reshape the very fast fiber output beam (96% of the output optical power is contained in an $f/2.8$ beam) into a slower $f/10$ matching the spectrograph collimator focal ratio (Fig. 1b).

2.3. Fiber Link for the ARGUS Mode

This basic construction was completely transposed in the fiber link for the ARGUS mode for sake of simplification and reduced costs and delays. At the input end, however, the 602 fibers of the bundle are assembled to form a rectangular aperture (20×13) arcsec^2 at the direct 0.67-arcsec sampling defined by the geometry of the microlens array cemented on the top of the fiber holder. Two other samplings (0.45 and 0.2 arcsec) are also provided by way of two focal enlargers arranged in front of the ARGUS fiber link in the incoming beam of the telescope.

Several prototypes for both modes of operation have been integrated in our laboratory in an effort to validate the proposed concept, to devise and develop the methods and tools for their delicate integration control and of course to make realistic predictions for our instrument performance. The following results are mainly supported by the extensive photometric tests and measurements completed on these laboratory prototypes.

3. Singularities and Advantages of the Design

It is important to emphasize that a fiber link such as the one described above is in many regards very different from a conventional fiber feed (i.e. a single fiber with no ancillary micro-optics) and it can also offer new interesting instrumental capabilities. Let us consider these differences and advantages.

3.1. Image Dissection

The first singularity concerns the potentiality offered by the coherent image dissection performed by the IFL seven-hexagon aperture. Owing to this particularity, it is now possible to determine the real positioning error of the fiber button with regard to its target star by simple analysis of the signal distribution across the global aperture. This information can then be used to the instrument's advantage to correct the position of the fiber buttons on the star plate after field assignation, or to balance the fiber transmission during data reduction to achieve accurate flux calibrations and perform adequate aperture photometry. The image-dissection capacity can also bring into play a new observational mode, the multi integral field mode. This can be done quite readily without any hardware development since the IFLs in the MEDUSA mode can indeed be considered as as many mini Integral Field units. Their corresponding field of view is of course very small, but with a few exposures taken through a step-by-step telescope drift it is possible to access reasonably-sized objects and engage in 2-D spectroscopic surveys in a multiplex mode (Fig. 2).

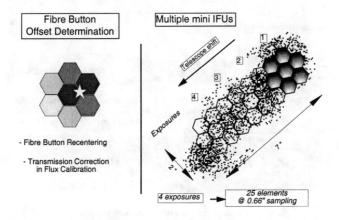

Figure 2. Image-dissection capability.

3.2. Pupil Imaging—Microlensing

A second singularity is introduced by the microlensing of the fiber tips. At the input end the lens working in a Fabry configuration creates an image of the telescope pupil on the fiber entrance face (Hill et al. 1983). This pupil imaging has two consequences. On practical grounds the fiber core can now be used effectively as a pupil stop for the whole system. In fact this is exactly what was done on FUEGOS since the instrument needed to be protected against a direct view of the sky taking place outside the useful $f/15$ Nasmyth beam due to incomplete telescope baffling. This protection was thus easily achieved by the microlens–fiber coupling designed with the appropriate characteristics which safely rejects the polluting light ring outside the fiber core (Fig. 3a).

More fundamentally, it is important to note that the pupil imaging radically transforms the injection conditions into the fibers. Indeed, instead of dealing geometric image. This image is in fact the pseudo-slit for the spectrograph that will be ultimately re-imaged on the CCD at the camera focus to yield a narrow-waist resolution element. As for the fiber near field, this is in turn re-imaged with a huge magnification (a few thousands) inside the spectrograph to form a flat collimated beam with a well defined sharp edge (Fig. 3b).

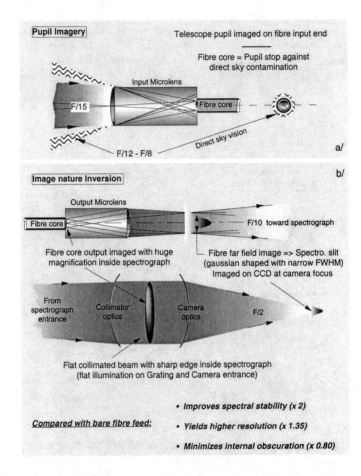

Figure 3. Microlensing interest and consequences.

These photometric features have important instrumental impact. Indeed, as was demonstrated through our comparative measurements, a spectrograph's spectral resolution is improved by about a factor of 1.35 when one is using a microlensed fiber link instead of a conventional one with the same equivalent aperture on the sky. We also showed that owing to the microlensing the level of spectral instability produced by the telescope injection fluctuations is reduced by a factor of 2, most certainly because of the conjugated effects of the flat

These photometric features have important instrumental impact. Indeed, as was demonstrated through our comparative measurements, a spectrograph's spectral resolution is improved by about a factor of 1.35 when one is using a microlensed fiber link instead of a conventional one with the same equivalent aperture on the sky. We also showed that owing to the microlensing the level of spectral instability produced by the telescope injection fluctuations is reduced by a factor of 2, most certainly because of the conjugated effects of the flat and stable illumination provided at the fiber entrance and on the grating. It is also worth noting that the flat distribution of light at the camera entrance can be a desirable characteristic when one is working with a spectrograph using a catadioptric camera with an internal detector (a near mandatory option with large-pupil and fast-output instruments when dioptric in-line solutions become problematic or very costly). With the microlensed fiber feed the photometric internal shadow is strictly contained within its geometric proportion with regard to the spectrograph pupil area. During our comparative measurements we determined that in the case of FUEGOS a conventional fiber injection would have increased the detector central obscuration by 20%.

4. Transmission Efficiency

The first and foremost property of a fiber link for a multi-purpose astronomical spectrograph is its ability to transfer the light collected at the telescope focus toward the entrance of the spectrograph. To assess this characteristic on FUEGOS as a first step we have analyzed every possible source of light loss taking place along the optical chain of an IFL, including at one extremity the right-angle prism in the magnetic button and at the other end the injection into the spectrograph's $f/10$ focal ratio. Then, as a second step, we have undertaken a precise photometric measurement from which we determined for an IFL an overall transmission efficiency of 62%, a value in perfect agreement with the preliminary estimate. This value is for the visible between 550 and 750 nm, while blueward and redward of this band the throughput is slightly degraded in accordance with the fiber's intrinsic spectral transmission. For FUEGOS, after many comparative trials, we have selected the new product proposed by CeramOptec, a very low OH silica fiber with special H treatment. This fiber can indeed be considered as near perfect, and today it is important to report that after completion of thorough ageing tests on such a sample we are totally convinced by the product and its characteristic longevity.

5. IFL Collection Efficiency

In the MEDUSA mode the capacity of a fiber link for one object to collect light from star images at the telescope focus is an other important characteristic. In the case of FUEGOS we have to consider the seven-hexagon aperture formed by the microlens array at the IFL input end. Giving this aperture an equivalent diameter of 2 arcsec on the sky and considering the 0.7-arcsec median seeing of the VLT site at Paranal, we have calculated and then measured for this aperture a collection efficiency of 94%.

It must be noted though that this value is for the ideal case corresponding to a fiber button perfectly centered on its target star. However, it is certainly more correct also to take into account the overall performance of the instrument assignation system, which on FUEGOS is required to be kept within a 0.3 rms arcsec equivalent precision on the sky (atmospheric effects and astrometry included). Considering the statistical distribution of the positioning errors corresponding to this particular performance we determined a mean collection efficiency of 84% for the totality of the fiber buttons. This is of course a more realistic figure than the previous one with regard to the multiplex capacity of FUEGOS and its automated fiber-positioning system. This performance clearly justifies the rather conservative 2-arcsec equivalent aperture adopted for an IFL, a choice considered as the best trade-off between antagonist considerations, throughput efficiency on the one hand, and spectral resolution and sky subtraction on the other.

6. Fiber Photometric Stability

FUEGOS is also required to achieve precise and reliable sky subtraction to better than 1% even for very faint object spectroscopy programs, which of course is non trivial. Although every precaution will be taken and tje best techniques used in approaching that performance (Wyse & Gilmore 1992; Cuby & Mignoli 1994), it was equally important to ascertain that our fiber link could provide the required photometric stability, i.e. a shape of their spectral transmission stable over time to better than at least 0.5%. While the fiber-to-fiber calibration on FUEGOS in envisaged to be taken sequentially with the astronomical exposures and with the fibers kept in their assignation configuration, it is not possible, however, to avoid the movements induced during the exposure time on the sky by the rotation of the fiber star plate attached to the Nasmyth field rotator. Could such motions (mainly a torque in the middle of the fiber link) produce stressful flexures sufficient to introduce small modifications in the fiber photometric throughput and jeopardize the sky-subtraction capability? To answer that crucial question it was decided to transpose into a photometric experimental set-up the very motion that an IFL will see during the largest-amplitude rotation that could occur on FUEGOS during a 1-hr exposure. Ultimately, it was found that no sensitivity changes across the whole 370–900-nm domain could be detected within our experimental ± 0.5% peak-to-peak photometric precision. While it is foreseen to improve that precision and to pursue these tests, this first result is very satisfactory as it already gives a strong indication that fiber photometric instability cannot be a significant source of noise in sky-subtraction procedures.

7. Signal-to-Noise Degradation in Fiber-Fed Spectrographs

We have also been engaged in some investigations regarding small-scale variations observed on high S/N ratio spectra obtained with fiber-fed spectrographs. To describe the phenomenon, let us consider the signal of a continuous spectrum extracted from a CCD frame after standard flatfielding (through the same fiber feed) and summation across the order width. For a spectrum obtained with the fiber kept perfectly still during the experiment, the S/N measured along the spectral dispersion is, as expected, determined by the photon statistic. For the same operation, but with a stress or motion applied on the fiber between the spectrum and the flatfield exposures, one may, in certain circumstances, witness a significant degradation of the photometric performance, yielding an S/N no longer in agreement with the fundamental limit set by the photon noise. After some cautious preliminary tests aimed at eliminating any instrumental artifact cause, we were forced to acknowledge that indeed, in certain conditions, some degradation of the S/N performances may occur in fiber-fed spectrographic data. From there we started a systematic study of these phenomena and today we are able to propose the following explanation.

The origin of the S/N degradation in fiber-fed spectra is directly linked to a noise affecting the optical power transmitted through multimode fibers. This is a modal noise that is present whenever three conditions are fulfilled simultaneously in a given experiment (Rawson et al. 1980):

1) A sufficiently narrow source spectrum

2) Some form of spatial or angular filtering that limits the total number of modes that can be transported by the fiber

3) Some shift or movement of the fiber between consecutive signal recordings

It is clear that such a situation can be met on astronomical fiber spectrographs since:

On every pixel of the CCD at the camera focus (or more precisely across every resolution element) the subtended spectral width may be very narrow indeed, especially when working at high resolution power

Most usually, an angular limitation is proposed for the fiber light injection and some vignetting-obscuration may take place along the optical path in the spectrograph between the fiber output extremity and the CCD detector (this also includes the order width selected for signal extraction)

The fiber link in its observational environment is moving with the telescope and most often is not in the same position during sky and flatfield exposures.

As a consequence, an astronomical fiber-fed spectrum after flatfield correction may be affected along the dispersion direction by this modal noise. According to the modal theory (Daino et al. 1980), this noise is proportional to the recorded signal, S, and its value is given by $S/\sqrt{n}$, where n is the number of modes participating in the propagation of the optical power through the fiber length for a given spectral width.

Therefore, in order to correctly assess the S/N performance of a fiber spectrograph we proposed that this modal noise, $S/\sqrt{n}$, be added in quadrature to the other noise sources, which gives:

$$\frac{S}{N} = \frac{S}{\sqrt{\left(S + RON^2 + \frac{S^2}{n}\right)}},$$

where RON is the read-out noise of the CCD.

From this expression it is interesting to note that at very high signal the S/N tends towards an asymptotic limit, a constant $\sqrt{n}$ which sets the ultimate value attainable in one flatfielded exposure with a fiber-fed spectrograph. The determination of n is of course what matters now but it is clear that the number of modes remains very dependent of the instrumental conditions, as is well indicated by its expression for a given monochromatic wavelength λ:

$$n_\lambda = \frac{V^2}{2},$$

where V is the fiber normalized frequency defined as

$$V = \frac{2\pi}{\lambda} d(N.A),$$

d being the fiber core diameter and $(N.A)$ its numerical aperture at maximum acceptance.

The monochromatic case yields already quite a large number of modes (e.g. about 24000 at 630 nm for a 100-μm core fiber with a standard 0.22 N.A) and, as can be intuitively guessed, the figure will be larger still when one considers the near but not quite monochromatic spectrum across one resolved element. It is beyond our purpose and competence here to attempt such a calculation, and from a practical point of view it is clearly simpler and more relevent to determine the S/N limit experimentally in one's own particular conditions. This is what was done for FUEGOS with a laboratory experiment where one of the typical configurations of the instrument was reproduced on a fiber-fed bench spectrograph, such that:

$R = 19000$	$\lambda_{\text{central}} = 520$ nm
$d = 100$ μm	N.A $= 0.22$
Fiber length $= 10$ m	Injection into the fiber $= f/3.5$

In Fig. 4 we have plotted the experimental points representing the S/N versus signal of a series of flat spectra (Flat 1) divided by another flat spectrum of very high S/N (Flat 2).

The plain curve is obtained with absolutely no perturbation on the fiber position during the exposure sequence, while the dashed curve is related to the same experiment performed when applying to the fiber link the kind of motion and displacement that will occur typically on FUEGOS in the course of one field assignation. We note first that the dashed curve tends effectively towards an asymptotic limit, $\sqrt{n}$, which in these conditions levels off toward 500 (which incidentally indicates that the number of modes participating in the transport of the optical power contained in one resolution element is around 250×10^3). We also verify here that these experimental points can be nicely fitted by the analytical expression proposed above for the S/N of a fiber spectrum, a result that confirms the validity of our hypothesis. Finally it is important to point out that the dashed curve departs significantly from the plain one only toward S/N ≈ 300, which from a practical point of view indicates that these modal perturbations will scarcely be noticed in our future instrument.

As a conclusion of these investigations one can say that it has been possible to explain some unwanted effects related to the modal nature of light transmis-

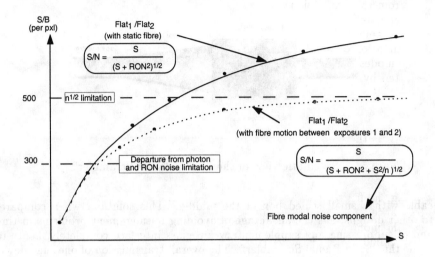

Figure 4. S/N degradation with modal noise in multimode fibers.

sion in multimode fibers. The magnitude of these effects is clearly dependent of the instrumental conditions but one should not be too alarmed since only experiments working at high spectral resolution may be concerned, and only in the presence of very high S/N data, a condition essential for bringing out the very faint modal features described here.

8. Solid Radial Scrambler

To conclude this work we shall present a new development undertaken slightly on the margin of our main FUEGOS activity, and which concerns more specifically the instrumentation dedicated to very precise radial-velocity studies.

It is well known that fiber-fed spectrographs already bring considerable improvement in this field with regard to conventional slit instruments owing to the inherent capacity of multimode fibers to scramble the fluctuations of light injection toward the spectrograph, fluctuations known as slit zonal errors and which are generated on a telescope mainly by seeing variations and guiding instabilities. But we are also aware, as has been stressed by some authors (Heacox 1986), that some memories of these injection fluctuations are still present in the fiber output beam because of the incomplete radial scrambling achieved by these fibers, and the more so when they happen to be of good-quality manufacture. So, when every other sources of spectral instability has been cautiously and expensively eliminated, this may prove to be the ultimate barrier to reaching the detection limit set by the photon noise. To further improve the situation some devices were proposed, all resorting to the same instrumental trick, i.e. exchange of the radial and azimuthal information, or in other words, in the swapping of the near and far fields of the fiber (Hunter & Ramsey 1992). On our side we

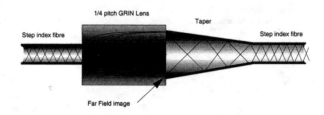

Figure 5. Schematic view of the "Solid Radial Scrambler".

cable with a small sealed box in the middle. This solution, when compared to other designs, has the advantage of avoiding misalignment problems in time and of suppressing any supplementary air–glass interface completely, hence its name the "Solid Radial Scrambler". The overall transmission of one such device prototype was measured on bench to be 65% of the efficiency of the plain fiber with the same core diameter. It has also been determined that the losses were mostly due to the taper, and that some manufacturing improvements should still be obtained here.

It is envisaged to test the "Solid Radial Scramble" in real conditions behind the 1.2-m Swiss telescope installed at the ESO La Silla site. It will feed the CORALIE spectrograph, an instrument specifically designed to access high-precision velocity studies dedicated to extra-solar planetary surveys.

9. Conclusion

In the course of the study phase of FUEGOS, the envisaged multi-aperture fiber spectrograph for the VLT, several prototypes of the conceptual solution chosen for the instrument fiber link have been assembled and thoroughly tested. Through this work it has been shown that the seven-fiber dissector is an elegant and simple way od performing the image slicing and anamorphosis required to reach the high spectral resolution domain. This solution also has the advantage of providing useful spatial information that can be used for fiber-button recentering, aperture photometry and multi integral field operations.

It was also demonstrated that ancillary matching optics at both the input and output ends of the fiber links, while providing a convenient solution for the instrument baffling and focal-ratio adaptations, are also generating favorable consequences such as a noticeable improvement in spectral resolution and spectral stability.

Through dedicated high photometry experiments we were able to prove that the spectral response of our fibers is not sensitive to instrumental changes of configuration. However, special attention should be given to modal features that may affect the S/N performances, especially when working in high spectral resolution mode.

It was also demonstrated that ancillary matching optics at both the input and output ends of the fiber links, while providing a convenient solution for the instrument baffling and focal-ratio adaptations, are also generating favorable consequences such as a noticeable improvement in spectral resolution and spectral stability.

Through dedicated high photometry experiments we were able to prove that the spectral response of our fibers is not sensitive to instrumental changes of configuration. However, special attention should be given to modal features that may affect the S/N performances, especially when working in high spectral resolution mode.

To complete this work a new kind of radial scrambler was also presented. The so called "Solid Radial Scrambler" should prove to be one of the mandatory components to equip a dedicated instrument for high-precision velocity studies.

Acknowledgments. We are very grateful to our colleagues J. Cretenet, A. Petitdemange, and R. Vitry. We know that nothing would have been achieved without their skilful collaboration. The authors also wish to thank Dr. G. Avila from ESO for his support during the scrambler developments.

References

Baudrand, J., Casse, M., Jocou, L., & Lemonnier, J.P. 1994, Proc. SPIE, 2198, 1071

Casse, M. 1995, Doctoral thesis, Université de Paris XI, Orsay

Cuby, J. G., & Mignoli, M. 1994, Proc. SPIE, 2198, 98

Daino, B., De Marchis, G., & Piazzola, S. 1980, Optica Acta, 27, 1151

Felenbok, P., Cuby, J. G., Lemonnier, J. P., & Baudrand, J. 1994, Proc. SPIE, 2198, 115

Heacox, W. D. 1986, AJ, 92, 219

Hill, J. M., Angel, J. R. P., & Richardson, E. H. 1983, Proc. Soc. Photo. Opt. Inst. Eng., 445, 85

Hunter, T., & Ramsey, L. 1992, PASP, 104, 1244

Rawson, E., Goodman, J., & Norton, R. 1980, J. Opt. Soc. Am., 70, No. 8 (August), 968

Wyse, R., & Gilmore, G. 1992, MNRAS, 257, 1

Results on Fiber Characterization at ESO

Gerardo Avila

European Southern Observatory. Karl-Schwarzschild-Str. 2 85748 Garching bei Muenchen Germany email: gavila@eso.org

Abstract. New fibers are being characterized in the ESO Fiber Laboratory to improve the performance of the fiber instruments at La Silla and Paranal Observatories. In particular, measurements of a new hydrogenated fiber optics are presented. They include transmission, focal-ratio degradation and spectrophotometric stability. Finally, a single-lens fiber scrambler design is proposed.

1. Introduction

CeramOptec has launched onto the market a new dry-hydrogenated fiber that in principle combines the features of the *dry* and *wet* fibers. We have measured the internal transmission of these fibers in different diameters as well as their focal-ratio degradation (FRD).

Working at high resolving power, Michael Pfeiffer from the Landsternwarte in Munich has found disturbing results on the spectral stability of fibers. The spectral response does not seem to be constant when the fiber is submitted to mechanical perturbations. Signal variation of up 5 % were reported (Pfeiffer 1994). We have measured the spectral stability of a single fiber when submitted to bendings.

A collaboration program with the Observatoire de Géneve has been undertaken to study the scrambling properties of fibers and develop dedicated fiber scramblers for two spectrographs: ELODIE at the Observatoire de Haute Provence and the Coudé Échelle Spectrograph (CES) linked with a fiber to the Cassegrain focus of the 3.6-m Telescope at La Silla. The results obtained with a *Solid Scrambler* using a graded index lens and a tapered fiber are presented by Baudrand et al. elsewhere in these proceedings. We propose here a scrambler made with a single lens.

2. Transmission

We use the traditional *cut-back* or *two-point technique* to measure the fiber transmission. A flux $P(L)$ at a given wavelength is first obtained on a fiber of length L. Leaving unchanged the launching conditions at the fiber input end, the fiber is cut back to a short distance l and the flux $P(l)$ at the same wavelength is measured. The ratio between these two measurements gives the internal transmission of the fiber corresponding to the length $L - l$.

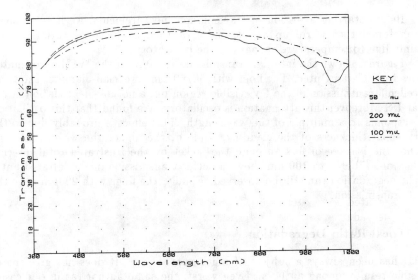

Figure 1. Internal transmission in 10 m of the hydrogenated CeramOptec fibers.

The measurements where performed with our *ESO General Purpose Spectrophotometer*. The instrument was adapted to launch a light $f/5$ beam into the fiber input end. The accuracy of the measurements has been estimated according to the error budget shown in Table 1.

Table 1. Spectrophotometer accuracy error budget

Polishing	< 0.08 %
Fiber-to-detector positioning	< 1.8 %
Spectrophotometer stability	< 0.6 % within 30 min
Fiber bending	< 0.5 %
Total	< 2 %

Most of the errors are related one other and it is very difficult to evaluate them separately. The polishing systematic error, for instance, is closely associated with all the other effects. We polished an output fiber end five times without dismounting it from its support and then we measured the output flux. This procedure reduced the influence of the positioning error in front of the detector.

We found the position error between the fiber and the detector (when the fiber has to be remounted onto the support after cut-off) to be the most critical one. To reduce this sensitivity, we designed a special fiber support to guarantee the positioning reproducibility of the fiber in front of the detector. The variations are due mainly to changes of reflected light from the fiber support generated by positioning inaccuracy of the support in front of the detector. Another important contribution was the rotation error of the fiber inside the connector.

Before starting a session of transmission measurements, we left the spectrophotometer to warm up and stabilize overnight. Residual variations are mainly due to temperature changes in the laboratory.

Figure 1 shows the internal transmission in 10 m of the WF hydrogenated series from CeramOptec. The fiber with 50/60-μm core/clad diameter shows an excellent transmission in the UV–visible region but a notable degradation in the near-IR. Moreover this attenuation is oscillatory. We found that this oscillation corresponds to a multiple of the wavelength. This effect is probably due to the very small thickness of the cladding. Indeed, the higher the wavelength, the higher the leakage of flux through the jacket by the frustrated total internal reflection. Fiber with 100 μm shows a better transmission in the infrared but a slight degradation can still be observed. Finally, the fiber with 200μm gave the best transmission.

3. Focal-Ratio Degradation

FRD has been always a subject of much controversy. It does not give reproducible results among authors. Even worst, the same author (as in our case!) sometimes does not find an acceptable reproducibility in his/her measurements. It is also common to find that FRD laboratory measurements on bare fibers are much more optimistic that the results obtained on the final prepared fiber. In reality, experience shows that FRD measurements are very sensitive to almost everything! The most relevant factors which increase the FRD are: 1) micro + macro bendings, 2) alignment, 3) perpendicularity between the fiber surface and fiber axis, 4) polishing errors, and 5) stability of the light source and movable mechanical parts.

We need to distinguish micro-bendings produced by the protection coating of the fiber (polyimide, acrylate, teflon, etc.) from those generated by the preparation of the fiber inside the cable and finished with connectors. We have compared the FRD in fibers with different coatings (Avila 1988). In particular, we have compared the FRD obtained with polymide and acrylate coatings and have not found any very significant differences. The more relevant microbendings are produced by the gluing of the fiber inside the ferrules or connectors. It is well known, for example, that fast-curing or very hard epoxies create intolerable FRD. Measurements of FRD in the laboratory may also be considerably affected by a strong or inappropriate squeezing of the fiber inside the holder. Another important source of micro-bending is the method of protection the fiber inside a cable. In order to avoid breakage, the fiber should be longer than the kevlar contained in the cable. When the cable is stretched, the stress is absorbed by the kevlar and not by the fiber. Of course, if the fiber is much longer than the cable, the fiber is strongly waved and therefore submitted to strong bendings.

The alignment of the fiber axis with respect to the telescope may generate a high FRD. For instance, if a telescope beam with an aperture $N =$f/# does not arrive perpendicular to the surface of the fiber, but at an angle β, the aperture of the exit beam will increase according to the following equation:

$$N_{\text{out}} = \frac{N_{\text{in}}}{1 + 2\beta N_{\text{in}}}. \tag{1}$$

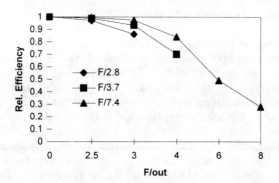

Figure 2. Focal-ratio degradation in CeramOptec fibers.

An $f/5$ beam arriving at only one degree from the axis of the fiber will emerge at $f/4.3$. If a collimator is opened to $f/5$, there will be a flux losses of 28%!

The same reasoning is applied when the fiber ends are not polished perpendicular to the axis of the fiber. This case is even more critical than the previous one. The degradation is given by:

$$N_{\text{out}} = \frac{N_{\text{in}}}{1 + 2n\beta N_{\text{in}}}, \qquad (2)$$

where β is the angle between the normal to the surface and the fiber axis, and n the refraction index of the fiber core.

As a numerical example, let us take an $f/5$ beam falling onto a surface whose normal is inclined by one degree with respect to the axis of the fiber, the output beam will be degraded to $f/4$! We assume that the output surface of the fiber is perpendicular to the fiber axis. In this case, the flux losses will be 35%!

Figure 2 shows the FRD for the CeramOptec fibers totally free of bending and submitted to a 30-mm diameter loop.

Table 2 shows the FRD comparison between CeramOptec and Polymicro fibers for the two specific configurations: FORS ($f/4.5 \rightarrow f/4.6$), and the link between the 3.6-m Telescope and CES ($f/3 \rightarrow f/3$).

Table 2. FRD between CeramOptec and Polymicro fibers

Aperture	WF 100		UV 102	
	Free	Bent	Free	Bent
$f/3 \rightarrow f/3$	85%	85%	94%	94%
$f/4.5 \rightarrow f/4.6$	67	65	83	80

Unfortunately the results obtained with the CeramOptec fiber (WF 100) are not suitable for the FEROS instrument.

4. Spectrophotometric Stability

Spectral variations of the fiber transmission when the fiber is submitted to mechanical stresses have been reported by Pfeiffer (this conference) and Baudrand et al. (these proceedings). This effect is especially important for spectrographs using fibers masked with real slits and working at high resolving powers. The highest variations were found when the fiber output beam was not centered with respect to the slit. A plausible explanation is the variation of the point spread function at the detector. The irradiance distribution of the flux at the fiber end shows radial symmetry. This distribution is very dependent of the stresses applied to the fiber. When the fiber end is not well alienated with respect the slit, the flux distribution on the slit is no longer symmetrical and therefore the irradiance of the image of the slit on the detector with the mechanical stresses on the fiber.

At ESO we wanted to confirm that this effect was negligible for bare fibers without a physical slit in front of them. We used a bench echelle spectrograph able to reach 30000 resolving power with a fiber of 100 μm. An interference filter centered at 620 nm was used to separate the relevant order. First, we took five flat-fields without fiber to produce a *master flat-field* to get the pixel-to-pixel variations. We worked at an S/N of 100. For this, we used a slit with a width slightly bigger that the core diameter of the fiber to be tested. We then took a second batch of flat-fields with the fiber in front of the slit, but *not* obstructed by the slit itself. During the exposures the fiber was not perturbed at all. Finally we took a third batch of five flat-fields but in this case the fiber was submitted to constant bending. After removal of the pixel-to-pixel variations with the master flat-field, we computed the S/N of the spectra of the fiber with and without bending and found a variation of the S/N below 0.5 %.

5. Single-Lens Fiber Scrambler

The optical fibers do not remove all spatial information of the incoming flux distribution at the input of the fiber. This *memory* is mainly kept in a form of radial symmetry. This effect limits the accuracy of the radial velocity measurement in a spectrograph because the point spread function changes with the positioning of the image of the star on the fiber input end. A way of reducing this effect is to cut the fiber in the middle and introduce a relay lens to conjugate the pupil of the output beam into the input of the second fiber. Figure 3a shows the principle of operation with a single lens. The fiber ends must lie in the focal plane of the lens. In the real case of the ELODIE spectrograph at the Observatoire de Haute Provence, a microlens of 0.55 mm focal distance would be needed to conjugate the pupil into the second fiber. The input and output beams are $f/8$ and $f/6$, respectively.

It is now possible to obtain commercially sapphire spherical lenses with focal distances of 0.574 mm ($\phi = 1$ mm). Figure 3b shows the optical design coupling two fibers of 100 μm.

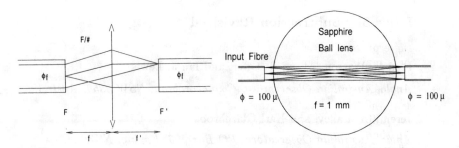

Figure 3. a) Single lens fiber scrambler. b) Scrambler with a sapphire ball lens.

6. Conclusion

The transmission of the hydrogenated fibers from CeramOptec is almost the same as the wet fibers in the blue region and does not show absorption bands in the near IR. However, the whole transmission degrades in the near infrared for small core diameters. The 200-μm fiber showed the best transmission over the whole spectral range (330 to 1000 nm). The FRD measured on bare fibers is not as good as that for the Polymicro fibers and the CeramOptec hydrogenated fibers would not be recommended for applications working with beams slower than $f/4$.

We measured the spectrophotometric stability of a bare fiber at 30000 resolving power and found negligible variations in the S/N (at S/N = 100).

We will test a fiber scrambler prototype made with a single microlens and we expect a coupling efficiency of around 70 %.

Acknowledgments. The author thanks B. Buzzoni, and M. Casse for their help in the transmission measurements; A. Kauffer, who basically made the measurements on the FRD, and finally J. Baudrand for fruitful discussions.

References

Avila, G. 1988, PASP, 3, 63
Pfeiffer, M. 1994, private communication

Fiber Sky-Subtraction Revisited

Fred G. Watson, Alison R. Offer, and Ian J. Lewis

Anglo–Australian Observatory, Coonabarabran, NSW 2357, Australia

Jeremy A. Bailey, and Karl Glazebrook

Anglo–Australian Observatory, PO Box 296, Epping, NSW 2121, Australia

Abstract. Although multi-fiber spectroscopy has been an established technique for a decade and a half now, there are still uncertainties as to the ultimate accuracy with which sky subtraction can be carried out. For objects fainter than a magnitude of about 20, the technique is frequently compared unfavorably with multi-slit spectroscopy. This paper reviews the recent work on fiber sky subtraction and examines some instrumental aspects currently limiting sky subtraction with the 2dF system on the Anglo–Australian Telescope.

For some special purposes, entirely novel approaches to fiber sky-subtraction are possible. One such scheme, using the optical properties of fibers themselves, is also presented.

1. Introduction

Over the fifteen or so years that multi-object spectroscopy with optical fibers has been routinely carried out, a "standard" method of sky subtraction has evolved. Depending on the particular multi-fiber system used, this yields an accuracy of ~ 2 per cent in the sky-subtracted sky. When the immediately adjacent sky is sampled with a near-identical instrumental path, as in a slit or multi-slit system, accuracies of considerably better than 1 per cent are typical. This comparison leads to the commonly held view that spectroscopy with multi-fiber systems is impractical for the faintest classes of object ($B \sim 21$ mag or fainter).

1.1. The Basic Sky-Subtraction Technique

The "standard" procedure involves the use of several fibers dedicated to sky observations (usually 10 to 20 per cent of the total, but see Wyse & Gilmore 1992). Ideally, these are spread uniformly over the fields of both the telescope and the spectrograph (i.e. the detector), but often this condition is not met.

Flat-field observations, either of the sky or a synthetic flat-field, are used to determine the relative transmission of the complete set of fibers. This transmission profile is often (incorrectly) referred to as the "vignetting function" by analogy with long-slit spectroscopy. Crucially, all the fibers are assumed to have identical spectral transmission. The justification for this is that they have nearly always been cut from the same original length of fiber, but it will be shown in this paper that other factors can modify the relative spectral transmission.

The sky correction is applied by taking a mean of the (wavelength-calibrated) sky samples, scaling it by the vignetting function, and subtracting it from the object spectra. Typically, the accuracy of sky subtraction in the continuum is much better than that in atmospheric emission lines. Thus, if observations are being made in the red, the large number of OH lines makes accurate sky subtraction unreliable for faint objects.

1.2. Reasons for the Shortcomings

A number of deficiencies have been cited for indifferent sky subtraction with multi-fiber systems (e.g. Barden et al. 1993). They include:

- spatial variations in the sky background;
- spatial variations in the spectral efficiency of the detector;
- flexure-induced variations in the broad-band and spectral transmission of the fibers;
- field-dependent aberrations in the spectrograph optics;
- flexure in the spectrograph, and
- inadequate sampling of the image profile of the spectrograph by the detector.

Additional limitations might result from the presence of scattered and reflected light in the spectrograph, and fiber-to-fiber cross-talk caused by scattering (or poor charge-transfer efficiency in the detector). The scattered-light problems are likely to depend on the exact design of the spectrograph. For example, an autocollimating Baranne design like WYFFOS (Bingham et al. 1994) will be more susceptible to scattering than a conventional fiber spectrograph such as 2dF (e.g. Gray et al. 1993).

Other problems could arise from non-flatness of the flat-field and, under certain circumstances, temporal variations of the sky background.

2. Improved Methods

A number of astronomers, frustrated by the limitations of the "standard" method, have explored techniques for improving the level of sky subtraction without modifying the basic instrumental set-up. In general they have shown that greater precision than the canonical 2 per cent is possible.

2.1. High-Precision Calibration

Parry & Carrasco (1990) demonstrated by means of computer simulations that the S/N in sky-subtracted spectra is critically dependent not only on the relative instrumental response of sky and object fibers, but also on correct sampling in the wavelength direction at the sub-pixel level. Their results indicated that systematic errors in flat-field and wavelength calibrations of 10 per cent and 0.5 pixels, respectively, will reduce the S/N by a factor of three (for a 22nd magnitude object observed with a 4-m telescope).

This result has self-evident implications for the calibration of fiber spectra.

2.2. Beam-Switching

In conventional slit (and multi-slit) spectroscopy, switching of an object between two apertures on the slit is carried out in such a way that the object spends exactly half its time in each aperture. Assuming the two apertures are sufficiently close together that there are no spatial variations in the sky background signal, the two object/sky pairs thus obtained can be combined to eliminate the sky signal (and/or any difference in instrumental efficiency between the two apertures).

The extension of this technique to multi-fiber systems by configuring the fibers in equally-offset pairs was suggested more than a decade ago (e.g. Watson, 1987), but has had limited application in practice. It is not simply that astronomers are reluctant to devote half the available fibers to sky. Often the disposition of fibers in the field makes the positioning of fibers in pairs difficult, while some fields are so crowded that it is impossible to avoid contamination by stars.

The technique has, however, been used with success by a few authors. Barden et al. (1993), using a method developed by Elston, eliminated the sky completely in the continuum, and reduced the residuals in bright atmospheric lines to under 1 per cent for galaxies with $R \sim 21$, using the Mayall Telescope. (An earlier, widely-cited report by these authors [Elston & Barden 1989] described a sky-chopping technique that used offset sky frames obtained with identical exposure times to the object frames. Thus, the sky sample for each object spectrum was obtained through the same fiber as the object. This, too, produced significant gains over the standard method. There is potential for further improvement of this technique by introducing charge-shuffling on the CCD chip, as proposed for multi-slit observation by Glazebrook et al. [1998]. Here, the sky frames are effectively obtained concurrently with the object frames rather than sequentially.)

Cuby & Mignoli (1994) tested a number of sky-subtraction methods, including classical beam-switching. They found that for objects with $B \sim 22$ observed with a 4-m class telescope, there is little to choose between the standard technique carefully applied (which they call the "mean-sky" method) and beam-switching. In evaluating the various methods, the authors took the pragmatic view that the effectiveness with which redshifts can be determined (i.e. S/N in the continuum) is paramount, rather than the effectiveness with which atmospheric lines can be subtracted. They also commented that particular care was taken with scattered light.

2.3. Modeling the Instrumental Background

In a seminal paper on sky subtraction with fibers, Wyse & Gilmore (1992) identified the need to account for the "instrumental sky" as well as the real night sky. This consists of scattered light within the spectrograph itelf (in addition to dark and bias backgrounds).

Two components of the instrumental sky are identified. Localized scattered light is the contamination of a spectrum by light from an adjacent or nearby spectrum. Uniform scattered light, on the other hand, is locally proportional to the total incident light. The constant of proportionality may vary with instrumental parameters such as grating angle.

Wyse and Gilmore stress the need to make scattered-light corrections to the flat-field frames, as well as the object/sky frames. Modeling of the localized scattered light is carried out using bright sources sparsely placed in the spectrograph field, while the level of uniform scattered light comes from unilluminated areas of the detector (such as the gaps corresponding to broken fibers).

In evaluating the success of a sky-subtraction process, it is not enough to determine the residuals in the sky-subtracted sky spectra. Wyse and Gilmore comment that these residuals do not necessarily measure the reliability of the sky subtraction. Only when the residuals from unilluminated areas of the detector are also small is the scattered light being correctly accounted for. The authors give an example of 0.1 per cent mean residuals (standard deviation ~ 3 per cent) in the sky-subtracted sky using their modification of the standard method.

2.4. Using Night-Sky Lines

When the spectral region of an observation includes a night-sky emission line (in particular the [O I] line at 5577.4 Å), the possibility arises of using it to define the relative transmission of each fiber. This method has been explored by Lissandrini et al. (1994). By replacing a standard flat-field with a flat-field derived from Gaussian-fits to the [O I] line (in the wavelength direction), these authors improved the rms sky-subtraction accuracy in low-dispersion (9Å per pixel) 3.6-m telescope spectra from ~ 4 per cent to ~ 1.4 per cent. (Account was also taken of both instrumental background effects and wavelength calibration effects due to distortion in the spectrograph.)

There appear to be three drawbacks to this technique. One is that the spectrograph image profile needs to be over-sampled for accurate Gaussian fitting. The spectra used by Lissandrini et al. were 7 pixels wide. The second is that, as the authors themselves point out, the [O I] line-intensities reflect not only the relative transmission of the fibers, but also spatial variations in the sky. There is no reason to think that spatial variations in the [O I] line intensity will be correlated with spatial variations in the continuum background. Finally, the technique is not applicable blueward of the 5577-Å line.

3. Sky-Subtraction with 2dF

We are currently investigating ways of improving the sky subtraction routinely obtained with 2dF on the Anglo–Australian Telescope. With its double-buffered, 400-fiber capability, 2dF will be the AAT's flagship instrument for the next five to ten years (see Lewis et al., these proceedings). In order for it to remain competitive in the 8-m telescope era, 2dF must deliver spectra of the highest possible quality, and sky subtraction is an important element of this.

3.1. Sky-Subtraction Procedure

To handle the 200-fiber frames that are read out from each of the two spectrographs at the end of an exposure, a fully automatic data-reduction system, 2dfdr, has been written by Jeremy Bailey (Bailey & Glazebrook 1998). Along with the target-object frames, 2dfdr requires (a) a fiber flat-field exposure (which need not be photometrically flat), (b) at least one arc exposure, and (c) at least one offset sky exposure (but preferably three or more at different positions, to allow for starlight accidentally entering the fibers).

The (curved) spectra are extracted by means of a tram-line map, which is generated from a model of the spectrograph's optical imaging. The offset and rotation required to fit this to the real data are determined from the fiber flat-field. The extracted spectra are then wavelength calibrated in the usual way.

The starting point for the sky subtraction is the 200 extracted, wavelength-calibrated spectra from each frame. Each spectrum is multiplied by an associated scalar throughput factor (determined from the median of the offset-sky exposures—the "vignetting function" of the standard method). Spectra from the sky fibers in the field are then combined into a median sky, which is subtracted from each of the 200 spectra. (2dfdr also has an option to determine the relative fiber transmissions using night-sky lines (see Section 2.4). It does not appear to work as well as the offset-sky method, possibly because the latter gives a better approximation to the object spectra in terms of instrumental scattered light.)

3.2. Typical Results

The general appearance of a typical sky-subtracted sky frame from 2dfdr is that residual sky is not apparent except in the strong night-sky lines. Here, residuals of up to 10 per cent of the line peak intensity are seen.

In Table 1, mean residual sky values are given together with standard deviations for four regions of a sky-subtracted sky frame. The shorter-wavelength region is dominated by continuum, while the longer is dominated by night-sky lines. The division into low and high fiber numbers was made because most of the 12 sky spectra in this frame are in the first two dozen or so spectra; the effects of this are obvious in the mean residuals. The overall sky-subtraction accuracy is consistently 3–4 per cent.

Table 1. Residual sky in a sky-subtracted sky frame from 2dfdr

	4800–5400 Å	7600–8000 Å
Low fiber numbers	$M = 0.6\%, \sigma = 3.7\%$	$M = 0.9\%, \sigma = 4.2\%$
High fiber numbers	$M = 3.2\%, \sigma = 2.8\%$	$M = -2.8\%, \sigma = 3.0\%$

The concentration of sky fibers in one area of the fiber slit in this frame is a reflection of the concentration of sky fibers in one area of the field (since 2dF fibers are ordered around the field and along the slit similarly). This is probably a consequence of the distribution of target objects in the field, the configuration software allocating sky fibers where there is the lowest concentration of targets.

3.3. Strategies for Improvement

The imperfect sky subtraction results from a number of effects, many of which were considered in the investigations described in the previous section. Methods of dealing with the most important of these are being investigated in our program.

Spectrum Extraction The current extraction of spectra from the data frame is by means of a top-hat function. This is non-photometric in the sense that not all the dispersed light from a given fiber is included in the spectrum. The faint wings of the spectrum profile are lost. A further complication arises because of the narrow inter-spectral region, where the wings of adjacent spectra overlap.

This imperfect extraction could give better sky subtraction if the sky-chopping technique of Elston & Barden (1989) were used, when light from each object and its sky sample would follow exactly the same path in the spectrograph. However, the full-length offset-sky exposures required are considered unacceptable in the large-scale survey work that is 2dF's main rôle.

The current approach at AAO is to attempt to define a spectrum profile that will better model the energy distribution in the spectra. Test data have been obtained using $\sim$ 20 fibers arranged on the field plate in a "sky cross", and sparsely distributed on the detector to allow the investigation of both the spectrum profile and the instrumental scattered light. A double Gaussian is one of the model profiles that have been tested, but the results cannot be applied satisfactorily to the real-life situation with 200 fibers because of the overlap problem. One of the difficulties with such profile fitting is that the projected fiber diameter in the spatial direction is only 2.8 pixels.

In practice, the full 200-fiber flat field is best modeled with a single Gaussian whose width is constrained to vary slowly across the field. This still has problems of slow fitting speed and uncertainties in the blue where the flat-field lamps have low output, but there is scope for progress on both these issues.

Wavelength Calibration It has been noted by other investigators that the wavelength calibration must be essentially perfect to allow the standard method to work properly. In the case of 2dF, the rebinning of the data onto a uniform wavelength scale is again compromised by rather poor sampling. At some grating angles, the fibers project to less than two pixels in the dispersion direction. Nevertheless, 2dfdr is still capable of centroiding to $\sim$0.1 pixel.

Another requirement for good wavelength calibration is uniformity of the spectrograph PSF over its entire field. Subtraction of the night-sky lines is particularly sensitive to PSF variations. Except where anomalous curvature of the detector itself is responsible, improvements can be brought about by optical adjustment of the spectrograph. Very considerable progress has been made in this area since the early days of 2dF.

Fiber-to-fiber Color Variation Like all versions of the standard method, 2dfdr assumes the fibers have identical spectral transmission. When unmodified lengths of identical fiber are used, as in a plug-plate system for example, this is probably true to a very high degree of precision. All AUTOFIB-type systems, however, have a prism cemented to the input face of each fiber, and it is possible for this to introduce variations in the fiber-to-fiber spectral transmission.

In flat-field observations normalized to the mean intensity of all the fibers, we have found fringes of equal chromatic order in a few fibers. These fringes fsattisfy the condition

$$2d = m\lambda,$$

where $d = nt$ is the optical path length in a medium of refractive index, n, and thickness, t, m is an integer, and λ is the wavelength of an intensity peak.

The most likely source of these fringes is an air-gap in the cement holding the prism onto the fiber end. The fringes have an amplitude of up to ~ 10 per cent in their lowest orders, and a range of spacings depending on the thickness of the air-gap. Two well-defined examples had d-values of 1.1 and 3.9 μm, which are consistent with expectations of the thickness of the cement interlayer.

These fringe patterns are very stable, and could be flat-fielded out. However, it would clearly compromise sky subtraction if one or more affected fibers were used to obtain the sky samples in data frames. The best remedy is re-cementing of the prisms, and this is being carried out.

4. An Optical Approach to Improved Sky Subtraction

With increasing interest in the far-red and near-infrared (J, H) spectral regions, and with the development of fiber-coupled instruments for 8-m class telescopes, the need for more exact fiber sky-subtraction techniques is greater than ever. In the far red and beyond, night-sky lines are highly variable, both spatially and with time, and derivatives of the standard method for multi-fiber spectroscopy are inadequate.

The detectors used in the near infrared differ from CCDs in that non-destructive readout can be used. They lend themselves to the development of novel digital methods of sky subtraction analogous to the charge-shuffling technique mentioned earlier in this paper. In the optical, the trend towards near diffraction-limited imaging suggests that mini-IFUs (integral-field units) will increasingly play a rôle in multi-object spectroscopy. This, too, will allow new approaches to sky subtraction to be adopted.

There is another possibility for sky subtraction with large telescopes in the optical and near infrared, and that is to mimic slit spectroscopy by using one fiber per object to transfer both the object signal and sky samples. As noted earlier, the great strength of slit (or multi-slit) spectroscopy for faint objects is that the adjacency of the object and sky samples is preserved both on the sky and at the detector. Also, the instrumental path to the detector for both signals is very nearly identical. Both these aspects can be simulated with modern, high-quality fibers.

4.1. Pseudo-Slit Spectroscopy using Single Fibers

The concept of using a microlens in front of the input end of a fiber to image the pupil of the telescope onto the fiber input face is well-known. In this set-up, the microlens is not a field lens: it is positioned so that the image surface of the telescope is in its first focal plane, and the fiber input face is in its second. The arrangement was first suggested by Hill et al. (1984), who also pointed out that the solid angle of the beam entering the fiber from the object (here assumed to be a point source) is proportional to the diameter of the seeing disk.

In a fiber with perfect focal-ratio degradation (FRD) characteristics, this solid angle would be preserved along the length of the fiber. With an identical pupil-imaging microlens at the other end (which images the fiber output face onto the collimator pupil of the spectrograph), the seeing disk would be reconstructed in the slit-plane of the spectrograph so that the spectral resolution would be determined by the seeing.

In fact, under these circumstances, *everything* within the field of the input-end microlens (defined by the lens focal length and the fiber diameter) would be photometrically re-imaged at the output end, but with perfect circular (azimuthal) scrambling. Thus, a point source centered on the axis of the fiber would be transferred to the focus of the collimator on the fiber axis at the output end. Likewise, the sky surrounding the object would reappear in the collimator focus.

In this simple situation (centered point object and surrounding featureless sky), the azimuthal scrambling is of no consequence, and the fiber is effectively transferring a small portion of the focal surface of the telescope to the spectrograph. By positioning a slitlet in the collimator focus, spectra of the object and samples of the adjacent sky can be formed on the detector with sufficient photometric accuracy to yield very high-quality sky subtraction.

4.2. Practical Realization of the Technique

For such a scheme to work, fibers with very high modal stability (i.e. ultra-low FRD) are required. In such fibers, the propagation modes carrying light from the object plus sky, and light from the sky alone will remain well separated. Clearly, fibers with perfect FRD characteristics do not exist, but the very best fibers come close to it (see, e.g. Worswick et al. 1994; Watson & Terry 1995).

The advantages of the set-up compared with the usual multi-fiber arrangement are:

1. for compact objects or point sources, it provides a sky sample that is a circularly-averaged mean of the sky immediately surrounding the object, and images it adjacent to the object on the detector;

2. only one fiber per object is required, and no additional sky fibers are needed;

3. the object and sky signals go down the same fiber (however, because they do not propagate at the same focal ratio, the light-paths they follow are not strictly identical); and

4. the full numerical aperture of the fiber is utilized.

There are, however, several drawbacks that could limit the usefulness of the scheme:

1. the telescope must be telecentric;

2. the effect of FRD (which will be most acute in the low-order modes carrying the object signal) is to degrade spectral resolution or, by virtue of the output-end slitlet, reduce throughput;

3. any centering errors in the positioning of the fibers or acquisition of the field will have the same effect;

4. any faint background object within the field of the input-end microlens will become azimuthally scrambled and contaminate the sky sample; and

5. to avoid underfilling the detector, the output ends of the fibers must be staggered along the slit.

In this method, there is a trade-off between the number of object/sky pairs that can be fitted onto the detector and the S/N in the sky spectrum associated with each object. The spatial extent of the sky sample obtained with each object (and hence the sky S/N) is determined by the field of the input- and output-end microlenses and the geometry of the slitlet. Sampling the sky out to a relatively large radius from the object is costly in terms of the number of object/sky pairs that can be imaged onto the detector. Conversely, that number can be maximized by limiting each sky sample to be the same width (in pixels) as the spectrum of the object. Here, the S/N in the sky spectrum is less than in the mean sky of the standard method, but it is of higher quality, being local to the object.

Further work on this method of sky subtraction awaits a demonstration (probably with a single fiber) that it is capable of providing the required photometric accuracy with real fibers.

5. Conclusions

Despite its coming of age, sky subtraction with multi-fiber spectroscopy systems remains a black art that produces varied results. It is clearly possible to carry out sky subtraction of very high quality for faint objects, but some of the most effective methods (beam-switching, and sky-chopping) are costly in terms of fiber numbers or exposure time. In general, there remain questions about the consistency with which high-quality sky subtraction can be done.

This is particularly relevant in the case of 2dF, which is now producing huge quantities of data that must be quickly and reliably reduced with the minimum of human intervention. The work described in §3 of this paper is directed towards that goal, and is expected to bring about improvements in the sky-subtraction accuracy routinely achieved.

Finally, the new generation of multi-fiber instruments now being planned for 8-m class telescopes provides opportunities for instrument designers to consider novel methods for accurate sky subtraction, such as the one presented in this paper.

Acknowledgments. This paper owes much to discussions with Ian Parry (IoA), Keith Taylor (AAO), Peter Browne (Macquarie University), and Quentin Parker (AAO), and these colleagues are warmly thanked. We are grateful to the co-editors of these proceedings for their forbearance concerning the late submission of the paper.

References

Bailey, J., & Glazebrook, K, 1998. 2dF User Manual, Anglo-Australian Observatory

Barden, S. C., Elston, R., Armandroff, T. & Pryor, C. P. 1993, in ASP Conf. Ser. Vol. 37, Fiber Optics in Astronomy II, ed. P. M. Gray (San Francisco: ASP), 223.

Bingham, R. G., Gellatly, D. W., Jenkins, C. R. & Worswick, S. P. 1994, Proc. SPIE, 2198, 56

Cuby, J.-G., & Mignoli, M. 1994, Proc. SPIE, 2198, 98

Elston, R. & Barden, S. 1989, NOAO Newsletter, 19, 21

Glazebrook, K., Bland-Hawthorn, J., Taylor, K., Farrell, T. J., Waller, L. G., & Lankshear, A. 1998. AAO Newsletter, 84, 9

Gray, P., Taylor, K., Parry, I., Lewis, I., & Sharples, R. 1993, in ASP Conf. Ser. Vol. 37, Fiber Optics in Astronomy II, ed. P. M. Gray (San Francisco: ASP), 145

Hill, J. M., Angel, J. R. P., & Richardson, E. H. 1984, Proc. SPIE, 445, 85

Lissandrini, C., Cristiani, S., & La Franca, F. 1994. PASP, 106, 1157

Parry, I. R., & Carrasco, E. 1990, Proc. SPIE, 1235, 702

Watson, F. G. 1987. PhD Thesis, University of Edinburgh, Chapter 5

Watson, F. G., & Terry, P. 1995, Proc. SPIE, 2476, 10

Worswick, S. P., Gellatly, D. W., Ferneyhough, N. K., Terry, P., Weise, A. J., Bingham, R. G., Jenkins, C. R., & Watson, F. G. 1994, Proc. SPIE, 2198, 44

Wyse, R. F. G., & Gilmore, G. 1992, MNRAS, 257, 1

Review of Fiber-Optic Instrumentation at NOAO

Samuel C. Barden

National Optical Astronomy Observatories, PO Box 26732, Tucson, AZ 85726-6732, USA

Tom E. Ingerson

Cerro Tololo InterAmerican Observatory, La Serena, Chile

Abstract. We give an overview of the current instruments in use and in development at NOAO which utilize fiber optics to transport the light from the telescope to the spectrograph. Two new instruments include a Hydra multi-fiber instrument for the Blanco 4 meter at CTIO and a simple 90-fiber IFU for the WIYN telescope at KPNO.

1. Multi-Object Spectrographs

1.1. ARGUS

ARGUS is the current fiber optic, multi-object spectrograph on the Blanco 4-meter telescope at CTIO (Ingerson 1993). It is based upon the MX style of fiber positioner (Hill et al. 1982) with 24 fiber positioners, each with an object and sky fiber, surrounding a 50-arcmin field of view. Each fiber subtends 1.8 arcsec on the sky. Reconfiguration of the target field takes approximately 90 s.

The 32-m long fibers, which can be disconnected from the telescope via a low-loss multi-fiber connector, feed a bench spectrograph for spectral observations in the 370- to 1000-nm spectral window.

ARGUS was operational in early 1989 and has been used about 15% of the time. This useful instrument will be decommissioned in 1998 as it will be replaced by the Hydra/CTIO fiber positioner. The instrument will be moved to the El Leoncito 2.4-m telescope in San Juan, Argentina.

1.2. Hydra

Hydra/WIYN Hydra/WIYN is currently in use at the WIYN telescope at KPNO (Barden et al. 1994). This instrument is based upon the Autofib style (Parry & Gray 1986) where the fibers are each mounted onto a magnetic button and are positioned by a single robotic positioner, one at a time. It is permanently installed at a Nasmyth focus and can be used at any time. Hydra utilizes a three-fingered gripper riding on an H-gantry X, Y, and Z stage. The fibers are placed onto a flat focal plate which, after all fibers have been positioned, is warped to a spherical shape by drawing a vacuum on the plate's backside and deflecting it against a spherical backstop. The instrument can accommodate up to 288 fibers mounted around a 1-degree field of view. At present the fiber positions are populated with 12 seven-fiber bundles used for acquisition, focusing, and guiding; 98 fibers with 2-arcsec apertures and a 450–1800-nm spectral window; and 96 fibers with 3-arcsec apertures and a 300–700-nm spectral window. Reconfiguration time is an average of 13 s per fiber, or typically about 25 min per target field.

The 25-m long fiber cables feed a bench spectrograph which has a variable camera–collimator angle (11 to 45 deg) in order to accommodate both low-order and echelle gratings. There is currently a choice of six gratings giving a range in resolving power from 1000 to 40000. Two cameras are available, a red camera which is all transmissive and a blue camera which is a Mangin-Mirror Maksutov. The blue camera has a significant central obstruction, but provides better efficiency than the red camera blueward of about 380 nm. The spectra are imaged onto a SITe 2048, 24-μm pixel CCD.

Hydra was initially fabricated for use at the Cassegrain focus on the KPNO Mayall 4-m telescope and was used on it about 20 to 30% of the time during the period from 1992 to 1995. The instrument was decommissioned from the 4-m, upgraded, and moved to the 3.5-m WIYN telescope at KPNO. Since 1996, the instrument has been used 50% of the time by both NOAO and the university partners at WIYN.

It is likely that Hydra/WIYN will be upgraded with motors and controllers similar to those used in Hydra/CTIO in order to have commonality in software and electronics between the two instruments and to reduce the configuration overhead in Hydra/WIYN by about a factor of two. Further reduction in overheads can be attained by replacing the X and Y stages, but that would require a relatively significant modification to the current instrument while the motors and controllers are effectively drop-in replacements.

Hydra/CTIO A second Hydra fiber positioner is under construction for implementation on the Cassegrain focus of the Blanco 4-m telescope. In general, the instrument is basically the same as that at WIYN, but has been enhanced in some aspects. It has a three-fingered gripper riding on an X, Y, and Z gantry to position the fibers one by one. Circling the 40-arcmin field-of-view focal plate are 288 fibers, of which 12 are acquisition and guide probes, 138 are 2-arcsec fibers, and 138 are 1.3-arcsec diameter fibers. The focal plate is warpable to align the fibers onto the curved focal surface and to tilt them into close alignment with the telescope exit pupil.

A number of upgrades in the instrument were driven somewhat by the fact that identical components on Hydra/WIYN are now obsolete and cannot be purchased off the shelf. This forced the project to purchase new state-of-the-art motors, controllers, and linear stages. Being more powerful, more compact, and more programmable, the instrument will be able to minimize the fiber-positioning overhead from 13 s per fiber down to about 3 s per fiber or better. Another difference between this instrument and that at WIYN is the inclusion of a multi-fiber connector to allow easy removal of the instrument from the Blanco telescope. This connector is built upon the experience at CTIO with the connector on the fiber cables for ARGUS. In addition, the fiber type for Hydra/CTIO will be the new STU-1.1 fiber which covers a 325–1800-nm spectral window (Lu et al., these proceedings).

The 30-m long fibers will feed a new spectrograph that will replace the ARGUS spectrograph. The spectrograph will have similar characteristics to the Bench Spectrograph at WIYN, but will utilize Schmidt collimators and cameras.

Hydra/CTIO is currently scheduled to be commissioned at CTIO between October and December of 1998 and should be available for shared-risk observations starting in early 1999.

2. Integral-Field Units

2.1. WIYN DensePak

A 7 by 13 fiber array was constructed for use at the WIYN telescope from the red fiber cable left over from the move of Hydra from the Mayall to the WIYN telescope (Barden et al. 1998). The lensless array provides 3-arcsec apertures with a center-to-center spacing of about 4 arcsec. It also has 4 fibers dedicated to sky, each about 1 arcmin from the center of the array. The fibers cover a spectral window from 370 to 1100 nm.

The array mounts onto the WIYN Indiana Fiber Optic Echelle telescope interface which replaces the WIYN imager when in use. The 25-m long fiber cable feeds to the Hydra Bench Spectrograph.

The array was fabricated at the end of 1996 in time for observations of Comet Hale-Bopp. It has been utilized by the WIYN university partners about 10% of the time for a variety of programs in addition to the comet studies. The array is also now available to the NOAO community.

This array primarily serves as a prototype for future IFU efforts on the WIYN telescope.

2.2. CTIO IFU

Plans are underway to fabricate a lensless, 37-fiber IFU with 1.3-arcsec apertures and 11-arcsec diameter sky coverage for use on the Blanco telescope at CTIO. Six additional fibers will monitor the sky. The fibers will be made from the STU-1.1 fiber material and will feed the Hydra/CTIO spectrograph.

An option is also underway to include a microlens array which will convert the sampling to 0.5-arcsec apertures with a 3.5-arcsec diameter total coverage on the sky.

This device is planned for implementation in late 1998 or 1999 and will serve as a prototype for future IFU development and implementation at CTIO.

3. Single-Object Spectrographs

3.1. CTIO Bench-Mounted Echelle

The CTIO Bench Mounted Echelle (BME) is a cross-dispersed, fiber-fed echelle spectrograph for single object use on the CTIO 1.5-m telescope. Two fiber diameters are available, 1.8 arcsec and 3.6 arcsec. The 16-m long fibers transmit light in the 320- to 880-nm spectral window. Resolving powers of from 15K up to 100K are possible with a variable entrance slit mounted directly behind the fiber input on the spectrograph.

This instrument has been used about 10% of the time since it was installed in the mid-1980s.

3.2. KPNO Coudé Spectrograph

There is a feasibility study under way to move the 2.1-m coudé spectrograph at KPNO over to the Mayall 4-meter Telescope where it will be fiber coupled. This will allow the closure of the coudé feed telescope while providing some enhancement in scientific capability for high-resolution spectroscopy at KPNO.

The fibers are intended to be installed on the telescope at the position of the #4 coudé mirror, making a quasi-Nasmyth focus. This would allow usage of the spectrograph when other instruments are mounted at either the Cassegrain or prime-focus locations.

The intent is to preserve the current resolving power of 250K with an echelle grating while maintaining relatively high throughput with a large core fiber (subtending between 0.8 to 1.0 arcsec on the sky). Image slicing is a must and produces a challenge in efficient implementation (about eight slices will be required). The ability to cross-disperse the spectra also becomes a design challenge due to the presence of the fairly large number of image slices.

If approved, it is hoped that this project will be implemented in 1999 or soon after.

References

Barden, S. C., Armandroff, T., Muller, G., Rudeen, A. C., Lewis, J., & Groves, L. 1994, Proc. SPIE, 2198, 87

Barden, S. C., Honeycutt, R. K., & Sawyer, D. G. 1998, Proc. SPIE, 3355, in press

Hill, J. M., Angel, J. R. P., Scott, J. S., Lindley, D., & Hintzen, P. 1982, Proc. SPIE, 331, 279

Ingerson, T. E. 1993, in ASP Conference Series, Vol. 37, Fiber Optics in Astronomy II, ed. Peter M. Gray, p. 76

Parry, I. R., & Gray, P. M. 1986, Proc. SPIE, 627, 118

Focal-Ratio Degradation Optimization for PMAS

J. Schmoll, E. Popow, and M. M. Roth

Astrophysikalisches Institut Potsdam, An der Sternwarte 16, D-14482 Potsdam, Germany

Abstract. The results of extensive focal-ratio degradation (FRD) tests of fibers for use in PMAS, the Potsdam Multiaperture Spectrophotometer, are described. In particular, we report the effects of oil immersion on FRD. From the results of differential stability measurements we derive an explanation for the source of the so far unexplained noise as observed in some fiber-coupled echelle spectrographs.

1. Method

The measurements were performed on the AIP photometric test bench under remote control. We have used the projected image of a 100-μm pinhole to illuminate the fiber input with a well-defined input f-ratio at λ = 550 nm (FWHM 10 nm). The light emerging from the pinhole is collimated with a 154-mm lens. After passing an iris diaphragm which is used to limit the diameter of the parallel beam, the light is imaged onto the fiber end by means of a 109-mm lens, thus creating a 70-μm spot on the fiber input face. The diaphragm allows for the selection of f-ratios between $f/2$ and $f/10$. The fiber output is attached to the dewar window of a cryogenic CCD camera using a thinned SITe TK1024 chip. The CCD acts as a screen, which is illuminated by the output cone at a certain distance from the source. This arrangement allows the the fiber output pattern to be registered as a function of angle.

In order to check the behavior of the fibers when used in immersion, we added a thin plane–parallel glass plate on the input end of the fiber and applied Cargille-index matching gel no. 0607 between the fiber and the plate. The same was done between the output end and the cryostat window.

Data reduction of the resulting FITS images was done with IDL routines that make a growth-curve analysis. First, output f-ratios were calibrated in terms of linear dimensions on the chip by directly illuminating the CCD with the fiber input projector which was focused onto the dewar window. Using this calibration, circles around the center of gravity of a given fiber image were associated with different output f-ratios. The fraction of light inside the circle of nominal input f-ratio as compared to the total is a measure of focal-ratio degradation. The program computes a curve of growth, giving the integral of flux within a circle of a certain radius as a function of radius (which translates into an f-ratio). The resulting values are normalized to the total flux on the chip which is assumed to be identical with the total output of the fiber.

2. Results

Different fiber types: We investigated fibers of 100-μm core diameter. Measurements were done with Ceram Optec Optran UV and Polymicro FHP/FVP fibers, and several individual pieces for each type in order to obtain some information on the statistical errors of the tests. We used 2 m as a standard length except for the stability test, which was carried out with a 21m FVP fiber. The comparison between five individual Optran UV 100/140/170 and FVP 100/120/140 fibers yielded only small differences. The scatter of performance among the Optran UV fibers appeared to be somewhat, but not significantly, higher.

Index matching: Immersion is used to increase the total throughput of an optical system where neighboring glass–air interfaces can be avoided by applying a layer of oil. During our experiments it became obvious that immersion also has the effect of compensating imperfections of the fiber end faces. In Fig. 1 the effect of immersion on FRD is visible. An important consequence is that the *intrinsic* FRD performance of a fiber due to stress or microbending may be seemingly decreased by improper treatment of the end faces.

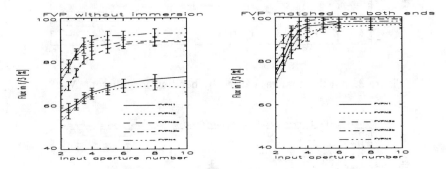

Figure 1. FRD behavior of five different FVP fibers without and with index matching; the fiber names are arbitrary.

In addition to an improvement in throughput, an improvement in FRD performance is notable: the scatter of the different fibers nearly disappears for input numbers larger than $N_{in} \approx 5$. Furthermore, fibers of particularly poor performance without immersion were "healed" out to get relative FRD performance increases from <70 % up to >95 %, at $N_{in} = 6$. The obvious advantages of relative independence from details of the fiber end faces, and of eliminating reflection losses convinced us to ask for a design change of the fiber spectrograph in order to have the fiber coupling in oil immersion.

Stress: Fibers are known to be sensitive in terms of mechanical stress and bending. In order to obtain more objective figures we made several FRD tests under the application of stress. These results will be used to constrain the mechanical design of the fiber-bundle device.

FRD behavior was investigated as a function of *bending, radial stress*, and *axial stress*. We used 2-m FVP fibers for these tests.

Figure 2 shows the effect of *bending* at various radii. The raw data falsely suggest a better performance at smaller radii. However, this effect is due to mode leakage of the higher modes. To take the light loss into account, the FRD curve was weighted by the quotient of the detected total light when bent, $F_{\text{bent}}(r)$, and the total light detected while the fiber was not bent, F_{unbent}. After correction, it becomes clear that bending actually decreases fiber performance at radii lower than ≈ 20 mm. Due to the weighting process, the error bars are larger. This kind of weighting was also used for the remaining stress plots.

The effect of *pressing* a fiber between the plastic sheets of a vice is displayed qualitatively in Figure 3. The following situations are shown: Fiber without manipulation (*free*), fiber slightly fixed between the vice's even plastic sheets (*clamp*), fiber pressed strongly (*clamp2*), fiber free again (*free2*), then fiber slightly clamped between grooved plastic sheets (*mbend*, fiber perpendicular to grooves), and finally the free fiber again (*free3*).

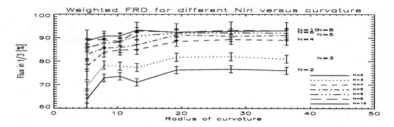

Figure 2.   FRD changes caused by bending for different N_{in}.

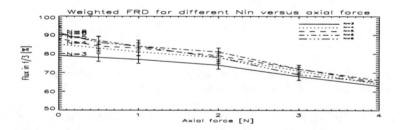

Figure 3.   FRD variations due to radial pressure (qualitatively, for different N_{in}).

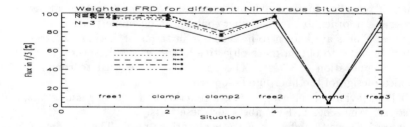

Figure 4.   FRD changes due to pulling axially for different N_{in}.

Finally, fiber behavior as a function of an axially *pulling* force was investigated. Forces between zero and 4 N were applied. Since there is no obvious way of introducing force along the fiber without causing secondary effects that are related to the fastening, we must treat the results with caution. At any rate, real fibers must be supported in some sensible way involving fastenings that may be similar to the experiment. We conclude as a working hypothesis that axial forces exceeding 2 N have to be avoided (see Fig. 4).

Stability: To investigate the long-term behavior of the output beam pattern and to study the effects of fiber motion and bending, simulating operation at the telescope, we used the FRD setup as before, however with a 21-m FVP fiber that was suspended free of stress throughout the lab and guided to a test place where the fiber was manipulated in various ways. The main rationale behind this test was to find arguments for or against the proposition of mounting the fiber spectrograph to Cassegrain and using short fibers instead of having a geometrically fixed position on the floor. We conducted several series of FRD measurements and compared each beam pattern as was projected on the CCD to a common reference (usually the first frame of a series). This differential approach is sensitive to any variation in the distribution of light within the output cone, showing up as intensity contrast in the differential image for each measurement. The general finding is that in fact small drifts of order 10^{-3} of the total flux redistributed over timescales of 2 hours are observed. Interestingly, the manual manipulation of the fiber which was performed qualitatively by lifting the fiber and more or less firmly pressing it between two fingers, resulted in rather strong shifts in intensity. This effect is illustrated in Fig. 5 (the contrast in the reference frame is only due to different levels of S/N within or outside the beam). The level of flux redistribution is now of the order of several per cent of the total flux, which indeed would be critical for the stability issue. Although total flux is preserved, the effect translates into an apparent flux variation on the detector when we take into account *vignetting* in the optical train of the spectrograph, e.g. the central obscuration of a Schmidt camera. We believe that this observation is able to explain the occurrence of "fiber noise" that happens to limit the maximum S/N in some fiber-fed echelle spectrographs (Pfeiffer, this conference). We will further investigate the problem with narrow-band illumination using a monochromator instead of our previous 10-nm bandwith measurements.

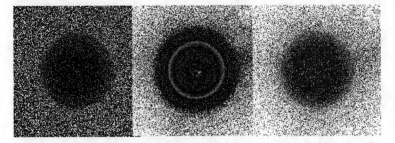

Figure 5. Three stages of stability checks, plotted as quotients relative to the first, undisturbed, image. *Left:* lightly held between two fingers. *Middle:* pressed slightly. it *Right:* pressed firmly.

Part 2: Multi-Object Spectroscopy

The Anglo–Australian Observatory 2dF Project: Current Status and the First Year of Science

Ian J. Lewis

Anglo–Australian Observatory, Coonabarabran, NSW 2357, Australia

Karl Glazebrook, and Keith Taylor

Anglo–Australian Observatory, PO Box 296, Epping, NSW 2121, Australia

Abstract. The 2dF Project will provide the Anglo–Australian Telescope with a unique ability to perform multi-object fiber spectroscopy of 400 objects within a 2-degree field of view. The 2dF project has now reached a stage where telescope time is being allocated to use this new facility on a regular basis for science observations. We review the current status of the 2dF project in terms of the performance and operation of the instrument. We also discuss the wide variety of science already being tackled with the 2dF facility.

1. Introduction

1.1. 2dF Project History

The 2dF Project was initiated in 1990 with the aim of providing the 3.9-m Anglo–Australian Telescope with a 2-deg field of view at prime focus. To access this focal surface 400 optical fibers would be positioned by a single robot positioner. These fibers would feed two identical spectrographs allowing simultaneous spectroscopy of up to 400 objects. A double buffering system would allow uninterrupted observations through one set of 400 fibers while the robot positioned a second set of fibers in readiness for the next set of observations.

Taylor et al. (1996b) give a summary of the progress on the 2dF Project just as the first science data had been obtained. The instrument as it then existed had fewer than 200 fibers and a single spectrograph. This paper will give an up-to-date description of the current status of the 2dF project and its capabilities and describe some of the science already undertaken using the 2dF facility.

1.2. A Brief Description of the Hardware

The 2dF facility is built on a new dedicated top-end ring for the AAT. In the center of the ring is the heart of 2dF, a four-element prime-focus optical corrector (Jones 1994) giving an unvignetted 2-deg field with atmospheric dispersion compensation (ADC) built into the first two elements. See Fig. 1 for a diagrammatic representation of the major subsystems of 2dF.

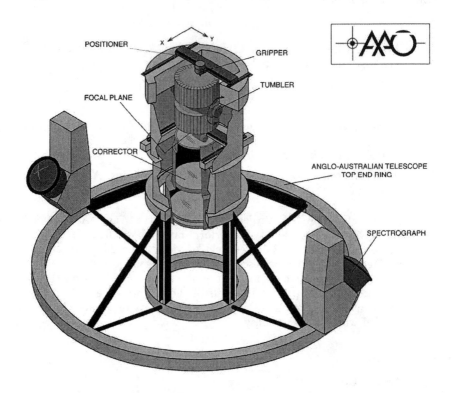

Figure 1. A diagrammatic view of the 2dF top-end ring showing all of the main hardware components.

Above the optical corrector are two fieldplates 440 mm in diameter which are mounted on a tumbler so that either plate can be brought into the focal plane. Each fieldplate has 400 optical fibers arranged around its periphery. The fibers are housed in retractor units in banks of ten fibers (40 for each fieldplate). The retractor units maintain tension on each of the fibers as they are moved in the focal plane by the robot positioner. The use of bare fiber (rather than fibers enclosed in stainless steel tubes) allows fiber–fiber crossovers, which increases the percentage of fibers assigned to objects in critical cases by the configuration software. The retractor units have been described by Parry et al. (1993).

Between the rear element of the corrector and the set of fibers in the focal plane a Peltier-cooled CCD may be brought into the optical beam on an XY gantry (the focal-plane imager system) for astrometric calibrations or to acquire a target field.

Above the second set of fibers, at the top of the prime-focus central obstruction lies the fiber positioner. This consists of a single robotic gripper on an XY gantry accessing the whole fieldplate. The gripper unit has two servo-controlled jaws and a video CCD robot-vision system to enable placement of fibers to 10-

µm accuracy. The fibers are positioned relative to a set of illuminated fiducial marks in the fieldplate to remove any telescope attitude and flexure effects.

The 800 fibers from both fieldplates are divided into two and pass out of the tumbler along the rotation axis and down two opposite spider vanes to the spectrographs located on the top-end ring. The fiber run is about 8 m in length.

The spectrograph optical design consists of an off-axis Maksutov $f/3.15$ collimator and an $f/1.0$ Schmidt camera with helium closed-cycle coolers. The spectrographs have a 150-mm collimated beam and use the standard AAO gratings.

Each spectrograph receives two sets of 200 fibers, only one of which is used for observing at any one time. The two sets of fibers are alternately switched into the optical beam synchronously with the fiber tumbler. Banks of LEDs are used to back-illuminate the set of fibers not being used for observations to enable the robot-vision system of the fiber positioner to accurately position the fibers. These LEDs must be very well shielded to avoid stray light within the spectrograph.

2. Current Status

For the period between first science data (1996 June) and 1997 August the 2dF facility was used with two sets of 200 fibers and a single spectrograph. Reconfiguration of the fibers took place at an average speed of 35 s per fiber. During the latter months of this period all available effort went into the integration of the control-system electronics and software in readiness for the arrival of the second spectrograph.

During 1997 August and September 2dF was upgraded to the full complement of two sets of 400 fibers and two spectrographs. First science observations with 400 fibers took place towards the end of September 1997.

2.1. The Fiber Positioner

The fiber positioner handles the movement of all 400 fibers directly from one target field to another. An average of 575 fiber movements are required to reconfigure the fibers from one field to the next. The positioner currently does this task at an average speed of 14 s per fiber. Reliability is at the level of one positioner failure every 1000 fiber movements. The current performance is still just over a factor of 2 slower than acceptable and we hope to reduce the failure rate by at least another factor of 10.

The tumbler is fully commissioned and requires a total of about one minute to change between fieldplates. This mechanism is fully interlocked with both gantries to avoid collisions.

The fieldplates are designed to have a small amount of controlled rotation to enable improved tracking of refraction effects. Currently this is disabled and awaiting commissioning.

2.2. The Spectrographs

Both spectrographs are partially commissioned with their optics in place and most operations under full computer control. The slit unit is the most complicated area of the spectrograph and consists of the two slit units each with 20 slitlets of 10 fibers, the fiber back-illumination LEDs within a lightproof shield, the slit interchange mechanism and two filter wheels which intercept the optical beam as it emerges from the fibers. The spectrograph shutter is a blade type immediately in front of the fiber slit.

The spectrographs use the standard AAO 150-mm by 200-mm gratings, which yield resolutions of 1.1 Å to 4.3 Å per pixel in first order. Second order will be available when the filter wheels are commissioned. Duplicates of several gratings have been obtained to allow identical dispersion options in both spectrographs. In the near future all of the AAO gratings will be fitted with magnetic bar-codes to enable automatic identification of the gratings and blaze direction for incorporation into the data header information.

The spectrograph cameras have an internal focus with the detector package (a Tek1024 CCD) mounted on an internal cold finger which is connected to an externally mounted closed-cycle cooler. Focus of the spectrograph is achieved by moving the detector package within the cryogenic camera using three motorized micrometer units with 0.06-μm resolution.

Currently, all of the spectrograph mechanisms are commissioned with the exception of the grating bar-codes and filter wheels.

2.3. The Control System

Low-level control of DC motors and encoders is performed by PMAC motion control boards with dedicated software for each mechanism. Mid-level control software is written in C and runs on VME 68030 processors using the VxWorks operating system. This mid-level software performs integration of the various mechanisms into logical units, for example the fiber positioner, the tumbler, the focal plane imager, the ADC, and each of the spectrographs.

Overall control is performed by the main 2dF control task, which provides the user interface to the mid-level tasks and also integrates the 2dF software with the control of the two CCD detectors and the telescope. The high level software runs on a Sun UltraSparc running Solaris.

All software (apart from the dedicated PMAC software) runs under the DRAMA environment (Farrell et al. 1995) which provides for transparent communications between tasks across the various operating systems and platforms.

2.4. Field Configuration

Preparation for observing takes place off line using specially written configuration software to take a target catalog and produce a valid configuration for all 400 fibers. This process is a blend of automated procedure and interactive tweaking. The fibers are allowed to be extended just past the center of the field and may deviate from the non-radial direction by up to 10 deg. Crossing of the fibers is unlimited, but once the user is happy with the configuration the software attempts to uncross as many fibers as possible without losing any of the target objects in order to reduce the number of fiber movements (and hence the reconfiguration time) to move directly between target fields.

2.5. Data Reduction

A new data-reduction package (2dFdr) has been written (Taylor et al. 1996a) specifically for the reduction of 2dF data. 2dFdr relies heavily on the header information contained within the data files to identify the type of data (flat, offset sky, arc, or object data) and the instrument configuration. The header information also contains all of the fiber information which is propagated all the way from the original target catalog via the configuration and control software to the data files. The data-reduction software then automatically associates each spectrum with its object label.

The data reduction software has recently had its first general release now that the data files contain full header information. In the long term it is planned to run two copies of the data reduction software in a parallel pipeline for automated data reduction with observers leaving the telescope with reduced data.

3. The First Year of Science

During the early commissioning phases of 2dF scientific data was obtained when the hardware status and weather conditions allowed. A broad range of projects obtained test data for evaluation purposes and to enable planning for future time applications with 2dF. Various calibrations of throughput were also made with the aim of providing a World Wide Web signal-to-noise calculator (http://www.aao.gov.au/cgi-bin/2dfsn.cgi).

3.1. The Redshift Surveys

Much of the design specification of 2dF is based around the scientific requirements for the combined 2dF galaxy and 2dF quasar redshift surveys. The former is a UK–Australian collaboration to measure the redshifts of 250000 galaxies from the APM survey (Maddox et al. 1990), and the latter is a UK project to identify and measure the redshifts of up to 30000 quasars in a subset of the area of the APM survey. The two surveys have been combined after a suggestion by the time allocation committees with an overall saving of 20% in the total amount of time required.

During the early testing of 2dF only 200 fibers were available, which, together with a very slow reconfiguration time, severely limited the efficiency of observations for the redshift surveys, which really require the full complement of 400 fibers and only require a 60-min exposure time. However approximately 20 fields of 200 objects were observed under differing instrument configurations with varying success rates (measured by completeness of identification and measurement of redshift).

After the upgrade of 2dF to 400 fibers with two spectrographs and an improvement in configuration speed by a factor of 2, the redshift surveys have had a total of four clear nights and obtained spectra of over 5000 targets with a completeness approaching 90%. See Fig. 2 for examples of the data obtained up to 1997 November and Fig. 3 for an RA slice showing the galaxy redshift information.

With the sheer amount of data being obtained for the redshift surveys it is imperative that a highly automated method for the measurement of redshifts

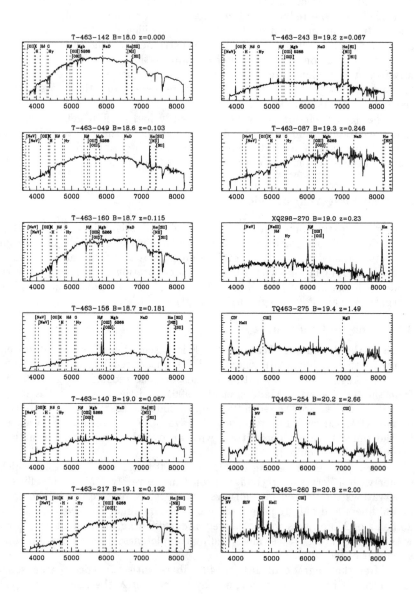

Figure 2. Example data from the galaxy and quasar redshift surveys.

The Anglo–Australian Observatory 2dF Project

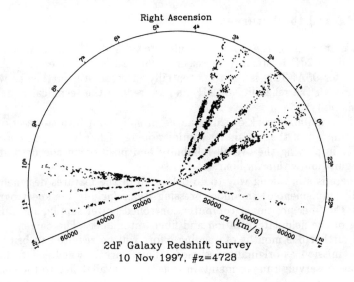

Figure 3. Redshift slice showing all of the data obtained to 1997 November.

from spectra is developed. The early test data has been manually assigned redshifts and used to test automated methods, for example a new one based on principle component analysis (Glazebrook et al. 1998). It is currently anticipated that the redshift surveys will be completed in about the year 2000.

3.2. Other Astronomy with 2dF

Of the time currently scheduled with 2dF over half is currently being used for projects other than the redshift surveys. Examples of these projects may be split into extragalactic and Galactic astronomy.

As well as the redshift surveys, other extragalactic surveys include surveys of the Fornax and Coma Clusters, dynamics of clusters of galaxies, searches for extragalactic globular clusters and the dynamics of nearby galaxies using stellar radial velocities.

Galactic astronomy is represented by a large Galactic structure survey, dynamics of globular clusters and planetary nebula searches. Stellar astronomy is also well represented with work on carbon stars and stellar abundances.

Novel uses for 2dF are constantly appearing before the time assignment committees; many of these proposals were not even conceived of during the design stage, and we are aconfident that this will ensure that 2dF has a useful l ife well after the completion of the redshift surveys.

4. 2dF and the Future

A quick analysis of the AAT schedule shows that the amount of time being allocated to 2dF is steadily increasing from 22 % in semester 97A to 28 % in semester 98A. This is set to potentially increase still further as we become happier with the reliability of 2dF and we reduce the technical restrictions on its availability.

The remaining work on 2dF can be broadly split into two sections: firstly completing all of the remaining commissioning of hardware and software, and secondly improving the overall reliability and positioning speed to allow a full reconfiguration within an hour.

There is still a long way to go before all of the features designed into 2dF are fully implemented; the more pressing items include the spectrograph filter wheels, the fiber positioner computer controlled power supply, and full commissioning of the fieldplate rotation and fiber autoguider. The current plans show that a further 6–9 months of ongoing effort will be required to bring 2dF into full commission as originally designed. This work is scheduled to fit between telescope observing runs to maintain instrument availability to the general user.

There is an ongoing program in place for the continuing refurbishment of the optical fibers to keep the complement of usable fibers as close as possible to 400. A total of ten spare fiber retractor units (100 fibers) are planned to enable swapping of defective units with a minimum of down time.

Much effort has been invested in the positioner hardware and software to determine the real causes of the slow fiber-positioning performance. We are confident that a series of minor hardware improvements to the gripper unit and a better understanding of the positioning process will allow the target of reconfiguring all 400 fibers in under one hour to be realized.

Acknowledgments. The commissioning of 2dF has only been made possible by the dedicated staff of the AAO. In particular, Chris McCowage, Allan Lankshear, Lew Waller, Tony Farrell, John Stevenson, Keith Shortridge, Greg Smith, Denis Whittard, John Straede, Llew Denning, and the night assistants and afternoon-shift technicians that have made the many commissioning runs a success. We also thank the 2dF Galaxy and Quasar Redshift Survey teams for permission to reproduce example data and some early results.

References

Farrell, T. J., Bailey, J. A., & Shortridge, K. 1995, ASP Conf. Ser. Vol. 77, Astronomical Data Analysis Software and Systems IV, ed. R. A. Shaw, H. E. Payne, & J. J. I. Hayes (San Francisco: ASP), 113

Jones, D. J. A. 1994, Appl. Opt., 33, 7362

Glazebrook, K., Offer, A. R., & Deeley, K. 1998, ApJ, 492, 98

Maddox, S. J., Sutherland, W. J., Efstathiou, G., & Loveday, J. 1990, MNRAS, 243, 692

Parry, I. R., Sharples, R. M., Lewis, I. J., & Gray, P. M. 1993, in ASP Conf. Ser. Vol. 37, Fiber Optics in Astronomy II, ed. P. M. Gray (San Francisco: ASP), 36

Taylor, K., Bailey, J. A., Wilkins, T., Shortridge, K., & Glazebrook, K. 1996a, in ASP Conf. Ser. Vol. 101, Astronomical DAta Analysis Software and Systems, ed. G. H. Jacoby & J. Barnes (San Francisco: ASP), 195

Taylor, K., Cannon, R. D., & Watson, F. G. 1996b, Proc. SPIE, 2871, 145

6dF: An Automated Multi-Object Fiber-Spectroscopy System for the UKST

Q. A. Parker, and F. G. Watson

Anglo–Australian Observatory, Coonabarabran, NSW 2357, Australia

S. Miziarski

Anglo-Australian Observatory, Epping, NSW 2121, Australia

Abstract. The FLAIR multi-fiber spectroscopy system on the Anglo–Australian Observatory's 1.2-m UK Schmidt Telescope (UKST) feeds an optically efficient all-Schmidt spectrograph with a thinned CCD. However, positioning of the 92 available fibers within the 40-deg^2 UKST field is only semi-automated and can take four to six hours. Typical observations of sufficient signal-to-noise usually take much less than this (e.g. about an hour for galaxy redshifts to $B \sim 17$ mag). Clearly the system is working well under its potential efficiency.

To address this imbalance, a fully-automated, off-telescope, pick-place fiber-positioning system known as 6dF has been proposed in which 150 fibers can be reconfigured in under an hour across a 6-degree circular field. Three field plates will be available with a 10–15 minute field plate changeover anticipated. The resulting factor 10 improvement in observing efficiency would deliver, for the first time, an effective means of tackling major, full-hemisphere, spectroscopic surveys. An all southern sky near-infrared selected galaxy redshift survey is one high-priority example.

The estimated cost of 6dF is $A450k. A design study has been completed and substantial funding is already in place to build the instrument over a 2-year timescale.

1. Introduction

The UKST is operated in two interchangeable modes: wide-field photographic imaging and fiber-coupled multi-object spectroscopy with the FLAIR system. These can combine to form a self-contained, strategic observing system in which photography provides input catalogs for follow-up spectroscopy of objects with $B \leq 18.5$ mag. In the five years in which FLAIR has been a common-user service, over 22000 object spectra have been obtained.

Despite the advent of powerful multi-object spectroscopy systems like 2dF (Taylor et al. 1997; Lewis, these proceedings) and SDSS (Kron 1997), the existing FLAIR II system on the UKST has remained competitive for certain types of projects due to its combination of very wide field and available fiber numbers

(Parker & Watson 1995). For these reasons, demand for FLAIR has remained high, despite the tedious semi-manual fiber set-up procedure.

At the same time, the UKST is entering a crucial new phase in its life as the major all-sky broad-band photographic surveys are nearing completion, and the future role of the UKST is debated. The UKST offers an exceptional wide field of view, which can still be effectively exploited by appropriate modern instrumentation. One way to remain relevant and productive in the future, whilst still utilizing the UKST's unique wide field, is by more fully exploiting the scientific potential of a wide-field MOS system via an automated fiber-positioning facility. Such a proposal, known as 6dF, has found favor among the astronomical community. Equipped with 150 fibers, the 6dF system would offer a factor 10 increase in efficiency over FLAIR for performing large-scale spectroscopic surveys. For modest cost, it would provide the UKST with a unique facility among 1-m class instruments as 2dF on the AAT is among 4-m telescopes.

In this paper we concentrate on the basic technical detail and the scientific benefits that such a 6dF system would deliver. The new facility would enable major new "full-hemisphere" scientific programs to be undertaken for the first time over realistic timescales. 6dF would maintain the UKST as a world-class facility well into the next century.

2. The Current FLAIR System and Its Drawbacks

Although the current FLAIR system has continued to evolve, with a much improved version being commissioned in 1992, the same basic principles employed by the original prototype FLAIR in 1985 are retained (Watson 1986; Watson et al. 1993). Serious drawbacks with the current semi-manual fiber-positioning system still exist and are briefly summarized below.

- FLAIR's semi-manual fiber-positioning system is laborious, time-consuming, messy, and needs regular maintenance

- A fully trained fiber can take > 4 hr to affix ~ 100 fibers. Visiting astronomers can take considerably longer (e.g. 6–8 hr)

- The system uses UV curing cement to affix fibers, which has health and safety implications: the UV curing cement is toxic; the UV lamp is high intensity - goggles must be worn during the fibering process; the UV lamp creates ozone, which must be safely ducted away.

3. The FLAIR Interim Upgrade

To alleviate many of these shortcomings an "interim" FLAIR upgrade has been undertaken and is currently being commissioned. It is seen as a half-way house between the old FLAIR and new 6dF. Apart from various spectrograph operational improvements such as automated grating rotation, Hartmann shutters, and focus adjustment (which will remain in place for 6dF), the interim upgrade makes use of magnetic buttons. An exhaustive demonstration of the viability of magnetic-button fiber positioning in the UKST environment was carried out.

This is a prerequisite before any 6dF magnetic button system could be considered. Unlike other 'AUTOFIB'-type systems, the FLAIR fibers are subject to vibration and shock after positioning when the fiber plateholder is loaded into the telescope. The interim-upgrade tests show that appropriately designed magnetic buttons will not present a problem for 6dF. The semi-manual fiber-positioning system of FLAIR was retained with the newly commissioned magnetic-button plateholders and should lead to reduced fibering time from 4–6 to 3–4 hr, mainly because the UV cement application and curing time is largely eliminated.

4. 6dF

Because of significant gains in FLAIR system efficiency, principally due to the commissioning of a thin, back-illuminated CCD in 1996 (Parker 1997), fiber set-up time now exceeds typical exposure times by a factor of up to three, even with the benefits of the FLAIR interim upgrade. The proposed fully automated 6dF positioner described here addresses this imbalance and will deliver a wide-field, Schmidt-based fiber system of unprecedented power and versatility. Specifically, a fully automated off-telescope, pick-place, fiber positioning engine would offer:

- Elimination of semi-manual fiber positioning, and reduced fibering time from 3–4 hr to ≤ 1 hr

- Greater multiplex advantage—an increase from 90 to 150 objects per field (the maximum that can be imaged with the existing CCD)

- Greater observational efficiency and flexibility with faster turn-around between fibered fields

- More fields per night can be observed, e.g. from the current 1 to 3–5

- More opportunistic observing possibilities with three field plates and a rapid fiber-positioning system to take advantage of non-photometric conditions at short notice

- Considerable savings over the long-term in manpower costs associated with the existing fibering system both in house and for visiting astronomers

- Anticipated significant reduction in preventative maintenance and repairs

4.1. 6dF Basic Technical Scientific Requirements

The technical scientific requirements for 6dF are described in detail by Parker & Watson (1997) but are summarized briefly below:

- Three interchangeable fiber field plates (i.e. fiber plateholders). Each would be configured off telescope via a robotic R–theta positioner before being loaded into the telescope

- 150 fibers per field plate, terminated with magnetic buttons and SF5 microprisms and arranged in a circular field of diameter 6 deg. Each fiber bundle terminates with a standard slit unit that would locate in the existing floor-mounted intermediate-dispersion spectrograph

- Fiber core diameter of 100 μm (6.7 arcsec), a reasonable match for galaxies at $B_j \leq 17.5$ mag, the anticipated limit of galaxy redshift surveys

- Fiber-positioning accuracy of better than 10 μm (0.7 arcsec)

- Fiber placement time (including any intermediate park-move necessary) of 24 s or less, allowing a complete reconfiguration in about an hour

- Minimum fiber–fiber proximity no greater than 4.5 mm (5 arcmin)

- Fiber placement over a full 6-deg diameter field

- Up to 10-arcmin field rotation to be provided to compensate for atmospheric refraction/polar axis elevation offsets (Watson 1984).

- Upgrade path for integral-field spectroscopy to be provided

4.2. The Proposed 6dF R–Theta Robotic Fiber Positioner

The principal component that differentiates 6dF from FLAIR is the automated R–theta magnetic-button fiber positioner. It will be designed to reconfigure a field of 150 fibers entirely automatically in about an hour. The fiber positioner will be located in its own enclosure in the dome, ensuring the fibers are positioned at the same ambient temperature as the observations are made while minimizing plateholder handling and transport.

The positioner will operate directly on a curved rigid field plate matching the focal surface of the UKST. This surface is a segment of a sphere of radius 3070 mm, so the R-motion will be along a curved track. Support and guidance of both R and theta motions will be by air bearings for minimum friction and no-stiction operation. This allows the radial motion to follow a curvature not covered by standard bearings.

A pneumatic fiber gripper with a piston-shaft assembly freely rotating in air bearings to lift and place the fiber buttons will move over the field plate under computer control, positioning each fiber in turn with an accuracy of 10 μm (0.67 arcsec). As with 2dF, machine vision with a CCD camera looking at the back-lit button through the center of the gripper's piston shaft assembly will ensure that this accuracy is achieved by iteration if necessary. Unlike 2dF, the time constraint is not critical, with up to 24 s being permitted for each fiber move. Figure 1 shows a side view of the complete R–theta and field-plate assembly whilst Fig. 2 gives the scale drawing of the gripper assembly.

4.3. Fiber Retractors and Field Plate

Unlike FLAIR, 6dF fibers will self-retract into the body of the field plate rather than requiring to be pushed in manually. Space restrictions in the Schmidt's focal surface mean that innovative solutions have been adopted for the retractor design; experimental work in this area has proved successful. Fibers will retract by means of small-diameter helical close-coiled springs wrapped around the fibers. The bending force of the spring provides the tension for self-retraction.

The rigid, curved field plate will allow a 10-arcmin rotation adjustment and the field-plate assembly, with 150 fiber retractors, will fit within the physical size of the existing Schmidt photographic plateholder envelope. To reduce

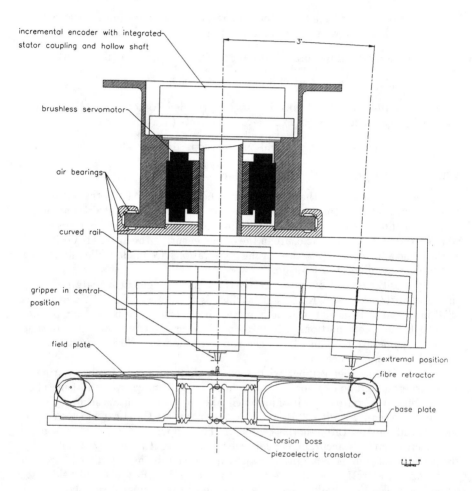

Figure 1. Side view of complete R–theta and field plate assembly.

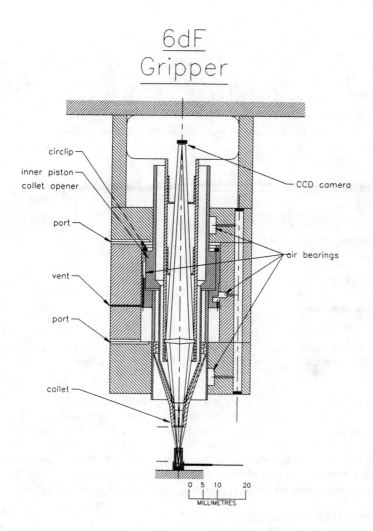

Figure 2.   Scale drawing of gripper assembly.

6dF Fibre Retractors

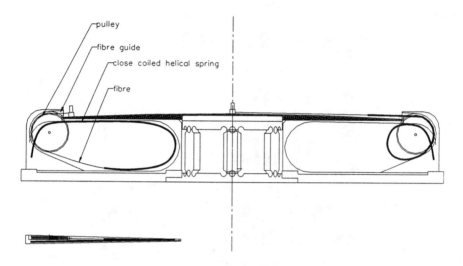

Figure 3. Schematic side view of the proposed retractor design within the field plate and a top view of one of the retractor segments.

weight and eliminate vibrations (by a choice of disparate resonant frequencies) a steel/composite sandwich is proposed for the field plate. Any thermal distortions will be within acceptable limits. The total mass will be low enough to prevent deflection of the flexible boss used for field-rotation adjustment. Figure 3 shows a schematic side view of the proposed retractor design within the field plate and a top view of one of the retractor segments. Figure 4 shows the circularly arranged position of the retractors under the field plate.

4.4. Magnetic-Button Design

The fiber buttons will be constructed as hollow cylinders incorporating SF5 glass prisms and rare-earth NdFeB magnets at the base. Figure 5 gives a scale drawing of the proposed button design. The prisms are extended in the vertical direction to allow the full incoming $f/2.48$ UKST beam to be accepted while retaining a reasonable length of tube for the gripper to handle. The practical constraints on gripper and button design limit adjacent button proximity to about 4.5 mm (5 arcmin). However, since the 6dF buttons will be circularly symmetric (except for the trailing fiber), the minimum separation is the same in all directions. This contrasts well with the existing system, which has a highly elongated rectangular footprint making positioning difficult for compact groups of more than three objects.

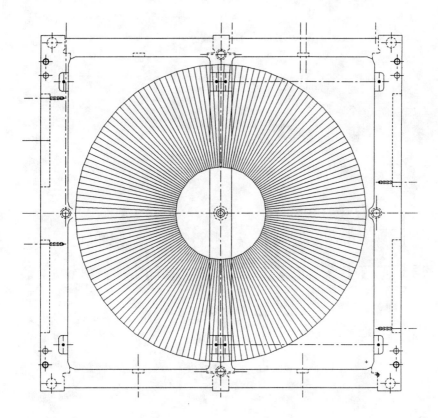

Figure 4. Schematic view of the circularly arranged retractors under the field plate.

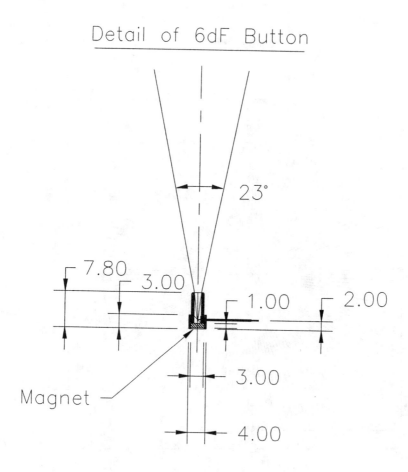

Figure 5. Scale drawing of proposed magnetic-button design.

4.5. Electronics and Software

The 6dF electronics design will follow the principles established for 2dF but with much simplified hardware with a PC communicating with the R–theta positioner via a PMAC motion controller. As many components as possible will be standard "off-the-shelf" items.

Unlike FLAIR, which relies on direct fiber positioning over a copy film of the target field, 6dF requires accurate object-input catalogs for the target observations (as provided by SuperCOSMOS for example), and so software will be necessary to transform object positions to the established current field coordinate grid. Due to the similarity in fiber arrangement between 2dF and 6dF, the 6dF field-configuration software may be ported from the 2dF regime before 6dF specific modification. Both the on- and off-line 2dF software may prove suitable for porting. For data reduction the 2dfdr pipeline software may also prove suitable, although the IRAF dofibers software could also be used.

4.6. Spectrograph Improvements

A robotic 6dF system could be commissioned without the need to do anything to change the existing CCD or spectrograph, apart from rotating the current 420×578-pixel EEV CCD02-06 CCD by 90 deg to allow all 150 fibers to be imaged. However, better sampling of the fiber beams and higher S/N could also be achieved with a new all-transmissive spectrograph design. It would incorporate a new CCD with smaller pixels and of larger format that is thinned and coated, and with a much lower readout noise. For example, a CCD with 1024×1024 10-μm pixels would be of physically similar dimension to the existing CCD, but the projected fibers would be much better sampled. This could be particularly valuable with the smaller-diameter fibers that may be adopted for use in the planned IFU field plate.

5. 6dF Operation

The provision of several 6dF field plates offers a "double-buffer" capability analogous to that used with 2dF. While one field plate is being used in the telescope, a second can be reconfigured. It is the shortest exposure time envisaged with 6dF (about an hour including overheads) that determines the minimum fiber set-up time. The point of having a third field plate is to allow repeated observations of the same field over a period of time (e.g. in time-resolved spectroscopic programs), or to have a back-up field for changing weather conditions. Field plates can be interchanged in the telescope in a matter of a few minutes with FLAIR; the same will be true of 6dF.

5.1. Field Acquisition

In 6dF the offset between the field-plate center and the telescope axis will be fixed, unlike the present system which involves "floating" deformable field plates. This eliminates the need for the 1.1-mm diameter coherent fiber bundle currently used to help acquire the field. 6dF field acquisition will be by means of the CCD autoguider and the existing intensified CCD and microscope assembly imaging

7 hexagonally arranged fibers in each of 4–6 fiducial bundles positioned over bright stars in the observed field.

6. Achievable 6dF Science Projects

Several major new large-scale projects have been put forward to take advantage of the UKST's unique capabilities for wide-field multi-object spectroscopy. However, without 6dF, none would be really practical in terms of timescales, efficiency and cost-effectiveness. A few such projects are:

- An all southern sky galaxy redshift survey to $B \leq 17$ mag (100000 galaxies)
- A J- or K-selected southern sky galaxy-redshift survey (e.g. from DENIS or 2MASS data) of a well defined volume of the local Universe
- Galaxy velocity dispersions—kinematics of galaxy clusters, distance estimates. 6dF could make a major impact in this area as fewer than 3000 galaxies currently have optical velocity-dispersion estimates
- The bright end ($B \leq 19.0$ mag) of the quasar luminosity function
- Possibilities for many large-scale stellar population and kinematical studies in the Galaxy and Magellanic Clouds based on DENIS/2MASS data, the new UKST H-alpha survey, and other surveys
- Follow-up observations to all-sky surveys performed in other wavebands where the surface density of objects are a few per square degree at the optical magnitude limits of 6dF (e.g. from satellite missions such as *ROSAT*, *IRAS*, and *FAUST*)

6.1. Illustrative 6dF Project Example

Let us consider three magnetic-button 6dF field plates being available with 150 fibers each and undertaking an all southern sky galaxy-redshift survey to $B \simeq 17$ mag (~ 100000 galaxies).

- Only 1.5–2-hr observation per field maximum would be required due to the high efficiency of the current thinned & back-illuminated CCD.
- Only 30 min turn-around between fields/plateholders (includes overheads—arcs, flats, etc.) are needed
- Typically 3–5 fields per night could be observed during the year giving 700+ redshifts/observations per night. This amounts to $\simeq 18\%$ of what 2dF would do on the AAT in strictly numerical terms.
- $\simeq 160$-degA^2 coverage per night could be achieved (c.f. 20 deg^2 for 2dF)

Thus, assuming: 365×0.7 (bright of moon fraction) $\times 0.66$ (spectroscopic observing fraction) equivalent to 170 suitable nights per year, we could observe $\simeq 680$ non-Galactic plane UKST standard survey fields per year. This is basically the entire extragalactic sky! If we simply maintained the existing FLAIR load on the telescope ($\sim 35\%$ of the available observing time) then the whole extragalactic southern sky could still be observed in 3 yr.

6.2. Timescales, Costs, and Funding

Although 6dF is currently only partially funded, a full Phase-A study has been carried out within the AAO. Several key elements of the design, including the fiber retractors, have already been successfully tested.

Following acceptance of the phase-A study, the fiber positioner could be completed within two years. An accurate cost estimate has emerged from the Phase-A study and is in the region of $450k. The additional costs for a new spectrograph are less certain but could amount to $100K + labour. This path would be followed only once full funding for 6dF had been obtained.

The funding to build 6dF is likely to come from a combination of AAO sources and third parties, for whom appropriate levels of guaranteed time on the instrument will be negotiated on a pro rata basis.

7. Summary

The scientific potential of a fully automated replacement for FLAIR is considered high. The time is right to exploit the niche that 6dF can offer, building as it does on the existing strengths and expertise of the AAO's fiber-spectroscopy program. A Phase-A study has just been completed which successfully demonstrates several key components of 6dF via working prototypes. Once 6dF is built, the UKST will be able to tackle full-hemisphere spectroscopic surveys for the first time over realistic timescales.

Acknowledgments. We thank Peter Gillingham, Greg Smith, Llew Waller, and Keith Shortridge for their input to the 6dF Phase-A study, and Matthew Colless, Gary Mamon, and Will Saunders for input to the science case. Q. Parker is on special leave from PPARC to work at the AAO.

References

Kron, R 1997, Proc. 2nd Conf. IAU Comm. 9, Wide-Field Spectroscopy, ed E. Kontizas et al. (Dordrecht: Kluwer), 41

Parker, Q. A., & Watson, F. G. 1995, Proc. SPIE, 2476, 34

Parker, Q. A. 1997, Proc. 2nd Conf. IAU Comm. 9, Wide-Field Spectroscopy, ed E. Kontizas et al. (Dordrecht: Kluwer), 25

Parker, Q. A., & Watson, F. G. 1997, AAO Internal Document: 'Scientific technical specifications for 6dF'

Taylor, K. Cannon, R. D., & Parker, Q. A. 1997, New Horizons from Multi-Wavelength Sky Surveys, IAU Symp. 179, ed. B. J. McLean et al. (Dordrecht: Kluwer), 135

Watson, F. G. 1986, Proc. SPIE, 331, 289

Watson, F. G. 1984 MNRAS, 206, 661

Watson, F. G. Gray, P. M. Oates, A. P. Lankshear, A., & Dean, R. G. 1993, ASP Conf. Ser. Vol. 37, Fiber Optics in Astronomy II, ed. P. M. Gray (San Francisco, ASP), 171

Performance of the Fiber-Positioning System for the Sloan Digital Sky Survey

Walter A. Siegmund, Russell E. Owen, Jessica Granderson, R. French Leger, Edward J. Mannery, and Patrick Waddell

ASTRONOMY Box 351580, University of Washington, Seattle, WA 98195, USA

Charles L. Hull

Observatories of the Carnegie Institute of Washington, 813 Santa Barbara St., Pasadena, CA 91101, USA

Abstract. A plug-plate system for multiple-object spectroscopy has been developed and tested. The holes are drilled on a milling machine in one set-up for the best accuracy. Good matches to the focal and telecentric surfaces are achieved with 3.2-mm thick aluminum plug-plates that are deformed elastically both for drilling and in the telescope.

1. Introduction

The Sloan Digital Sky Survey (SDSS) is a project by a diverse international collaboration to build and operate a facility to perform a very large imaging and spectrographic survey (Knapp 1997). The fiber-positioning system of the SDSS project is based on that of Schectman (1993). The plug-plates locate the optical-fiber plugs spatially and define the plug tilt with respect to the surface of best focus. The aluminum alloy 6061 plates are ϕ795 mm and 3.2 mm thick. (A number preceded by ϕ indicates a diameter.) Approximately 670 holes are drilled in each plate. These include 640 science holes, 10 guide holes and 9 quality-assurance holes. The balance are locating holes, light-trap holes, and a sky-brightness monitor hole.

For drilling, the plate is deformed elastically by a drilling fixture so that its upper surface is slightly convex. The holes are drilled parallel to one another. In the telescope, the plate is deformed to match the concave surface of best focus. When this is done, the hole axes are normal to the telecentric surface (Limmongkol et al. 1993). The telecentric surface is defined over the field of view such that its normal is centered on the bundle of rays from the telescope pupil.

Once the plug-plate is assembled and installed on the telescope, guide stars imaged onto coherent fiber-optic bundles are used to determine the errors in telescope pointing, image scale and rotator angle. A low-bandwidth control system acts to minimize the position errors of the guide stars on the guide fiber bundles. The telescope scale is adjusted by moving both the primary and secondary mirrors axially. The guide holes are drilled intermingled with the science holes to minimize any offset between the two hole patterns.

2. Hole Locations

Two plates, uw0111 and uw0112, were drilled at the University of Washington (UW) Physics Instrument Shop. A vertical milling machine, a Dahlih MCV-2100, was used. Plates ke0111 and ke0112 were drilled by a commercial vendor, Karsten Engineering, Inc., Phoenix, AZ. A horizontal milling machine, a Dixi 420TPA, was used. A new precision carbide spade drill bit was used to drill each plate. Drill run-out was measured prior to drilling and was 6.4 μm or less. A program written in C converted the hole list file to the computer numerically controlled (CNC) program used to control the milling machines.

The plug-plates were measured at Fermi National Accelerator Laboratory (FNAL) using a coordinate-measuring machine (CMM). The locations, x and y, of each hole, were measured at $z = -1.58$ mm (middle of the plate). The desired hole locations were subtracted from these values to get hole location errors.

During operation, guide stars on $\phi 8.25''$ coherent fiber-optic bundles will be used to determine scale, a_1, rotation, b_1, and pointing errors, δ_x and δ_y. The telescope image scale, pointing and rotator angle will be adjusted accordingly. To remove these effects, the functions $f(x,y) = \delta_x + b_1 y + (a_1 + a_3 r^2 + a_5 r^4)x$ and $g(x,y) = \delta_y - b_1 x + (a_1 + a_3 r^2 + a_5 r^4)y$ were fit to the x and y errors, respectively. Once the coefficients were determined, the residual errors were calculated (Table 1).

The coefficients a_3 and a_5 account for higher-order effects due to the drilling fixture. Unlike the other effects, these cannot be corrected by the telescope and must be anticipated in the drilling process. These coefficients were set to the mean of the coefficients found from separate least-squares solutions for uw0111 and uw0112.

Table 1. Hole location error (μm RMS)

Plate	x	y
uw0111	8.4	4.5
uw0112	10.6	5.3
ke0111	7.2	11.6
ke0112	6.7	11.0

For the four plates, the residual errors in one axis are dominated by a linear trend with hole number. With this removed, the residual error in x is 4.4 μm root mean square (RMS) for both uw0111 and uw0112 (Fig. 1). The error in y is 6.2 and 6.5 μm RMS for ke0111 and ke0112, respectively. The residual errors in the orthogonal axis do not show this correlation.

In the case of the UW machine, the spindle housing is heated by bearing friction. The resulting thermal expansion of the spindle housing increases the separation of the spindle and the vertical ways of the machine. Direct measurements of the spindle housing temperature and the displacement of the spindle housing with respect to the table are consistent with this explanation.

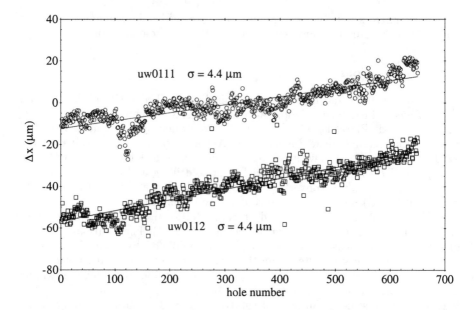

Figure 1. Hole location error in x. Standard deviations after trend removal are given. For clarity, 40 μm has been subtracted from the data for uw0112.

3. Hole Tilts and Diameters

As described in the introduction, the holes are drilled tilted with respect to the normal to the plate surface. Hole tilts were determined on six plates drilled prior to plate uw0111. For these plates, the holes were measured at three different depths. The differences of top and bottom hole location measurements were calculated. This and the separation of the two measurements were used to calculate the tilt of each hole. The radial component of the tilt as a function of radius was compared to the tilt of the telecentric surface calculated from the optical design (Fig. 2). The hole tilt error with respect to the optical design ranged from 2.1 to 3.3 mrad rms for the six plates.

The plug-hole clearance must be specified so that nearly all holes can be plugged given the distribution of hole and plug diameters. However, excessive plug-hole clearance increases fiber location error unnecessarily. The moment applied to the plug by the protective fiber tubing causes the plug to tilt so that the plug tip is pushed against one side of the hole. This decenters the fiber tip in the hole and causes misalignment of the fiber axis with the hole. Consequently, it is desirable that the hole diameters be as uniform as possible so that the plug-hole clearance can be small.

The holes in four plates were gauged with pin gauges (ϕ2.1666, ϕ2.1692, ϕ2.1717 and ϕ2.1742 mm). Each hole was assigned the diameter of the largest gauge that could be inserted in the hole. The project goal is $\phi 2.167^{+0.010}_{-0.000}$ mm and was satisfied by nearly all the holes (Fig. 3).

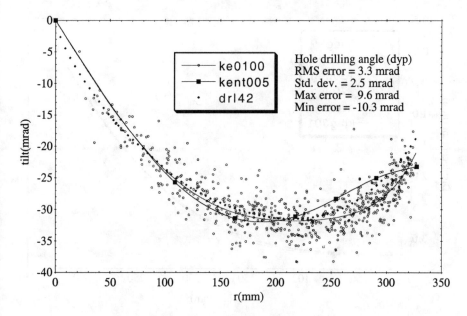

Figure 2. Measured hole tilts for ke0100 are compared with the tilts calculated from the optical design (kent005) and the finite-element model (drl42). The rms error is calculated with respect to the optical design.

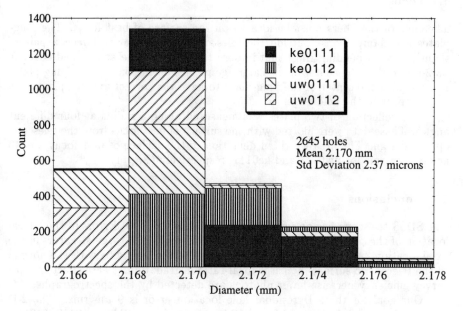

Figure 3. Hole diameter distribution. If the fit of the largest pin gauge was judged loose, it was assigned to the right-most bin.

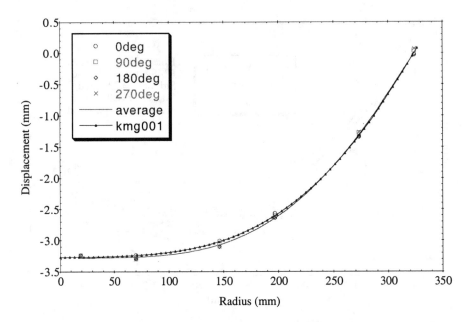

Figure 4. Measurements along four radii of plate uw0111 compared with the surface of best focus from optical design kmg001.

4. Focus

The ends of the fibers must be located on the surface of best focus. The plug-plates are clamped just outside the telescope field of view between a pair of slightly conical bending rings that impose the proper slope at the field edge. A ϕ3.2-mm rod at the center of the cartridge pushes on the back of the plug-plate. It constrains the center of the plug-plate to have the correct axial displacement with respect to the field edge.

The deflection of two plates was measured along a radius at four different angles. These data were plotted with the surface of best focus from the telescope optical design. The area-weighted deviation of the surface of best focus was 43 and 37 μm rms for uw0111 and ke0111, respectively (Fig. 4).

5. Conclusions

The SDSS telescope has an image scale of 60 μm/″ at $f/5$. Errors in the lateral location of the ϕ180-μm science fibers cause decentering of targets. Errors in the axial location cause image blur. Misalignment of the fiber axis with the normal to the telecentric surface increases focal-ratio degradation. These effects reduce survey efficiency because fewer photons are detected by the spectrographs.

Our goal for the 2-D residual hole location error is 9 μm rms. The 2-D errors measured, 9.5, 11.9, 13.6 and 12.9 μm rms for uw0111, uw0112, ke0111, and ke0112, respectively, are not quite at this level. They would be if the linear trend in hole number of the residual error were corrected or compensated. One option, under consideration for the Dahlih machine, is to implement an available temperature-control system for the spindle lubricant.

The hole tilt measurements indicate excellent performance. The holes on the worst plate have an rms error of 3.3 mrad, much better than the 6 mrad allotted.

The measurements of hole diameters suggest a large-diameter tail in the distribution that exceeds our specified range. Since only 1 or 2% of the holes are affected, this appears to be inconsequential.

The plates match the surface of best focus to 43 and 37 μm area-weighted rms for uw0111 and ke0111, respectively. These do not meet our goal of 25 μm rms. It appears that the error is largely due to inadequate bending ring flatness. Correcting this appears to be straightforward.

Acknowledgments. We are grateful to Paul Mantsch, Robert Riley, Barb Sizemore, and Charles Matthews at FNAL for their assistance with various aspects of plug-plate drilling and measurement. Jim Gunn at Princeton University provided inspiration and encouragement. It is a pleasure to thank Ron Musgrave, Larry Stark, Dan Skow, Siriluk Limmongkol, and Jeff Morgan at the UW for their interest and comments.

References

Knapp, G. R. 1997, S&T, 94, 40

Limmongkol, S., Owen, R. E., Siegmund, W. A., & Hull, C. L. 1993, in ASP Conf. Ser. Vol. 37, Fiber Optics in Astronomy II, ed. P. Gray (San Francisco: ASP), 127

Schectman, S. 1993, in ASP Conf. Ser. Vol. 37, Fiber Optics in Astronomy II, ed. P. Gray (San Francisco: ASP), 26

Fiber Optics in Astronomy III
ASP Conference Series, Vol. 152, 1998
S. Arribas, E. Mediavilla, and F. Watson, eds.

Status of the Fiber Feed for the Sloan Digital Sky Survey

Russell E. Owen, Matthew J. Buffaloe, R. French Leger, Edward J. Mannery, Walter A. Siegmund, and Patrick Waddell

University of Washington, Astronomy Dept., PO Box 351580, Seattle WA 98195, USA

Charles L. Hull

The Observatories of the Carnegie Institute of Washington, 813 Santa Barbara St., Pasadena, CA 91101, USA

Abstract. The Sloan Digital Sky Survey fiber feed consists of nine interchangeable plug-plate cartridges that will be plugged during the day and exchanged at night. Each cartridge contains 640 science fibers distributed between two spectrograph fiber slits, one coherent fiber bundle for sky-brightness measurement and ten coherent fiber bundles for guiding. One plug-plate cartridge has been assembled. We present throughput measurements of 5200 science fibers.

1. Introduction

The Sloan Digital Sky Survey is a project with the primary goal to measure the redshifts of approximately one million galaxies and one hundred thousand quasars in the northern Galactic cap. The survey consists of two parts: digital imaging in five colors to provide data for object selection and astrometric position determination, and spectroscopy of the chosen objects. The project will utilize a dedicated 2.5-m telescope with a 3° field of view, located at Apache Point Observatory in the Sacramento Mountains of New Mexico.

Spectra will be obtained using a pair of 320-object fiber-fed spectrographs mounted on the instrument rotator at the Cassegrain focus. Each spectrograph has a wavelength range of 390 to 910 nm with a resolution of 2000. This is achieved using a pair of cameras, red and blue, each with a 2048×2048-pixel Tektronix CCD.

The spectrograph fiber feed consists of nine interchangeable plug-plate cartridges. Each plug-plate cartridge contains 640 optical fibers for science targets and sky spectra plus one coherent fiber bundle for broad-band sky -brightness measurement and ten coherent fiber bundles for telescope guiding and image-quality monitoring. These cartridges will be manually plugged during the day and exchanged as needed at night. The integration time for each field will be approximately 45 minutes. For maximum efficiency, the interchange process has been highly optimized.

As of 1998 January, one plug-plate cartridge has been assembled (except for the coherent bundles) and the remaining cartridges are ready to be assembled.

Figure 1. Plug-plate cartridge.

All science fibers have been procured except for some spares. The optical fibers were terminated by a commercial vendor, C Technologies, Inc., in New Jersey. The vendor tested each fiber using a highly automated throughput tester we designed and built. We used a second identical tester to measure most of the fibers ourselves. The results of these measurements are presented below.

2. Plug-Plate Cartridge

Each plug-plate cartridge (see Fig. 1) consists of a cast aluminum housing supporting a plug-plate bending ring and two fiber slit assemblies. The cartridge contains 640 optical fibers and will contain 11 fiber-optic bundles for guiding and sky-background measurement. The plug-plate bending ring positions the plate and bends it to match the curve of the focal plane (Siegmund et al., these proceedings). Each slit assembly is spring-mounted to the housing and is automatically kinematically aligned to its spectrograph when the plug-plate cartridge is installed in the telescope.

3. Science Fibers

The optical fibers for science targets are Polymicro FHP all-silica UV-enhanced step-index fiber. The core diameter is 180 μm, or 3 arcsec on the sky. The fibers are only 1.8 m long, so absorption is minimal from 390 to 910 nm, the bandpass of the spectrographs. The telescope feeds the fibers an $f/5$ beam, and the spectrograph accepts $f/4$, resulting in negligible loss due to focal-ratio degradation. The fibers are uncoated. Predicted loss due to end reflections, focal-ratio degradation and absorption is 7 %.

The science fibers are packaged as harnesses of 20 fibers (see Fig. 2). For ease of manufacturing, testing and installation, the individual fibers are permanently mounted in their harness and are not individually replaceable.

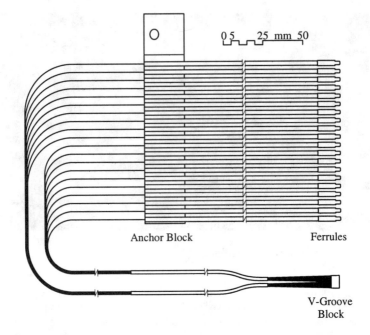

Figure 2. Harness of 20 science fibers.

Each harness is terminated at the fiber slit in an 8-mm wide stainless steel V-groove block. The exit face of this block is polished flat and has slightly fanned grooves to provide the correct alignment of the output beam from each fiber. This V-groove block is glued to a thin aluminum substrate to form the fiber slit. The resulting slit thickness is 3.5 mm, which results in a 3 % loss of light due to obstruction of the collimated beam. A high-precision steel positioning jig is used during assembly to align the V-groove blocks along the appropriate curve.

At the other end of the fiber harness the fibers are terminated in individual 3-mm diameter stainless steel ferrules. These ferrules are off-the-shelf components used in standard connectors, and as such are both very high precision and inexpensive. The high precision allows a very close fit between the ferrule and its hole in the plug-plate (Owen et al. 1994).

The fibers are protected in the plugging region using 3-mm diameter nylon tubing. The tubing terminates below the plug-plate (above it when the cartridge is turned over for plugging) in an aluminum anchor block. The tubing is permanently attached to this anchor block but the fiber is free to slide inside it, allowing for some expansion and stretching of the tubing without harming the optical fiber.

The fiber harness is attached to the plug-plate cartridge in only two places. At one end the V-groove block is glued to the slit plate using a strong, fast setting acrylate adhesive. At the other end, the anchor block is bolted to the plug-plate cartridge using a single bolt. This simple system permits easy installation and replacement of the science fibers. Mounting all 640 science fibers in our first plug-plate cartridge took less than two days.

4. Coherent Fiber Bundles for Guiding

Guiding and sky-background measurements are performed using a set of Sumitomo Corp. coherent fiber-optic bundles. At the plug-plate the guide bundles are terminated using the same ferrules as the science fibers (but with a larger-diameter central hole). This allows science-fiber holes and guide-bundle holes to be drilled in the same operation, for maximum precision. At the other end, the fibers are imaged onto a Photometrics 768×512-pixel SenSys thermoelectrically cooled CCD camera with 9-μm pixels.

One coherent bundle 1 mm (16 arcsec) in diameter will be used for sky-brightness measurements. Ten coherent bundles will be used for guiding, one with a diameter of 1 mm and nine with a diameter of 0.5 mm. The larger guide bundle will also be used for field acquisition, if necessary. Guide-star data from ten stars distributed across the 3° field of view will enable us to actively correct not only pointing but also the angle of the instrument rotator and the plate scale of the telescope. The plate scale is adjustable (over a fairly small range) by moving the primary and secondary mirrors. This allows the aluminum plug-plates to be drilled in advance and used over a range of operating temperatures.

The 11 coherent fiber bundles for one plug-plate cartridge will be assembled as a single harness. The overall design of this harness is similar to that of the science fibers. There will be one coherent bundle per anchor block, allowing the anchor blocks to be distributed across the plug-plate. The output end of the coherent bundles are terminated in a puck attached the slit head; this puck is then reimaged onto the guide camera. The re-imaging optics and camera are mounted to the spectrograph

5. Throughput of the Science Fibers

The fiber vendor was required to measure the throughput of each science fiber and to meet the following specifications: the 20 fibers in a harness must have a mean throughput of at least 90 %, and each fiber must have a throughput of at least 87 %. These specifications were originally determined by measuring the throughput of prototype fiber harnesses fabricated by several vendors during the development phase.

Fiber throughput is measured by feeding a uniform broad-band $f/5$ beam into the input end of the fiber and detecting all output light that falls within an $f/4$ cone. Absolute calibration is obtained by bringing the output detector to the input source to measure the source intensity. The light source is feedback-stabilized and as an added precaution the input intensity is measured both before and after measuring each harness. To simplify testing and reduce testing expenses we made the testers highly automated (Owen 1997).

We built two identical throughput testers, one for the vendor and one for in-house testing. The vendor has measured all of the 5880 fibers delivered and we have measured 5220 of these. The results of our measurements are shown in Figure 3. As can be seen, the fibers greatly exceed the throughput specifications.

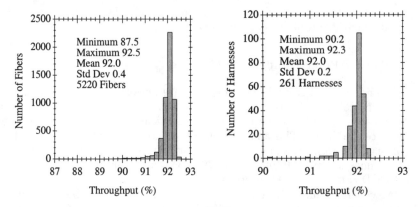

Figure 3. Throughput for each fiber and mean throughput for each harness.

We also compared throughput measurements made at the University of Washington with those reported by C Technologies. These data were acquired with different throughput testers of the same design. Earlier tests indicated that the system used by the vendor consistently measured lower throughputs by a few tenths of a per cent. In addition, the throughput measurements are not directly comparable due to the way the fibers were cleaned. Our intent at the University of Washington was to accurately characterize the fibers, so we tested various cleaning techniques and settled on reagent-grade isopropyl alcohol and cotton swabs. In contrast, C Technologies was only asked to verify that the fibers were acceptable, not to characterize them, so C Technologies used the more efficient cleaning technique of wiping the fibers with tissue.

Subtracting C Technologies throughput measurements from ours, we found a mean of 0.4 %, a standard deviation of 0.6 %, a minimum of −3.8 %, and a maximum of 4.7 % for the 5220 fibers we measured. The mean is consistent with the differences between the two testers and the different cleaning regimens discussed above.

Figure 4 shows a scatter plot of UW vs. C Technologies throughput measurements. The diagonal line shows the ideal case of equal throughput. Most of the points are clustered at the upper-right corner, indicating both excellent agreement and excellent fibers. The horizontal spreading of the main clump of data and the scattered points to the left of the clump are probably due to the different cleaning regimens discussed above. The few points scattered lower along the line of equal throughput appear to be genuinely inferior fibers. The very few points along the right-hand side and well below the line appear to be fibers we did not clean well. We tested this conjecture by carefully recleaning some of the worst examples in an earlier version of the graph and the corresponding throughputs did move above the line.

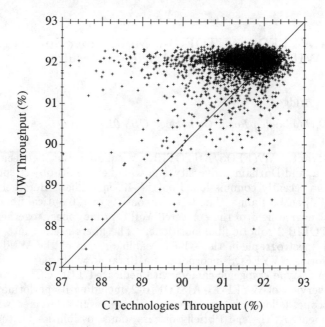

Figure 4. UW vs. C Technologies throughput measurements.

6. Conclusion

The design and manufacturing process for the Sloan Digital Sky Survey fiber feed has been very successful. The science-fiber vendor has delivered 5880 optical fibers of excellent quality. The fibers have high throughput and are very easy to install in the plug-plate cartridge. The plug-plate cartridge is robust, provides safe fiber routing, and offers generous room for plugging.

Throughputs of 5220 science fibers were measured by both the fiber vendor and ourselves using two automated fiber testers of the same design. The fibers have a mean throughput of 92.0 % and a standard deviation of 0.4 %. The two fiber testers agree to within 0.4 %.

Acknowledgments. We are grateful to C Technologies, Inc. for their excellent work manufacturing and testing the science fiber harnesses.

References

Owen, R. E., Siegmund, W. A., Limmongkol, S. & Hull, C. L. 1994, Proc. SPIE, 2198, 110

Owen, R. E. 1997, SDSS Telescope Technical Note 19970415

The WYFFOS/AUTOFIB-2 Multi-Fiber Spectrograph on the WHT: Description and Science Results

Terry Bridges

RGO, Madingley Road, Cambridge CB3 0EZ, UK

Abstract. WYFFOS/AUTOFIB-2, a collaboration between RGO, La Palma, and Durham University, is a wide-field multi-object spectroscopy system recently commissioned on the 4.2-m William Herschel Telescope (WHT) on La Palma. The top end consists of 126 optical fibers covering the 1-degree field of the corrected WHT prime focus, together with the AUTOFIB-2 robotic fiber positioner. The fibers are fed into the WYFFOS spectrograph in the GHRIL cabin on one of the WHT Nasmyth platforms. WYFFOS incorporates a Baranne white-pupil design, and a Schmidt-type cryogenic camera with dedicated TEK CCD. I first present an overview of WYFFOS/AUTOFIB-2, including its performance on the telescope, followed by some recent science results obtained with the instrument. At the end I briefly discuss some possibilities for the future.

1. Introduction

At the end of 1993, a prime-focus corrector was installed on the William Herschel Telescope (WHT). This corrector gives a 1-deg diameter field (40-arcmin diameter unvignetted), and includes both an atmospheric-dispersion compensator (ADC) and a rotating instrument platform. In 1994 and 1995 a wide-field multi-fiber spectroscopic system was commissioned at the WHT prime focus. This facility consists of a robotic fiber positioner (AUTOFIB-2, built by Durham University), the fiber feeds, and an intermediate-dispersion spectrograph (WYFFOS, built at RGO) mounted permanently in a cabin on one of the WHT Nasmyth platforms. WYFFOS/AUTOFIB-2 is now in routine operation on the WHT. I will first give a brief description of the facility, its performance on the telescope, and then give a flavor of the science that has been done with this instrument.

2. Description of the System

This section will be brief, since details of the system have already been published elsewhere (e.g. Watson 1995; see also the WYFFOS WWW page at http://www.ast.cam.ac.uk/RGO/wyffos/wyffos.html). Table 1 summarizes the main parameters of the facility.

Table 1. WYFFOS/AUTOFIB-2: main parameters

Field diameter	1 deg
Unvignetted field	40 arcmin
Plate scale	17.6 arcsec mm^{-1}
Focal ratio	$f/2.81$
Fiber positioner	AUTOFIB-2; capacity 160 fibers
Fiducial fibers	10×7 fiber guide bundles, res. 1 arcsec
Autoguider imageguides	1 fixed, 1 movable
Science fibers (module 1)	126×153 μm≡2.7 arcsec
Science fibers (module 2)	150×90 μm≡1.5 arcsec
Cycle time	~ 10 s per fiber
Positioning accuracy	~ 20 μm (absolute)
Science fibers length	26 m
Slit optics (module 1)	Sapphire microlenses
Spectrograph	WYFFOS: Baranne white-pupil type
Collimator	$f/8.2$; $f=820$ mm
Camera	$f/1.2$; $f=132$ mm (effective)
Dispersions (plane gratings)	11.5–0.8 Å pixel^{-1}
Dispersion (grism)	~6 Å pixel^{-1}
Resolving power (echelle)	~8000
Detector	Thinned Tektronix TK1024
Pixel size	24 μm square

2.1. AUTOFIB-2

AUTOFIB-2 is a robotic fiber positioner, similar in philosophy to AUTOFIB-1, and to the 2dF instrument; see Parry et al. (1994) for more details. The (x, y) carriage carries the gripper head and a mobile probe that can view and centroid on back-illuminated fibers. The mobile probe also has an image guide feeding the autoguider CCD camera, used to view star images (to a limiting magnitude limit of ~ 16 mag); this allows mapping of astrometric starfields and characterization of the prime-focus geometry. There is also a "fixed sky probe" which can be used as an off-axis autoguider. Finally, there are ten 7-fiber fiducial bundles which feed a TV viewing system (nine are currently in operation), used for field acquisition and manual guiding. There is a Vax-based observing system which allows communication with the VME-based machine control system that talks directly to the robot; see Lewis, Jones, & Parry (1994) for more details.

2.2. Fiber Feeds

The current fiber module, largely constructed at RGO, contains 126 × 26 m of high-OH ("wet") Polymicro fibers, in 14 individually-cabled bundles of nine fibers (only eight bundles of 117 fibers can be imaged onto the TEK detector at the present time). The fibers cover 2.7 arcsec on the sky, and have been used mainly for galaxy observations. Each of the bundles has a nine-way fiber connector on the top end of the WHT, to allow the lower part of the feed to remain permanently mounted on the telescope. The fiber input ends have microprisms to bend the light into the fibers, plus magnetic buttons to hold them onto the field plate. The fiber output ends are terminated by 2-mm sapphire microlenses which convert the outgoing $f/2.5$ beam to an $f/8.2$ beam to feed the WYFFOS collimator. In order to accommodate the microlenses, each bundle of nine fibers is brought into a skewed 3×3 array; thus the spectrograph sees three staggered fiber slits. See Worswick et al. (1994) for further details.

A second fiber module, with 150×1.5 arcsec fibers will be commissioned in 1998. Given the lower levels of sky admitted by these smaller fibers, they will be much better for observations of point sources. The fibers are again high-OH Polymicro fibers, but there will not be connectors used with this module.

2.3. WYFFOS

WYFFOS is permanently mounted in the GHRIL cabin on one of the WHT Nasmyth platforms. It uses Baranne's white-pupil design, in which a dioptric collimator is used in double-pass (near-Littrow) mode with a reflection grating to form an intermediate spectrum close to a concave spherical relay-mirror. This intermediate spectrum is then re-imaged onto the detector inside the Schmidt-type camera. The WYFFOS gratings give resolutions between 2.5 and 20 Å, with the 2.7-arcsec fibers and the TEK CCD. There is also an "echelle" grating (actually a 632-line grating used at large angle and low orders [3–7]), which has a resolving power $R \sim 5000$ (the resolution in order 7, limited by the fibers, is ~ 0.75 Å). Finally, the spectrograph can also be used in transmission mode, where a second collimator and grism give a dispersion of ~ 6 Å pixel^{-1}. See Bingham et al. (1994) for more information about WYFFOS.

WYFFOS is controlled using EPICS/VsWorks on a VME micro, and motor-drives operate the slit translation and focus, filter slides, and the grating table. At present, wavelength calibration is performed by directing a beam from calibration lamps on the GHRIL table via the Nasmyth flat to the prime-focus corrector and then to the fibers. In 1998, a fully remotely-controlled calibration facility will be mounted at the WHT "broken-Cassegrain" focus.

2.4. Software

Too often software is forgotten when astronomical instrumentation is discussed. As well as the control software which runs WYFFOS and AUTOFIB-2, and the higher-level software that links these subsystems to the telescope, there are two more important pieces of software, both written by Jim Lewis (RGO). Af2_configure takes positions for the program and fiducial objects (with optional weights), and then optimally allocates fibers to objects; the user may then interactively "tweak" the configuration. Wyf_red is an IRAF package used to reduce WYFFOS data, based on Frank Valdes's dofibers task. Both packages plus documentation are available on request from Jim Lewis (jrl@ast.cam.ac.uk).

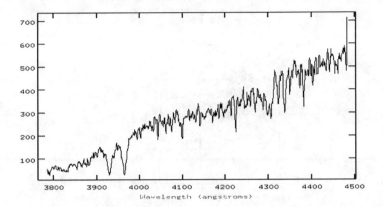

Figure 1. WYFFOS spectrum for M 31 globular cluster 158-213 ($V=14.7$ mag), at 2.5-Å resolution. Absorption lines from Ca H&K, Hδ, and the Mg I triplet are clearly visible.

2.5. Performance

After extensive re-engineering of AUTOFIB-2 at La Palma in the fall of 1997, the entire system is performing well. At the present time, it takes $\sim$ 25 min to position the 110 currently-enabled fibers ($\sim$ 13 s per fiber), with further gains expected. The current absolute positioning tolerance, as determined from mapping of astrometric fields, is $\sim$ 0.35 arcsec ($\sim$ 20 μm). In 1997 December, observations of standard stars were taken to measure the system throughput, and show that a star with an AB mag of 16.5 mag gives 1 photon s^{-1} Å^{-1} for an *average* fiber (lower for the holographic 2400-line grating, and for the echelle grating). The limiting magnitude for absorption-line spectra is B=21.5–22.0 mag. The true performance of the system is best demonstrated by the excellent science that has already been carried out (next section): note that one program obtained $\sim$ 1000 spectra in two nights of observations!

3. Science Results

This is not meant to be an exhaustive list of the science programs that have been proposed and/or carried out with WYFFOS, but merely to give a flavor of the kinds of science that can be done with the instrument.

3.1. M 31 Globular Clusters

In 1996 November, Bridges, Carter, Irwin, Brodie and Huchra obtained spectra for $\sim$ 180 globular clusters in M 31 with 2.5-Å resolution, covering five 1-deg fields along the major axis, and complete to V=18; Figure 1 shows a spectrum for one of the brightest clusters (V=14.7 mag). The velocities (good to 10 km s^{-1} or better) and abundances as determined from absorption-line indices, will be used to study the kinematics and chemical-enrichment history of the clusters, and to determine the mass distribution in the M 31 halo.

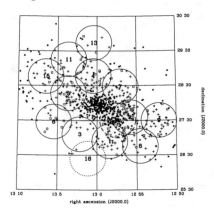

Figure 2.   Coma fields obtained with WYFFOS by M. van Haarlem.

3.2. Cluster Galaxies

Virgo   In 1996 March, Gorgas, Pedraz, Guzman, Cardiel, and Gonzalez obtained WYFFOS spectra for $\sim$ 25 dwarf ellipticals (dEs) in the Virgo Cluster. From these spectra they have derived central Mg_2, Fe, and $H\beta$ line-strengths, with the goal of determining the ages and abundances of these galaxies. They find that: (1) the dEs follow the Fe–Mg trend found for Galactic globular clusters, rather than that of more luminous ellipticals, and (2) most of their dEs are significantly younger than more luminous ellipticals and Galactic globular clusters.

Coma   Also in 1996 March, M. van Haarlem obtained spectra for just over 1000 objects in the Coma Cluster, of which 196 were confirmed to be Coma galaxies without previous velocities. These data will be used to study how the cluster interacts with its larger-scale environment (e.g. filaments and The Great Wall). Figure 2 shows the WYFFOS fields obtained by van Haarlem and the confirmed Coma Cluster galaxies; see van Haarlem (1998) for more information.

3.3. Lyman-Alpha Absorbers at Low Redshift

Bowen, Pettini, and Boyle used WYFFOS in 1995 August to obtain redshifts for 69 galaxies near the line of sight to the bright QSO 1821 643 (z=0.297). From a combined WYFFOS-HYDRA sample, only one galaxy within 500 kpc of the QSO sight-line is associated with a strong Lyman-α absorption line from the QSO. Bowen et al. (1998) conclude that strong Lyman-α absorbers are at most weakly correlated with galaxies at low redshift.

3.4. A UV-Selected Galaxy-Redshift Survey

In 1997 April, Treyer, Ellis, Milliard, Donas, and Bridges obtained WYFFOS spectra for $\sim$ 70 galaxies selected in the rest-frame UV from the FOCA balloon-borne experiment. Many of these galaxies show strong emission lines; see Figure 3. The spectra have been used to determine the galaxy redshift distribution and to derive a rest-frame UV luminosity function and star-formation rate at low

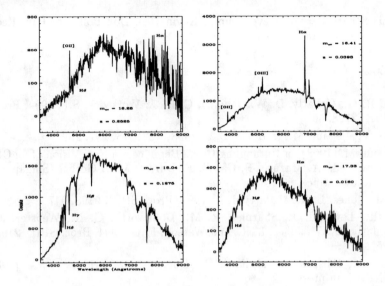

Figure 3. Representative spectra from the UV-selected galaxy-redshift survey of Treyer et al.

redshift. The data imply that the local abundance of star-forming galaxies has been underestimated, and that claims for strong evolution in the global star-formation rate over the range $0 < z < 1$ have been overstated. See Treyer et al. (1998) for more information.

4. The Future

WYFFOS/AUTOFIB-2 has already proven to be a powerful instrument for wide-field multi-object spectroscopy. During the next year we will commission the second module of 1.5-arcsec fibers, which will be used extensively for observations of point sources (stellar populations in Local Group galaxies, cluster galaxies at intermediate redshift, QSOs). Integral-field spectroscopy will become ever more important, and recently we have seen the commissioning of INTEGRAL (see paper by S. Arribas, this volume), which uses the WYFFOS spectrograph. Sue Worswick has an optical design for a longer focal-length camera which will allow better sampling of integral-field fibers (e.g. TEIFU, see paper by R. Haynes, this volume). There is also interest in building a multiple integral-fiber feed (MIFF) for the WHT, which would have something like 10 4×4 arcsec fiber bundles covering the 15-arcmin Nasmyth field, feeding the new long camera. Even more speculative is the idea of an IR multi-fiber system for the WHT.

Acknowledgments. Space precludes the naming of everyone who has been involved with WYFFOS/AUTOFIB-2. However, key players have been: (from the RGO) Richard Bingham, Nick Ferneyhough, Charles Jenkins, Dave King, Jim Lewis, Paddy Oates, Fred Watson (AAO), and Sue Worswick; (from Durham) Ian Lewis (AAO), Ian Parry (IoA), and Ray Sharples; and (from

La Palma) Stuart Barker, Steve Crump, Kevin Dee, Don Pollacco, Paul Rees (LJMU), and Bart van Venrooy.

References

Bingham, R. G., Gellatly, D. W., Jenkins, C. R., & Worswick, S.P. 1994, Proc. SPIE, 2198, 56
Bowen, D. V., Pettini, M., & Boyle, B. J. 1998, MNRAS, 297, 239
van Haarlem, M. (1998), in Untangling Coma Berenices: a New Vision of an Old Cluster, ed. A. Mazure, F. Casoli, F. Durret, & D. Gerbal (Singapore: World Scientific), p. 46
Lewis, I. J., Jones, L. R., & Parry, I. R. 1994, Proc. SPIE, 2198, 947
Parry, I. R., Lewis, I. J., Sharples, R. M., Dodsworth, G. N., Webster, J., Gellatly, D. W., Jones, L. R., and Watson, F. G. 1994, Proc. SPIE, 2198, 125
Treyer, M. A., Ellis, R. S., Milliard, B., Donas, J., & Bridges, T. J. 1998, MNRAS, in press
Watson, F. G. 1995, in 35th Herstmonceux Conference, Wide-Field Spectroscopy and the Distant Universe, p. 25
Worswich, S. P., Gellatly, D. W., Ferneyhough, N. K., Terry, P., Weise, A. J., Bingham, R. G., Jenkins, C. R., & Watson, F. G. 1994, Proc. SPIE, 2198, 44

Fiber Optics in Astronomy III
ASP Conference Series, Vol. 152, 1998
S. Arribas, E. Mediavilla, and F. Watson, eds.

Fiber-to-Object Allocation Algorithms for Fiber Positioners

F. Sourd

*DAEC - Observatoire de Paris, F-92195 Meudon Cedex,
Laboratoire d'Informatique de Paris 6, UPMC, 4 place Jussieu,
F-75252 Paris Cedex 05, France*

Abstract. In multi-object spectroscopy, the fiber-to object-allocation is of paramount importance in order to maximize the number of observations. We present new algorithms that compute automatically as many assignments as possible for fiber positioners that do not allow any fiber cross-overs. Thanks to them, we are able to observe up to 10% more objects than with previous published algorithms and in many examples it is proved that it is not possible to find better results. Then we carry out experiments and simulations to investigate the effect of different technical choices on the efficiency of the allocation process.

1. Introduction

To find the best possible assignment of fibers to objects in a target field is a difficult problem as well for a human being as for a computer. But the latter, helped by good algorithms, can find better solutions more quickly than a trained person. This paper presents new algorithms to perform automated assignments. These are not linked to an existing fiber positioner. The only supposed condition is that fibers are parked around the field and that cross-overs between fibers are not possible, which is the case of most instruments.

Some algorithms have already been published to solve the fiber-to-object allocation problem for different instruments. In order to compare these, we adapted them to a virtual instrument of 100 fibers and we tested them on fields of 100 objects.

Tournassoud & Vaillant (1988) describe an algorithm based on the Hungarian method (Kuhn 1955): when some mechanical constraints are relaxed, the problem is equivalent to a linear problem that can be solved easily; once this is done, the assignments that do not satisfy the relaxed constraints are deleted. Results are not very good when there are more than 50 fibers: in our tests, only 60 fibers are allocated. The heuristic presented by Donnelly et al. (1992) builds a solution thanks to evolution operators such as allocation and dis-allocation. Lewis et al. (1993) present the algorithm that further on we will call H_0. Experimental tests have shown that these last two algorithms have more or less the same efficiency: about 80 fibers are allocated, and the one by Lewis is faster.

Section 2 gives a mathematical model of the problem and its formal terms. Section 3 presents the algorithm and its efficiency. Then §4 shows simulations and analyzes the effect of technical parameters on the efficiency of the allocation process.

2. The Model

Let N be the number of fibers and n the number of objects in the target field. Let $\mathcal{O}$ be the set of objects. Because of mechanical constraints—length of the fiber, maximum non-radial angle of the fiber—a fiber F cannot be assigned to any object $o \in \mathcal{O}$. An assignement of F to o is denoted $\langle F, o \rangle$. The non-assignment of F is denoted by $\langle F, \emptyset \rangle$.

Two assignments are said to be *compatible* if the corresponding fibers do not intersect. A (partial) solution is a list of compatible assignments. We search for a solution containing as many assignments as possible.

3. The Allocation Algorithm

The allocation algorithm has a construction phase that gives a possible assignment and a post-optimization phase that improves this first solution.

The choice of the first fiber and of the clockwise or counterclockwise selection of the other fibers is not very important, so we suppose that the first allocated fiber is F_1, the second one is F_2, and so on until F_N.

3.1. H_0 Algorithm

This algorithm is based on Lewis et al. We will call the angle between the fiber and the line between the park position and the center of the target field the anti-radial angle of an assignment.

Let us denote by A_i a partial solution $(a_1, a_2, \cdots, a_i)$. The procedure $H_0(A_i)$ below is a greedy algorithm that extends A_i by successively computing the a_j's, $j \in \{i+1, \cdots, N\}$ as follows:

```
procedure H_0(A_i)
for j := i + 1 to N do (for all unallocated fibers)
begin
    R := {o ∈ O | A_{j-1} · ⟨F_j, o⟩ is a partial solution}
    for each o ∈ R let θ_o be the anti-radial angle of ⟨F_j, o⟩
        if R = ∅ (no possible allocation)
            then ω := ∅
            else
                if v(A_{j-1}) = 0 (allocation of the first fiber)
                    ttf then choose ω ∈ R such that |θ_ω| is minimum
                        else choose ω ∈ R such that θ_ω is minimum
        A_j := A_{j-1} · ⟨F_j, ω⟩
end
```

Note that A_0 is the empty list, and that $H_0(A_0)$ is the solution provided by the lowest-level heuristic H_0. In what follows, we also use the notation $H_0(A_i)$ to denote the solution provided by H_0 for the input A_i.

This algorithm gives quite good results within about one second, but the solution may often be improved while changing by hand the assignment of a few fibers.

3.2. H_1 and H_k Algorithms

For any partial solution A_j, we will denote its value by $v(A_j)$, i.e. its real number of assignments—the unallocated fibers are not counted. The procedure $H_1(A_i)$ below chooses at step j the assignment decision corresponding to the maximum-valued solution in the subset

$$\{H_0(A_{j-1} \cdot a) \mid A_{j-1} \cdot a \text{ is a partial solution}\}$$

procedure $H_1(A_i)$
for $j := i+1$ **to** N **do** *(for all unallocated fibers)*
begin
  1. $R := \{\omega \in \mathcal{O} \cup \{\emptyset\} \mid A_{j-1} \cdot \langle F_j, \omega \rangle \text{ is a partial solution}\}$
  2. **for each** $\omega \in R$ let g_ω be the value $v(H_0(A_{j-1} \cdot \langle F_j, \omega \rangle))$
  3. choose $\omega \in R$ such that g_ω is maximum
  4. $A_j := A_{j-1} \cdot \langle F_j, \omega \rangle$
end

If, at the third line of the loop, the maximum is obtained for more than one $\omega \in R$, the tie may be broken by selecting the object with the minimum anti-radial angle. Note also that R cannot be empty since $A_{j-1} \cdot \langle F_j, \emptyset \rangle$ is always a partial solution.

It is easy to generalize the construction of H_1. Hence H_2 is defined as H_1 but will call H_1 instead of H_0 in Step 2. More generally, we can define H_k in the same way: in Step 2, it simply calls H_{k-1} instead of H_0. This can be easily programmed using recursive procedures.

It can be shown that H_k gives a better result than H_{k-1}, but it is also much longer. If H_0 takes one second, H_1 needs about 40 seconds to complete and H_2 30 minutes... With current computers and more than 100 fibers H_1 seems the best algorithm considering the time–efficiency ratio.

This ratio can easily be improved by introducing a limit on the cardinality of R in Step 1. Instead of selecting all possible objects, we can choose the two or three (including $\emptyset$) with the lowest anti-radial angle, θ_ω.

3.3. Post-Optimization

The post-optimization makes changes on the solution found by H_1 in order to improve it. It consists more precisely of disallocations and re-allocations of fibers. We suppose that H_1 has allocated the fibers counterclockwise. The post-optimization phase will run clockwise as follows:

1. It selects a non-allocated fiber.

2. The d following fibers (when going clockwise) are disallocated.

3. They are reallocated using H_2.

4. If the new solution is at least as good as the previous one, it becomes the current solution.

5. Search the following unallocated fiber. If no stopping condition is met, go to step 2.

The typical stopping condition is that one lap of disallocations and reallocations has been completed. Two—or more—laps of post-optimization can be achieved. The process seems more efficient if the direction of selection of the fibers is changed after each lap, and if the start fiber is selected randomly. The efficiency of post-optimization is experimentally at its best when d is greater than 10 and less than 15.

3.4. Computational Efficiency

This algorithm gives better results than any previously published algorithm within 5 min CPU time on a 133-MHz Pentium PC. When there are far more fibers than objects or far more objects than fibers, the problem is quite easy, and H_0—with little or no post-optimization—often finds a near-optimal solution. H_k with $k \geq 1$ and post-optimization are required when the number of objects is close to the number of fibers. From now on we will only consider these kind of instances: let us suppose that $n = N$.

When $n = N = 100$, H_1, helped with post-optimization, finds on average 10 assignments more than H_0, and this difference of efficiency between these two algorithms increases with n (and N).

The optimal solution of an instance can be computed with branch-and-bound methods (Sourd 1997). It is too time consuming—about 24 hours with $n = N = 100$—to be used operationally. Anyway, to know the optimal solution is of great importance for reaching conclusions on the efficiency of our algorithm. In many instances, this exact algorithm does not find a better solution than those found by our approximation algorithm. We can conclude that this one often finds optimal solutions in a few minutes.

4. Experiments and Simulations

This configuration algorithm has been tested for:

- 100- to 400-fiber instruments
- FUEGOS and 2dF geometries
- Clustered and uniform target fields

In FUEGOS geometry, the fibers are held inside a tube of fixed length. The outer end-point of the tube is moving with the retracting system of the fiber. Its course is represented by one dotted segment in Figure 1.

In 2dF geometry, however, the retracting system is fixed, and the length of the *visible* fiber depends on the position of the allocated object.

The following two experiments do not aim to test or improve an existing instrument. They just show general features that can be useful for instrument designers.

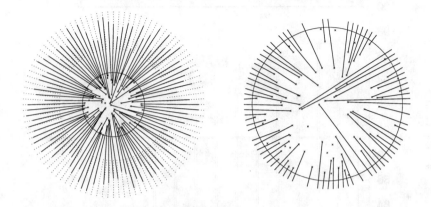

Figure 1. Two geometries for fiber positioners (on the left the FUEGOS geometry and on the right the 2dF geometry). The FUEGOS moving retracting system is larger than the 2dF fixed one.

4.1. Instrument Geometry

We tested the algorithm with a field radius of 452 mm, a button of 9 mm and a maximum possible angle of 8 deg. With uniform target fields there were on average 2 % more assignments for FUEGOS-type instruments but with clustered target fields there where about 2 % more assignments for 2dF-type instruments.

4.2. Button Size

In this simulation, the field radius was set at 452 mm for a FUEGOS-type instrument of 100 fibers. The algorithm was run on a uniform target field. Figure 2 shows how the number of assignments varies with the size of the button. In fact, a 1-mm larger button means one assignement less.

5. Conclusion

We presented efficient algorithms to maximize the number of observations in multi-object spectroscopy. Their implementation is quite simple. A C program and an OBJECT-PASCAL program performing this automated assignment are available to any interested party. Persons desiring a current copy should contact the author by mail or by electronic mail at **Francis.Sourd@lip6.fr**.

The presented algorithm can easily be modified in order to consider *weighed* stars. A number—e.g. between 1 and 10—is associated with each star and means the priority of observation of the star. The sum of the weights of observed stars has to be maximized.

Because of its geometrical approach, the algorithm can easily consider such additional technical constraints as constrained cross-overs, shadows on the target field, or sky fibers.

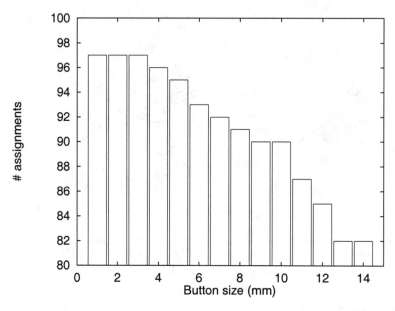

Figure 2. Effect of varying the size of the button on the number of assignments.

Acknowledgments. I would like to thank Paul Felenbok and Philippe Chrétienne for helpful discussions regarding technical modeling and combinatorial optimization.

References

Donnelly, R. H., Brodie, J. P., & Allen, S. L. 1992, PASP, 104, 752

Kuhn, H. W. 1955, Naval Research Logistic Quaternal, 3, 253

Lewis, I. J., Parry, I. R., Sharples, R. M., & Taylor, K. 1993, ASP Conf. Ser. Vol. 37, Fiber Optics in Astronomy II, ed. P. M. Gray (San Francisco: ASP), 249

Sourd, F. 1997, Assignation de fibers optiques pour la spectroscopy multi-objets, Rapport de DEA (1997), DAEC - Observatoire de Paris, F-92195 Meudon CEDEX

Tournassoud, P., & Vaillant, R. 1988, Des robots et des étoiles, Note, INRIA, Domaine de Voluceau, F-78153 Le Chesnay CEDEX

Multi-Fiber Spectroscopy at the Observatorio "Guillermo Haro"

B. E. Carrasco, and S. Vázquez

Instituto Nacional de Astrofísica, Optica y Electrónica, Luis Enrique Erro 1, C.P. 72840 Tonantzintla, Pue., Mexico

Deqing Ren & R.M. Sharples

Durham University, Science Laboratories, South Road, Durham DH1 3LE, UK

R. Langarica

Instituto de Astronomía, UNAM, Cd. Universitaria, D.F., Mexico

I. J. Lewis

Anglo–Australian Observatory, P.O.Box 296, Epping, NSW 2121, Australia

I. R. Parry

Institute of Astronomy, University of Cambridge, Madingley Road, Cambridge CB3 OHA, UK

Abstract. Within a collaborative program between INAOE and Durham University, we present a project to adapt the fiber-positioning system AUTOFIB-1.5 (Af-1.5) to the 2.1-m telescope at the Observatorio "Guillermo Haro" in Cananea, Son., Mexico. Af-1.5 is a robot that moves In the x, y & z directions to position 55 fibers across a field plate. It was built at Durham University as a prototype for the William Herschel Telescope (WHT) prime-focus fiber-positioning system AUTOFIB-2. Af-1.5 has been used on the WHT during two observing runs and its performance has been extensively evaluated in the laboratory. The 2.1-m Cananea telescope with a new corrector system will provide a 47.8-arcmin field of view. The corrector mounting is also the mechanical interface between the telescope and the fiber positioner. Af-1.5 fiber diameters are equivalent to 2.1 arcsec, the positioning accuracy to 0.2 arcsec and the minimum fiber separation to 16 arcsec. In the first stage the multi-fiber system will be used with a low-resolution fiber bench spectrograph to study the satellite dynamics around elliptical galaxies to determine the mass and extension of dark galactic halos.

1. Introduction

AUTOFIB-1.5 (Af-1.5) was designed and built at Durham University as a prototype for the prime-focus automated fiber positioner, AUTOFIB-2 (Af-2), for the 4.2-m WHT (Parry & Lewis 1990). Based on the original AUTOFIB system for the Anglo–Australian Observatory, Af-1.5 incorporated several new design features including a viewing system that allows a fiber and the object it is trying to acquire to be seen simultaneously, an improvement in positioning accuracy and in the minimum distance between fibers. Af-1.5 operates at the Cassegrain focus and was commissioned on the WHT telescope in 1989 February and 1990 April. The lessons learnt in building the prototype were folded into the plans for Af-2 which now is routinely used at the prime focus of the WHT.

Af-1.5 has 55 spectroscopic fibers and 8 guide fibers, which can be arranged on a field plate of 380 mm. However, the Cananea telescope field of view is 47.8 arcmin, which corresponds to a 360-mm diameter field. The fiber core diameter is 260 μm, and the fiber is of the polymide-coated type with an outer diameter of 315 μm. Polymicro high-OH all-silica fiber is used for the spectrograph fibers. At the input end a microprism of 1.2-mm side length is cemented onto the polished end face so that the light traveling normal to the field plate is reflected into the fiber, which is held parallel to the field-plate. The top half of the instrument is a pick-and-place robot, which places the fibers in the desired position one at a time. The robot consists of an $x-y$ positioning table which carries an electromagnetic gripper which can be moved in the z direction so that a fiber can be lifted above the others when it is moved from one place to another.

Given its characteristics, Af-1.5 is an instrument that can be used on a 2-m class telescope, where the scales of allocation time allow projects to be carried out that can fully exploit the multi-object capability. The goal of this project is to turn Af-1.5 into a stand-alone instrument routinely used at the Cananea telescope. To achieve this, building an interface for the 2.1-m telescope and a new spectrograph are required. The project can be divided into two main parts, the first regarding the fiber positioner and its interface with the telescope, and the second regarding the spectrograph. In the following section we present all the work done to achieve our goal.

2. Telescope Interface

The $f/12$ Ritchey–Chrétien telescope optics (Cornejo & Malacara 1973), with an $f/2.7$ primary, was built at the INAOE optical workshop. The telescope has a one component field-flattening system which gives a 30-arcmin field, however its back focal distance was too short to feed the fiber positioner. Furthermore, the image size at the edge of the field was bigger than the fiber's diameter.

A two-component system was required to flatten the field and to correct for astigmatism. The design, by S. Vázquez, also corrects for chromatic aberration and is formed by an LF5 meniscus-type lens, and a bi-concave lens. The two 360-mm diameter corrector components give a 47.8-arcmin corrected field with an image size of less than 115 μm, equivalent to less than 1 arcsec over the field and a 157-mm back focal distance, enough to feed Af-1.5.

With the new corrector, the field of view of Af-1.5 will be as large as the maximum field achievable given the telescopes rotator diameter. The 260-μm diameter fibers will be equivalent to 2.1 arcsec, the 30-μm positioning accuracy to 0.2 arcsec, and the fibers can be placed as close as 2 mm, equivalent to 16 arcsec.

Non-telecentricity is unvoidable in a wide field corrector with two elements and a back focal distance of about 150 mm. At the edge of the field the principal ray is at an angle of 6.46°. The half angle of the $f/12$ beam is 2.4°, so an extreme ray hits the focal plane at an angle of 8.86°, which is equivalent to $f/3.2$. A way to compensate this effect is to use prisms at the ends of the fibers which have a tilt of 6.46/2=3.23° built in so that they are 3.23° off at the center and the edge and correct half way out. This would make the extreme rays enter the fibers at 5.7°, which is equivalent to $f/5$. However, for the first stage of this project we will start with the fibers as they were built originally with right-angled prisms at the ends of the fibers. Changing the right-angled prisms to prisms with an appropriate tilt will be postponed for a future update.

The two elements of the corrector system were built at the INAOE optical workshop. To minimize losses due to reflection, each surface was coated with a broad-band anti-reflection coating (BBAR) by Denton Vacuum Inc. The LF5 transmittance of a witness sample with and without coating on both surfaces was measured by using a spectrophotometer. The results show that with the anti-reflection coating there is an improvement of more than 10%.

3. Mechanical Interface

The corrector mounting, designed by R. Langarica, is also the mechanical interface between the telescope and the fiber positioner. The mounting was aluminum cast and subsequently machined at the INAOE mechanical workshop. Finite Elements Analysis was carried out during the design stage to minimize flexure of the mounting and to estimate possible flexures of the fiber positioner. Considering the extreme case of the mounting and Af-1.5 at a horizontal position and with the boundary condition that Af-1.5 flange is fixed, the results show that in the worst case the flexure expected is very small. The mounting has not been tested at the telescope as we decided it would be better to test it in combination with Af-1.5, because testing requires a large $x-y$ system, which will be available with it.

4. Optical Interface

Two cameras are required for the optical interface: one for acquisition and guidance and a second for closed-loop fiber positioning. INAOE has two SBIG (ST6 and ST7) Peltier-cooled cameras with the corresponding control PCs available for this project. The very sensitive 765×510 pixel ST7 camera with 9×9 μm pixel size will be used for acquisition and guiding. While the 375×242 pixel ST6 camera with 23×27 μm pixel size will be used for closed-loop fiber positioning.

The coherent bundle will be projected onto one of the CCD camera via optical rails and relay lenses inside a light-tight box. Similarly, for the guide fibers an optical arrangement to relay the guide-fiber image is needed. While the spectroscopic fibers are 14 m long, the guide fibers are much shorter, so the CCD cameras and their optics must be located close to Af-1.5. The coherent bundle in particular is very short, which imposes severe limits on the location of the CCD systems and optics. Software enhancements are required to fully integrate these two new cameras to the fiber-positioner control system, although it is possible to operate Af-1.5 without making any changes to its software.

5. Spectrograph

At the first stage the multi-fiber system will be used with a low-resolution fiber bench spectrograph to study the satellite dynamics around elliptical galaxies to determine the mass and extension of dark galactic halos. For this project we need to measure relative velocities between the galaxy candidates to the primary galaxy of about 500 km s^{-1}, requiring a spectrograph with a resolution, $\Delta\lambda$, of about 1 nm.

The Af-1.5 fiber focal-ratio degradation performance has been measured in the optics laboratory (Carrasco 1992). For an $f/11.4$ input beam, the output beam for all the fibers will be within an $f/5$ beam. For the spectrograph design an $f/4$ fiber output beam was assumed. A Nikon $f = 135$ mm, $f/2$ lens will be use as a collimator and a Nikon $f = 85$ mm, $f/1.4$ camera lens will be use in combination with a 1200-line mm^{-1} grating, blazed at 600 nm with a blaze angle of 21° 06'. The angle between the collimator and the camera is 50° (45.95/-4.05 deg). The slit width is given by the fiber core diameter of 260 μm (2.1 arcsec). The detector will be a 1024×1024 pixel (24 μm × 24 μm) Tektronix chip. With these components each fiber will be projected onto 4.7 pixels in the spectral direction and 6.8 pixels in the spatial direction. Forty-five fibers can be arranged along a 17-mm slit without vignetting.

The spectrograph has been tested in the optics laboratory by Ren Deqing, using a Pixtor 416XT 768×512 pixel (9μm × 9μm) detector. The measured resolution is $\Delta\lambda = 1.1$ nm, the anamorphic factor is 1.46, and the free spectral range is 240 nm, centered at 540 nm. The spectral-line G-band (430 nm), H$_\beta$ (486.1 nm), Mg I (517.1nm), Na I (5889,5895 nm), and H$_\alpha$ (656.3 nm) are within this spectral range. The spectrograph resolution is limited by the fiber size, as expected.

References

Carrasco, B. E. 1992, PhD Thesis, Durham University

Cornejo, A., & Malacara D. Boletín del Instituto de Tonantzintla, 1, 35

Parry, I. R., & Lewis I. J. Proc. SPIE, 1235, 681

Part 3: Two-Dimensional Fiber Spectroscopy

Integral-Field Spectroscopy with Optical Fibers on Medium-Size (1.5–4-m) Telescopes

C. Vanderriest

DAEC, Observatoire de Paris-Meudon, 92195 Meudon Cedex, France

Abstract. Most of the new large (8–10-m) telescopes will be equipped with integral-field spectrographs. This does not mean that the same technique is unimportant for smaller telescopes. In fact, the combination of the optical parameters (scale, field of view, etc.) and availability of such telescopes allows spatial, spectral and temporal domains to be explored which are in many respects complementary to those covered with large light collectors.

We discuss the capabilities of some fiber-optic spectrographs associated with medium-size telescopes. Results on quasars and AGNs are presented that are representative of the good spatial sampling case.

1. Introduction: the Coming of Age of IFS

Standards in telescope making have changed rapidly in the past 20 years; this has induced some vocabulary shifts. The new generation of *large telescopes* have 8–10-m apertures, 4-m telescopes being henceforth considered as *medium size* while 1.5–2 m may be reduced to the status of *auxiliary instruments*. At the same time, modern detectors and multiplex techniques have largely increased the power of observation (in terms of photon and information gathering) for a telescope of any given diameter. This is obvious with the fast spread of multi-object spectrographs, a key factor in the exploration of large-scale structures of the universe. Integral-field spectrography (IFS) has the same multiple potentialities, with some *qualitative* advantages too. At first, its development was at a slower pace, but it is now coming of age, as can be judged from numerous communications at this meeting, and also from the fact that this kind of equipment is now considered indispensable for the new large telescopes. Hence, some naïve questions:

Q: Why do we need IFS?

A: Because of the bad behavior of astronomical objects, whose shapes (generally) do not match the nice slits of our classical spectrographs.

Q: Why does this imply special techniques?

A: Because of the bad behavior of our detectors, which (generally) offer only 2 dimensions for recording 3 variables (α, δ, λ).

Several solutions have been elaborated for resolving this geometrical puzzle. The principles are not very new, but we had to wait for technical improvements before practical realizations were possible. We present here some instruments based on optical fibers and discuss illustrative results.

2. IFS Techniques

Strictly speaking, the limitation of present detectors to two dimensions is not insurmountable. One may recall the attempts to record spectral variables in the thickness of a photographic emulsion (see Lindegren & Dravins 1978; Connes 1995, and references therein) and some possible tricks allowing the measurement of the energy of a photon, in addition to its coordinates, with electronic detectors. However this is still marginal or exploratory. Current systems are divided into two broad classes.

- Scanning systems that use time as a third dimension include:
- Long-slit scans (the "poor man's IFS")
- Scanning Fabry-Perot interferometers
- Fourier transform interferometers for the infrared
- IFS *stricto sensu* changes the geometry of the telescopic image (squeezing three dimensions into two) before entering an otherwise "classical" spectrograph.
- The oldest system used for the geometrical transformation is the mirror image-slicer. It was designed mostly for increasing the resolution of stellar coudé spectrographs and has limited capabilities.
- It was first suggested seriously 40 years ago (Kapany 1958) to replace these image-slicers by optical-fiber devices arranged in a "peacock's tail" geometry (that prompted the name ARGUS): in a bundle, the fibers form a two-dimensional array at the focal plane of the telescope and are realigned along a "pseudo-slit" at the entrance of the spectrograph.
- If an array of microlenses, instead, is put in the focal plane, it splits the field into an array of micropupils that may be used as entrance apertures of the spectrograph: this is the TIGER approach (Courtès et al. 1988).

Finally, combining the advantages of both ARGUS and TIGER devices into a single system is the optimal solution for the "integral-field units" (IFUs) of modern spectrographs (Courtès 1982). In this case, 100% coverage of the focal image is insured by the (square or hexagonal) microlenses, while the redistribution of the pupils by the fiber bundle ensures the best coverage of the detector area and a large spectral range.

An important advantage of IFS is that all the information is recorded simultaneously, i.e. in the same conditions of observation. Accurate centering in a slit is no longer needed. Differential atmospheric refraction does not ruin the spectrophotometric measurements (although it does not help for the data processing: the use of atmospheric dispersion compensators is recommended). The software for data processing is partly similar to the one used for long-slit data. So maps in flux, velocity, line ratios, etc., may be easily reconstructed. The spectra from fibers corresponding to the same physical region may be co-added, which gives a constant spectral resolution for areas of different extent.

The main limitation, inherent in the method, is the small field of view (typically 10″ diameter for sub-arcsecond sampling, no more than 1′ for 2″ sampling). This is enough, however, for many types of objects. Larger objects may be covered in several fields if variations in the observing conditions are tolerated.

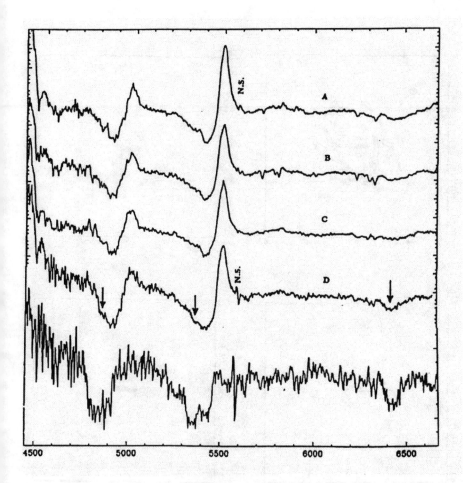

Figure 1. Spectra of the four images of H1413+117, from SILFID observations in 1989 march. For each image the spectrum has been summed over the maximum possible area while avoiding the "blended" fibers, i.e. fibers receiving light from more than one image. Note the $z = 1.44$ and 1.66 absorption systems, mostly visible in B, and the peculiar behavior of D (see text). The lower spectrum is the difference $(D-\langle S \rangle)$, where $\langle S \rangle$ is the average spectrum of A, B and C, normalized to the intensity of the broad emission lines. The "excess" of continuum and of some BAL components is interpreted as differential microlensing effects in D.

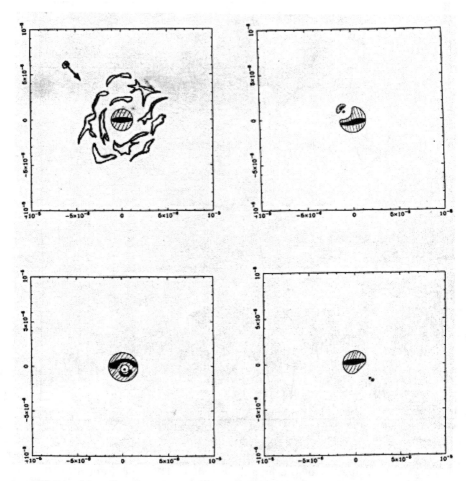

Figure 2. Simulation of microlensing effect on H1413+117. The quasar is modeled by a disk that emits the continuum ($2.5 \times 10^{-6}{''}$ diameter), and a perfectly absorbing cloud (ellipse with $2.5 \times 10^{-6}{''}$ and $0.5 \times 10^{-6}{''}$ major and minor axes). In the first frame, we have sketched the BLR clouds. We assume $Z_{\text{source}} = 2.54$ and $Z_{\text{lens}} = 1.5$; the microlens is a 1 $M_\odot$ star passing in front of the quasar along a 45° path.

3. Examples of Scientific Production from IFS

Scientific results obtained with various instruments will be presented at this conference: SILFID (Chatzichristou), MOS/ARGUS (Soucail), Albireo (Herpe et al.), PMAS (Roth), Hexaflex, 2D-FIS, and Integral (talks and posters by the IAC group). Here are some other examples of interesting findings that would not have been possible without IFS.

3.1. SILFID

Our first test of a fiber bundle (Vanderriest 1980) prompted us to build an adapted spectrograph (fast collimator accepting a long slit): SILFID (Vanderriest & Lemonnier 1988). Its first real scientific use on the sky was on comet P/Halley in 1985 (Malivoir et al. 1990). Mostly used at CFHT between 1986 and 1992, it is presently being modified in order to be adapted to the 2.3 m telescope (Vainu Bappu) at Kavalur, south India. Here we would seek to cover a large field of view with moderate spatial resolution.

In 1989, the "clover-leaf" (a gravitational mirage with four bright images of a BAL quasar) was observed at CFHT with SILFID in very good seeing conditions (see Angonin et al. 1990). The configuration is very tight, with a typical separation $\sim 1''$ between the images; we thus used the best available sampling, 0.33 arcsec per fiber. This allowed the spectra of individual images to be extracted with a minimum contamination (Fig. 1).

The spectrum of component D shows definite departures from the averaged spectrum of the other components; this may be explained by microlensing effects. It is known that the flux amplification by microlensing depends critically of the ratio between the size of the source and the Einstein radius of the microlens. With a stellar microlens ($m \sim 1 M_\odot$) and the typical size and distance of a quasar source, it is expected to observe differential amplification between the different structures of the quasar. This is indeed observed in several other cases of gravitational mirages (Q 0957+561, the "Einstein cross", etc.), as an excess of continuum with respect to the broad emission lines. Compared to the Einstein radius of the microlens, the size of the BLR is larger, and hence with a negligible amplification, while the size of the disk that emits the continuum is smaller. The slightly bluer color of this continuum for D can even be explained by a relatively larger amplification of the inner (smaller) parts of the disk.

Likewise, microlensing may further enhance some BAL components whose angular size is a fraction of that of the continuum disk. Figure 2 shows the result of a numerical simulation. When the amplification is close to the maximum (step 3 in the sequence), the observed spectrum is well reproduced. Microlensing is a transient phenomenon, whose time scale depends on the relative proper motion between the microlens and the source. If this interpretation is valid, the amplification should thus decrease with time and the spectrum of image D should become progressively similar to that of the other images. Monitoring the spectrum would give information about the structure of the BAL clouds on a micro-arcsecond scale. Further observations of the "clover leaf" in 1995 and 1997, although suffering from poorer seeing, definitely showed the same effect on image D, with the same enhanced BAL components. No significant variation of the amplification factors is observed, which means that the characteristic

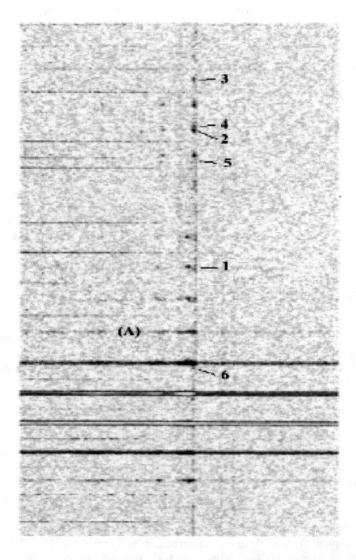

Figure 3. Raw MOS/ARGUS spectra of 3C 48 in the H_β-[O III] spectral region. Several clouds, with slightly different radial velocities, may be identified (the numbers point to the maximum intensity of each cloud). Note the velocity gradient and resolved lines in cloud 1, the strong emission of cloud 2 and the double structure of the [O III] lines in the nuclear region.

crossing-time for the microlens could be a few tens of years. It could take that long for obtaining a natural "cut" of the BAL structure from the microlens effect. However, since the structure of BAL quasars is poorly understood, a patient monitoring by IFS is worth the effort.

3.2. MOS-ARGUS

In 1993 MOS/ARGUS has replaced SILFID at CFHT. The trick was to insert a fiber bundle into the existing MOS spectrograph (see Vanderriest 1995 for a description). This was a cheap way for doing integral-field spectroscopy with the set of grisms and the detectors that MOS accommodates. The field of view ($8'' \times 12''$) and sampling (0.4 arcsec per fiber) are well adapted to the study of low-redshift quasars.

Figure 3 shows the raw data from an observation with MOS/ARGUS of the allegedly "well-known" quasar 3C 48. The aim was to study the kinematics and ionization state of the emission clouds detected by Stockton & MacKenty (1987), and to ascertain the status of the faint secondary nucleus 3C 48A (Stockton & Ridgway 1991; Hook et al. 1994). It is believed that this is the remnant core of a merging galaxy that reactivated the quasar.

In the raw data, it is already possible to recognize the different clumps of gas, showing some velocity differences. The secondary nucleus appears as a featureless red continuum (A). On the reconstructed images (Fig. 4), the pattern of gas clouds appears clearly north of the quasar nucleus. Clouds 2–6 present narrow [O II], H$_\beta$, and [O III] emission lines with intensity ratios compatible with photoionization from the quasar nucleus. Region 1 is more complex, with a resolved blue (280 km s^{-1}) component showing an obvious velocity gradient, and a fainter red component. It is likely that we see in this region the projection of two distinct clouds. It should be noted that this is the region where the radio jet of 3C 48 interacts strongly with some gaseous material, the interstellar medium of the galaxy that harbors the quasar and/or of the merging companion.

A previously unknown feature is the splitting of the NLR emission lines into two distinct components. On poorer quality spectra, Gelderman & Whittle (1994) noticed that 3C 48 has broad and "flat top" [O III] lines. Here, these lines may be nicely fitted by two Gaussian components separated by 570 km s^{-1} (Table 1). The red component is also spatially resolved ($\sim 0.8''$), and its centroid does not coincide with that of the blue component, hence the variable ratio from fiber to fiber (Fig. 4). This complex structure is likely to be related, directly or indirectly, to the merging process. Either both components are produced by a bowshock driven by the motion of the radio-emitting plasma through the interstellar medium, or the blue component comes from material belonging to the merging companion.

A more detailed discussion may be found elsewhere (Chatzichristou et al. 1998). I just wanted to stress with this example the invaluable advantages of IFS for studying such structures close to the spatial-resolution limit.

3.3. The Canarian IFS

Up to now, CFHT is the observatory that has devoted the largest number of nights to IFS in its different flavors (fibers with SILFID and MOS/ARGUS, micro-lenses with TIGER, various devices with OASIS) and for various subjects

Table 1. Line splitting in the NLR of 3C 48 (the spectra have been co-added over the nuclear region (15 fibers), then the lines have been fitted with two Gaussian components denoted "b" [blue] and "r" [red])

Line	Z_b	$FWHM_b$ km s^{-1}	Z_r	$FWHM_r$ km s^{-1}	ΔV km s^{-1}
[Ne V] 3426	0.3667	1130 ± 200	0.3694	480 ± 120	590 ± 100
[O II] 3727	0.3668	1000 ± 180	0.3696	390 ± 100	610 ± 80
[Ne III] 3869	0.3670	980 ± 170	0.3695	400 ± 100	550 ± 80
[O III] 4959	0.3668	1015 ± 90	0.3694	370 ± 80	570 ± 50
[O III] 5007	0.3669	1010 ± 80	0.3694	380 ± 60	550 ± 40
weighted mean→	0.36685 ±0.00005	1010 ± 60	0.36945 ±0.00005	380 ± 50	570 ± 30

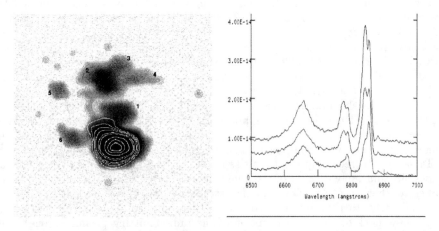

Figure 4. A *(left)* Image reconstruction of 3C 48 in [O III]$_{5007}$ (gray levels) showing six main emission clouds, and in the continuum ($\lambda_{\rm obs}$ = 7000–8000 Å, isophotes) showing the secondary nucleus 3C 48A. For the emission-line picture, an automatic single-Gaussian fitting procedure was used. North is up, east to the left. B *(right)* Spectra from three adjacent fibers in the central region of the quasar showing the variations of the blue/red components (the separation between fibers is 0.4").

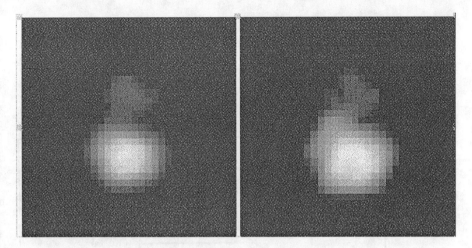

Figure 5. Images produced by fitting two Gaussians to the [O III]$_{5007}$ emission profiles. The blue image (left) is not significantly resolved, while the red one (right) has a less peaked distribution (FWHM $\geq$ 0.5″) and dominates toward the west and north. The offset between the centroids of the two components is $\sim$ 0.8″.

(see Vanderriest et al. 1994). The IAC has also a strong record of development and efficient use of IFS with fiber devices. The fiber bundles were realized at IAC itself and associated with various spectrographs and telescopes of the Roque de los Muchachos Observatory. This instrumentation is now used mostly for extragalactic astronomy.

- Hexaflex (*Hexa*gonal field, *flex*ible bundle) was first used with the William Herschel Telescope (WHT), and then adapted to the Nordic Optical Telescope (NOT). It has 125 fibers with a sampling of 0.7 arcsec per fiber.
- 2D-FIS is adapted to a double spectrograph (ISIS, with one blue and one red arm) at WHT. It has also 125 fibers with 0.9 arcsec per fiber plus 30 "sky" fibers in a circle of radius 38″.
- INTEGRAL, adapted to the WYFFOS spectrograph on the WHT, was commissioned in 1997 July. It is a versatile instrument giving several spatial samplings and coverages (see elsewhere in these proceedings for details).

An active program on active galaxy nuclei and their environment has been undertaken with these instruments. Interesting results have been obtained notably on NGC 1068, 3227, 3516, 4151, 5728, 7331, etc. As an example, Fig. 5 shows a reconstructed map of emitting clouds in the central region of the Seyfert galaxy NGC 3516. This is obtained from multiple Gaussian fitting of the [O III] lines. Several components are recognized, with different locations, relative velocities and line broadenings (Arribas et al. 1997). Such data give real access to the three-dimensional structure of the NLR. It allows a choice to be made among the different models proposed for explaining the morphology and kinematics of the ionized gas in the inner region of the galaxy.

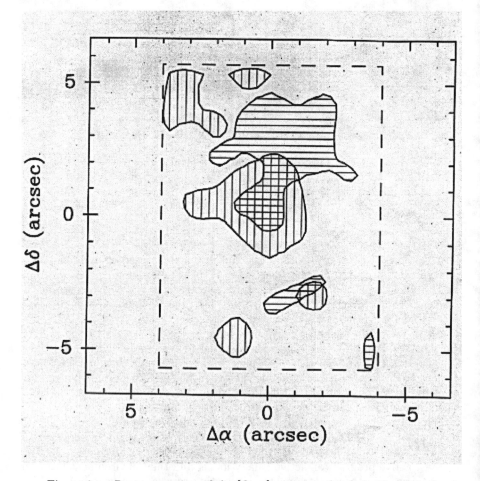

Figure 6. Reconstruction of the [O III] emission clouds in the central region of NGC 3516 (from Arribas et al. 1997). In addition to the two kinematically distinct components shown, a third redshifted one is present over the entire field.

4. Conclusion

IFS with optical fibers is now developing at a fast pace and is more and more "user friendly". Most of the new very large telescopes will provide this mode of observation as a general tool. It could also improve the efficiency of medium size (1.5–4-m) telescopes in several ways.

1) Some observational "niches" may be filled more easily with medium size telescopes rather than very large ones; for instance covering large fields ($\sim 1''$) with moderate spatial sampling ($\sim 2''$). In that case, the image scale may be more convenient compared to the pixel size of current detectors.

2) It is also possible to organize systematic observations for some complete samples of not-too-faint objects. In this case, easier availability of the telescope is the important point.

3) For the same reason, IFS on medium-size telescopes could help to select candidates for further, more detailed observations on large telescopes.

4) Spectrographic monitoring of variable extended or complex objects is also a job for that class of telescopes. With existing instruments, it may already be possible to organize a continuous (24h/24h) monitoring in IFS, by coordinated observations in several sites. The targets I can identify are objects of the solar system (planetary surfaces or atmospheres, and comets), as well as transient events in crowded stellar fields.

IFS with large telescopes is most efficient for objects that need the highest possible spatial resolution and *when this resolution is indeed achievable*, i.e. at very good sites and/or with adaptive-optics techniques. The situation is quite different when spatial resolution is not a crucial factor. The number of photons from an object with a surface brightness B, observed with a telescope of diameter D, during a time t, if we assume a spectrograph with N fibers, each corresponding to an angular diameter L (arcsec) on the sky, is proportional to:

$$B \cdot D^2 \cdot t \cdot N \cdot L^2.$$

If everything else (i.e. including L) is kept constant, this is proportional to D^2 and an 8-m telescope is 16 times more efficient than a 2 meter. But $L = d/F$, where d is the linear diameter of the fibers (μm) and F the focal length of the telescope, so the above expression is also:

$$B \cdot (D/F)^2 \cdot t \cdot N \cdot d^2 = B \cdot (D/F)^2 \cdot t \cdot S,$$

S being the total area (in μm^2) of the "slit" built with the fibers. Considering a given fiber device, an object with constant surface brightness and homothetic telescopes (same F/D), the efficiency no longer depends on the diameter of the telescope. Of course, the resolved spatial elements will be $0.3''$, for instance, with the 8-m telescope and only $1.2''$ with the 2-m but for a lot of objects this could be sufficient. Also, it is probably easier to obtain a longer total observation time, t, on the small telescope.

References

Angonin, M.-C., Remy, M., Surdej, J., & Vanderriest, C. 1990, A&A, 233, L5
Arribas, S., Mediavilla, E., Garcia-Lorenzo, B., & del Burgo, C. 1997, ApJ, 490, 227
Chatzichristou, E., Vanderriest, C., & Jaffe, W. 1998, in press
Connes, P. 1995, in ASP Conf. Ser. Vol. 71, Tridimensional Optical Spectroscopic Methods in Astrophysics, ed. G. Comte & M. Marcelin (San Francisco: ASP), 38
Courtès, G., 1982, in Instrumentation for Astronomy with Large Optical Telescopes, ed. C. Humphries (Dordrecht: Reidel), 123.
Courtès, G., Georgelin, Y. P., Bacon, R., Monnet, G., & Boulesteix, J. 1988, in Instrumentation for Ground-Based Optical Telescopes, ed. L. Robinson (New York: Springer), 266.
Gelderman, R. & Whittle, M. 1994, ApJS, 91, 491
Hook, R., Lucy, N., Stockton, A., & Ridgway, S. 1994, ST-ECF Newsletter, 21, 16
Kapany, N. 1958, in Concepts of Classical Optics, ed. J. Strong (San Francisco: Freeman)
Lindegren, L., & Dravins, D. 1978, A&A, 67, 241
Malivoir, C., Encrenaz, T., Vanderriest, C., Lemonnier J.-P., & Kohl-Moreira, J.-L. 1990, Icarus, 87, 412
Stockton, A., & MacKenty, J. 1987, ApJ, 316, 584
Stockton, A., & Ridgway, S. 1991, AJ, 102, 488
Vanderriest, C. 1980, PASP, 92, 858
Vanderriest, C., & Lemonnier, J.-P. 1988, in Instrumentation for Ground-Based Optical Telescopes, ed. L. Robinson (New York: Springer), 304
Vanderriest, C., Bacon, R., Georgelin, Y., LeCoarer, E., & Monnet, G., 1994, Proc. SPIE, 2198, 1376
Vanderriest, C. 1995, in ASP Conf. Ser. Vol. 71, Tridimensional Optical Spectroscopic Methods in Astrophysics, ed. G. Comte & M. Marcelin (San Francisco: ASP), 209

The ARGUS Mode of the ALBIREO Spectrograph: Evaluation of Its Performance and First Results

G. Herpe, and C. Vanderriest

DAEC, Observatoire de Paris-Meudon, 92195 Meudon, France

J. Sánchez del Rìo, and A. del Olmo

Instituto de Astrofisica de Andalucia, 18080 Granada, Spain

Abstract. Albireo is a multi-function spectrograph for the 1.5-m telescope of the Sierra Nevada Observatory (Spain). Its "spatial stage" allows a choice between 3 spectrographic modes and an imaging mode. The ARGUS mode for integral-field spectroscopy uses a bundle of 202 fibers (80-μm diameter = 1.5" sampling) covering an area of 16" × 30". The useful wavelength range is 350–1000 nm. This mode will be used mostly for the kinematical study of galaxies up to magnitude 18, and point-like objects up to 20 mag.

1. Introduction

A collaboration between the EROES team of the Laboratoire d'Astronomie Extragalactique et de Cosmologie (Observatoire de Paris-Meudon, France) and the Instituto de Astrofisica de Andalucia (Granada, Spain) led to the realization of a multi-purpose spectrograph, dubbed Albireo. This instrument is permanently mounted at one of the two Nasmyth foci ($f/8$, scale: 1"=58.7 μm) of the 1.52-m telescope at the Sierra Nevada Observatory, the other focus being devoted to direct imaging and photometry. Albireo was designed to be used for a wide variety of stellar and extragalactic programs. It has three spectroscopic modes: classical (long-slit), multi-object (MEDUSA) and integral field (ARGUS). We briefly describe this instrument and give more details on the ARGUS mode.

2. Structure of Albireo

The instrumental chain consists of three parts: (i) the *mechanical interface* with the telescope, (ii) the *spectrograph* itself, and (iii) the *detector*.

(i) The *rotator* is an interface between the telescope and the spectrograph. It allows a full rotation for the multi-object and long-slit modes.

(ii) The *spectrograph* (Herpe et al. 1995) includes three separate parts:
• The *bonnette* fulfils the functions of field acquisition, guiding and calibration. A high-sensitivity movable video camera insures acquisition and guiding for the different observing configurations. The bonnette also contains a wavelength-calibration unit.

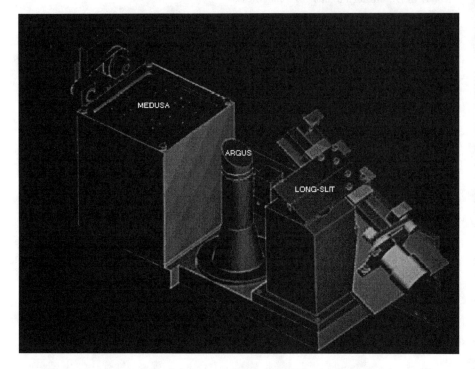

Figure 1. Drawing of the spatial stage (SAM), showing the three different observing modes in Albireo. For each mode, the optical assembly has its entrance aperture(s) in the telescope focal plane and its exit at the collimator focus.

- The *spatial stage* offers three spectrographic configurations with different spatial coverage: a classical slit with 6' length and adjustable width; multi-object spectroscopy (MEDUSA) with 40 fibers of 200μm (=3.3") diameter plugged into an aperture plate covering 20', and integral-field spectroscopy (ARGUS) with a bundle of 202 fibers (80-μm core diameter $\simeq$1.5") covering an area of 16" × 30". Direct imaging is possible, but with poor sampling, through the wide open slit; a wheel contains standard $UBVR$ filters. The slit, ARGUS, MEDUSA devices are mounted on the same sliding assembly, nicknamed SAM (Fig. 1). Changing the configuration (i.e. moving SAM and the acquisition/guiding unit) is easily done by remote control.
- The *spectrographic stage* is of classical design, with a dioptric collimator at $F/D = 5.6$ and camera at $F/D = 2$. Five reflection gratings are mounted on a wheel, which also holds a mirror for the imaging mode. The present set of gratings (300 l per mm to 1800 l per mm) gives dispersions of between, roughly, 200 Å mm^{-1} and 40 Å mm^{-1}, corresponding to resolved elements between 10 Å and 2 Å in ARGUS mode. The useful wavelength range is 3500–10000 Å.

(iii) The present *detector* is a thick CCD with 1152×770 (22.5-μm) pixels.

A full description of Albireo, its electronics, and command software is given elsewhere (Sánchez et al. 1998).

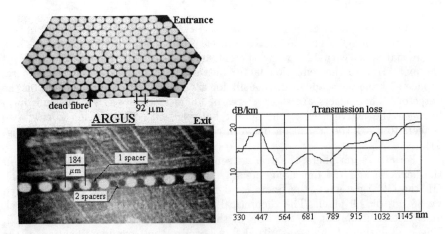

Figure 2. *Left:* Entrance and exit parts of the ARGUS device. *Right:* Transmission loss in the fluoride fibers.

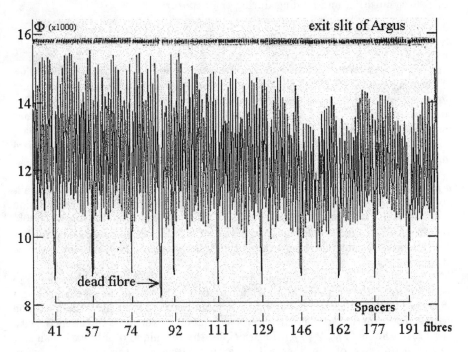

Figure 3. Image of the pseudo-slit and corresponding modulation of the light through the ARGUS device.

3. The ARGUS Mode

In this configuration, a fiber bundle is used for the geometrical transformation of the focal image. The 202 active optical fibers of the device, arranged in a hexagonal array with 12 rows at the entrance, are aligned into a pseudo-slit at the exit. Due to the reduction factor between the collimator and the camera, "spacers" have been added in this slit for a better separation of spectra on the detector. The spacers are short pieces of fibers inserted (at the slit-end only) between the active fibers: one spacer between each fiber within a row, two spacers to mark the transition between consecutive rows (Fig. 2 (*left*), and Fig. 3)).

For 80/88-μm fibers, the average separation between two fibers is 92 μm in the hexagon and 184 μm along the slit (the extra microns are due to glueing). On the CCD detector, the fibers (and hence the spectra) are then separated by 66 μm or 2.9 pixel. The theoretical diameter of a fiber image is 1.4-pixel, and, after convolution by the PSF of the optics, it is measured at $\sim$1.8 pixel. The modulation of the transmitted flux permits the clear identification of the different fibers (Fig. 3). Using a CCD with smaller pixels would help to reduce the blending between spectra in the spatial direction and slightly improve the resolution, up to $R = 5000$ with the highest-dispersion grating.

A tilted mirror around the entrance field allows the use of a video camera for field acquisition and guiding during the exposure.

4. Manufacture of the ARGUS Device

The device was built entirely in our laboratory. It necessitates careful handling of the tiny fibers. High mechanical precision is needed, as well as very good polishing at both ends of the fiber bundle (use of 3-μm emery paper). The principal difficulties arose from the very small parts, especially at the entrance because the guiding mirror should allow the closest possible access to the main field. The central hole in this mirror for the fibers and mechanical support is 4 mm.

The length of the optical fibers in the bundle being 150 mm, silica was not *a priori* an imposed choice. The criteria for choosing the material are the availability in 8-0μm diameter and thin cladding, while keeping a good transmission over the useful wavelength range ($T > 90\%$ from 4000 to 10000 Å and $T > 80\%$ at 3500 Å—see Fig. 2 (*right*)). It was also of interest to use a material with a dilatation coefficient matching that of the brass support. This led to the choice of fluoride glass.

The glueing of fibers is made row by row at the entrance. The fibers are loose between the entrance and exit for a better absorption of mechanical and thermal shocks. The whole assembly is mounted in a metallic cylinder. The geometrical ordering at the entrance is satisfactory over most of the field, with few broken fibers (Fig. 2 (*left*)). In principle, the technique of fabrication should keep the ordering of the fiber within a row. However, the relation between entrance and exit locations has been checked for each fiber, and a few inversions have been identified. This is taken into account in the reduction software (see Teyssandier, these proceedings).

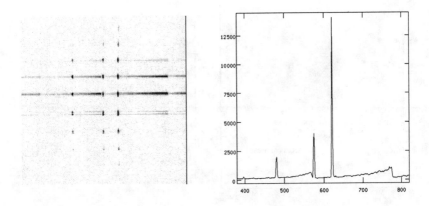

Figure 4. Left: Raw data for R Aqr with the 1800 l-mm^{-1} grating. Right: Spectrum of the brightest region, showing the H$_\beta$ and [O III] emission lines of the nebulosity, and the underlying M-star continuum of R Aqr itself.

5. Performance and First Results

From observations of spectrophotometric standards, we measure a practical limiting magnitude of $\simeq$ 17 mag (S/N=10) for a single 30-min exposure with a 300-l mm^{-1} grating. The total derived efficiency, 0.035 at 5500 Å, seems relatively low but could be improved in 2 ways: 1) by changing the present thick CCD for a thinned back-illuminated CCD, and 2) by performing the long-overdue aluminizing of the telescope mirrors. This would boost the efficiency above 0.10. Here are some typical observations with the present instrumental chain:

- R Aqr is a well known symbiotic star composed of a Mira (period = 386 days) and a faint hot companion. It is embedded in a large nebulosity. This component is well visible in the raw data (Fig. 4 [left], 20-min exposure).
- NGC 7674 (= HCG 96a) is an SBc galaxy belonging to a compact group (Fig. 5 [left]). This galaxy shows morphological as well as spectroscopic evidence of interaction with another member of the group (Verdes-Montenegro et al. 1997). A remarkable feature, probably related to the interaction, is the broad asymmetric structure of the [O III] emission lines seen on a 30-min exposure at 40 Å mm^{-1} (Fig. 5 [right]; compare with R Aqr, Fig. 4 [right]). Previously classified as Seyfert 2, it could be in fact an obscured Sy 1.
- Mrk 1014 (= IRAS 01572+0009) is a 16th mag quasar at z = 0.163. Only slit spectroscopy of the host galaxy was available (MacKenty & Stockton 1984). Here, on a 20-min exposure at 100Å mm^{-1} (not shown), the [O III] emission extends over 15″ around the nucleus. This shows that studying the environment of low-redshift quasars is possible with a "small" 1.5-m telescope.

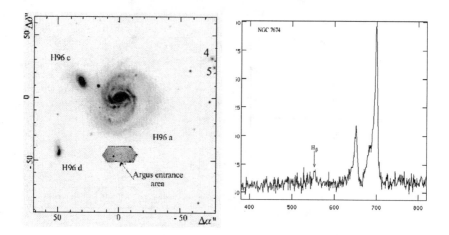

Figure 5. *Left:* Field of NGC 7674 (= Hickson 96a). The field of view of ARGUS is shown. *Right:* Spectrum of the nuclear region (integration over 6 fibers) in the H$_\beta$–[O III] region with the 1800-1 mm^{-1} grating.

6. Conclusion

The Albireo spectrograph is now commissioned and its ARGUS mode is beginning to give interesting results. With freshly aluminized mirrors and a more efficient CCD, the practical limiting magnitude for a 2-hr exposure would be of the order of 20 mag for a point-like object and 18 mag for a 30″ diameter galaxy.

This will open up a wide field of investigation, either for self-consistent survey work, or as a preparation for observations with larger telescopes.

Acknowledgments. We thank L. Ponzevera for his help in the making of the mechanical parts, as well as A. Ruy-Falco, J. Pereira, and S. Lelong for their helpful suggestions and support.

References

Herpe, G., Sanchez, J., Vanderriest, C., & Moles M. 1995, in ASP Conf. Ser. Vol. 71, IAU Coll. 149: Tridimensional Optical Spectroscopic Methods in Astrophysics, ed. G. Comte, & M. Marcelin (San Francisco: ASP), 223.

MacKenty, J., &Stockton, A. 1984, ApJ, 283, 64

Sanchez, J., Ruiz-falco, A., & Herpe, G. 1998, Rev. Sc. Instr., submitted

Verdes-Montenegro, I., del Olmo, A., Perea, J., Athanassoula, E., Marquez, I., & Augarde, R. 1997, A&A, 321, 409

Mapping the Structure and Dynamics of Late-Stage Mergers

E. T. Chatzichristou

Sterrewacht Leiden, Postbus 9513, 2300 RA Leiden, The Netherlands

Abstract. I present results obtained through detailed mapping of a number of interacting/merging candidates, using the technique of integral-field spectroscopy at the CFHT. Two-dimensional spectra, covering the full spectral range $\sim$ 4200–8300 Å with spatial resolution of 0.7–1.4″per fiber over a $(16'')^2$–$(32'')^2$ field of view, are used to analyze the kinematics of the gas component and the details of the ionization structure in the central 10–20 kpc regions of these systems.

1. Introduction

With this contribution I wish to demonstrate the power of the technique of integral-field spectroscopy (IFS) for addressing some of the most fascinating questions in current astronomical research. The main objective of the present work is to study the link between galaxy interactions/merging and the initiation of galactic activity. The observational evidence for interactions and mergers to trigger nuclear activity, although strong, remains circumstantial or qualitative. It is mainly based on statistical studies, often suffering from biased working samples and/or inappropriate control samples, which indicate only the probability that such a link exists.

This has motivated us to propose instead the *detailed* study of selected individual interacting systems in the hope that such dedicated studies could provide direct evidence for the mechanisms through which interaction-induced activity could be possible. In this context, the main questions that I will be addressing here are:
- What dominates the gas dynamics in the emitting regions of interacting/merging galaxies?
- Can we infer the energy-input mechanisms and how they relate to the "type" of galactic activity?

2. Sample Selection and Observations

We have used the technique of integral-field spectroscopy with fiber spectrographs in their various incarnations at the CFHT, taking advantage of the high spatial and spectral resolution and the large wavelength coverage that they provide, and of the sub-arcsecond conditions. We obtained two-dimensional spectra and complementary broad-band images, in order to study the detailed kinematics, ionization, and morphological properties of a selected number of Seyfert and starburst galaxies that are undergoing strong interactions. In this way we can begin to unravel the physics of the merger process and thus begin to answer the questions posed above. Two valid approaches are possible, so we have defined and carried out the following projects, in collaboration with C. Vanderriest:

(i) In an attempt to study the *evolutionary* sequence of galactic merging we defined a sample of objects selected to "represent" various interaction stages (Chatzichristou & Vanderriest 1995; Chatzichristou et al. 1997). Our sample galaxies were specifically chosen to have a range of IR color temperatures (probing different dust-heating mechanisms) and are all relatively nearby objects ($\langle z \rangle \sim 0.04$) to provide good spatial resolution. We have used the SILFID spectrograph (Vanderriest, these proceedings) with the 2048×2048×15μm Lick2 CCD on the CFHT in the ARGUS configuration designed for IFS. The entrance field consists of 397 fibers (100-μm diameter and 4-μm cladding) arranged in a compact hexagonal array. We used two grisms covering the spectral range 4200–8300 Å, with 100 Å mm^{-1} dispersion and resolution better than 5 Å at the center of the field. The spatial sampling and field size vary within 0.7–1.4″per fiber and $(16'')^2$–$(32'')^2$, depending on the configuration used. Results for some of the objects of this sample will be presented and discussed in the next section. The main drawback with this approach is that there is no unique description of what a merging sequence would be, due to the multiplicity of factors that can influence the evolution and thus the observable characteristics of an interacting system.

(ii) Alternatively, We attempt to isolate a particular stage of the interaction process and try to establish the characteristics that are *proper* to it. In this respect, we are collecting data for a sample of *advanced merger* systems that exhibit high IR and nuclear activity levels and for a control sample of single systems that match the merger sample in luminosity, redshift range, and IR properties. We have used the MOS/ARGUS spectrograph (Vanderriest, these proceedings) with the 2048×2048×21μm Stis CCD on the CFHT. A hexagonal arrangement of 594 active fibers gives a useful field of 12.8″× 7.8″with spatial sampling 0.4″per fiber. We have used two grisms covering the spectral range 3800–10000 Å, with resolution 2.2 and 5 Å at the center of the field. We have mapped the *whole* extent of each system (typically $\sim 20''$) using mosaics of 2-3 exposures.

Dedicated software (Teyssandier & Vanderriest 1997) is used to analyze the data and to produce intensity/velocity/dispersion maps at various wavelengths. The resulting data cubes are of excellent quality and, unlike slit spectra, provide a direct two-dimensional picture of the wavelength-dependent emission and absorption-line properties of these galaxies. To address the questions posed above:

(i) Mapping the velocity field of the gaseous and stellar component can provide evidence of gas motions tidally induced or associated with the nuclear activity

(ii) Line-emission diagnostics help to identify the ionization mechanisms and physical conditions throughout the emitting regions

(iii) The reconstructed intensity images coupled to the broad-band direct images are used to trace the continuum/ionized gas/stellar component/dust distribution and to identify additional components

(iv) Searching for relations between the kinematics and physical conditions in the emitting regions can help to understand what the energy-input mechanisms are and how they relate to the galaxy's nuclear activity.

3. Results and Discussion

In this section I will present the main results for three merger candidates of our first sample, Mrk 463, Mrk 789, and UGC 4085, that have a range of IR and nuclear activity. These objects are selected in order to indicate the power of IFS in analyzing and ultimately confirming or rejecting the merger nature of three apparently similar, but distinct as I will show, cases of "double nucleus" systems.

Mrk 463 is a double-nucleus system, both its nuclei being activated by the merger. Our detailed spectroscopy confirms the Seyfert-type east nucleus and the lower-ionization AGN in the west nucleus. The most activated eastern nucleus is the center of a triple radio source with approximately N–S orientation and the main contributor to the strong far-IR emission of this object ($L_{\rm FIR}$=1.8 $10^{11} L_\odot$), comparable to that of ULIRGs. We have studied this object in great detail, using the 1.4″per fiber configuration, with a seeing of $\sim 0.8''$, covering a $(32'')^2$ field and the full wavelength range 4200–8300 Å (Chatzichristou & Vanderriest 1995; Fig. 1).

A striking feature on the [O III]$_{5007}$ and H_α reconstructed images and on the $V-R$ colour map, is the extended emission ($\gtrsim 20''$) centered on the east nucleus and collimated along the radio axis. We have identified several emitting regions with distinct physical conditions, the most interesting being (i) a highly ionized knot of emission $\sim 14''$ in the direction of the southern radio jet, due to either beamed ionizing radiation from the east nucleus or "in situ" generated continuum, (ii) shock-like emission-line ratios, spatially coinciding with the path of the radio jets north and south of the east nucleus. Velocity maps constructed for various emission lines reveal a sharp (previously undetected) velocity gradient blueshifted by $\sim 450\pm50$ km s^{-1} within 2–4″north of the east nucleus, as well as asymmetric profiles with obvious sub-structure, phenomena that we attribute to bowshocks driven by the radio jets into the ambient gas. With the exception of this region, there is a remarkable absence of any radial component of motion throughout the whole system. Regions of star formation are rather compact and knotty, distributed further around the east nucleus and comparison of the blue and H_α luminosities with the IR properties of this object indicate that the AGN provides the main dust-heating mechanism. Consequently, we find that in Mrk 463 all kinematic and ionization properties are driven by the coupling between radio emission and ambient material, that is, dominated by hydrodynamic processes due to the powerful AGN that was triggered by the merger.

Mrk 789 is a recent merger product with a fuzzy appearance and multiple knots of emission and with IR properties that make it directly comparable to the class of powerful IR galaxies. We have studied this object with the 0.7″per fiber

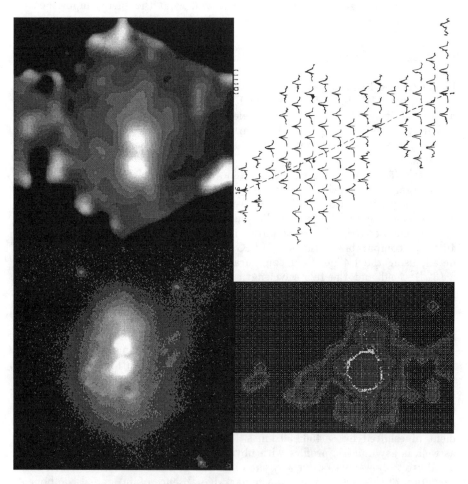

Figure 1. Mrk 463: $B+V+R$ fov$\sim(44'')^2$ (*upper left*), reconstructed continuum (*upper right*) and H_α emission (lower left) images with fov$\sim(32'')^2$ and [O III]$_{5007}$ line-profile map (*lower right*). The radio/emission axis is indicated as a dashed line, the two nuclei as asterisks and the blueshifted region is labeled "RS" (the radio source, as well).

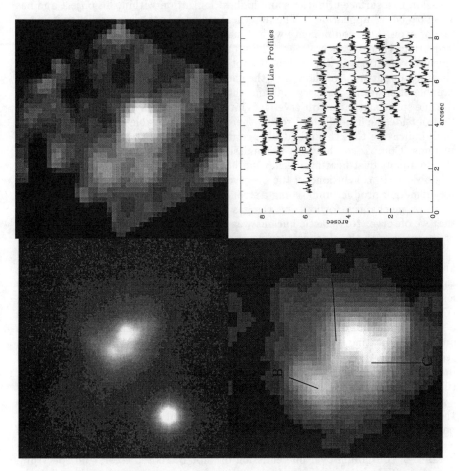

Figure 2. Mrk 789: $B+V+R$ fov$\sim(36'')^2$ (*upper left*), reconstructed continuum (*upper right*), and H_α emission (*lower left*) images with fov$\sim(16'')^2$ and [O III]$_{5007}$ line-profile map (lower right). The three main emission knots are labeled.

configuration, with a seeing of $\sim 1''$, covering a $(16'')^2$ field and the wavelength range 4200–6900 Å (Chatzichristou et al. 1997; Fig. 2). None of the emission knots seems to have the characteristics of a galactic nucleus, all being extended and showing ionization levels consistent with photoionization by hot stars. Of particular interest is a region of a few arcsec (~ 2.4 kpc) across (unidentified in previous studies), which we name knot C. This is the region of the highest emission-line surface brightness and highest ionization within this object and has a "sharp-edge" appearance. The main feature here is the line profile splitting in two or even three components with a maximum separation of 600–700 km s^{-1}, indicating violent gas motions. This one-to-one relation suggests anomalous physical conditions in the region of knot C. Using parameters derived from the observed quantities, we show that the energetics of the starburst in Mrk 789 are capable of powering a superwind, and that it has had sufficient time to generate the line splitting that we observe. It is interesting here, as in the case of Mrk 463, that outside the line-splitting region there is no appreciable line-of-sight velocity field, indicating that this might be a common characteristic in advanced merger systems. Comparison between the emission-line and IR properties indicate that the dominant dust-heating mechanism is due to the high star formation within this object. In conclusion, all the observed properties suggest that Mrk 789 is a *recent* merger product undergoing a strong burst of star formation, this being the common mechanism for ionizing the gas and producing the large-scale motions that we observe. No galactic nucleus was identified, either because it is obscured or confused with the starburst, or because it has not yet have the time to form.

UGC 4085 exhibits an asymmetric spiral structure with two giant emission knots within the central ~ 3 kpc, one of which is embedded in a bar-like structure (Fig. 3). We have used the 0.7"per fiber configuration for this object, with a seeing of $\sim 0.8''$, covering a $(16'')^2$ field and the full 4200–8300-Å spectral range (Chatzichristou et al. 1997). Although UGC 4085 was included in our sample as a merger candidate, we find no such evidence, all its properties extracted from our data indicating that this is a normal spiral galaxy with several giant H II regions: (i) both central knots are extended and resolved in our reconstructed images and show low ionization spectra that are characteristic of H II regions; (ii) the general velocity field, reconstructed for various emission lines, resembles a smooth retrograde-rotating disk with (the faintest optically) knot B close to the kinematic center; (iii) the emission-line luminosity and distribution are comparable to those for normal spirals, its IR properties indicating the dominance of a cold dust component, similar to that in isolated galaxies, and (iv) the preponderance of continuum over line emission of knot B. Its red colors and spectrum, as well as its location close to the kinematic center, indicate that this might be the location of the galactic nucleus that remains embedded in large amounts of dust.

4. Conclusions

We have undertaken a program for studying the central regions of interacting/merging galaxies, using the technique of IFS with fiber spectrographs. In this contribution I have presented a few examples of the large amount and diversity of information that such data can provide. I have shown that line-emission

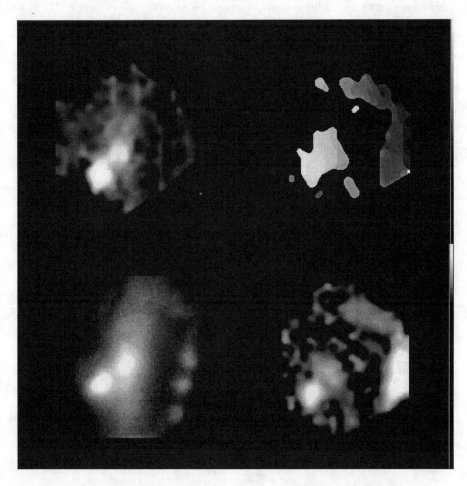

Figure 3. UGC 4085, the $(16.1'')^2$ ($\sim (7.4 \text{ kpc})^2$) region: $B+V+R$ band (*upper left*), reconstructed 5300–5500-Å continuum (*upper right*), reconstructed H_α emission (*lower left*), reconstructed H_α velocity (*lower right*) images.

diagnostics coupled with detailed kinematics are a powerful tool for identifying or rejecting putative nuclei or other components, as well as for understanding what the main energy-input mechanism is and how it relates to the activity "type" in each of these systems. In particular, I have shown that the gas dynamics in the central 10–20 kpc regions of late-stage mergers is likely to be dominated by the (more energetic) hydrodynamical processes ($\sim$ several 10^2–10^3 km s^{-1}), while the motions that a merger can generate result in much smaller velocities generally. Interestingly, a common feature of the *overall* velocity field in late-stage mergers seems to be the absence of an appreciable radial component of motion. We argue that this might be due to the presence of tidally induced motions perpendicular to the plane, that annul any signatures of rotational motion in the parent galaxies, giving to the systems their patchy appearance.

References

Chatzichristou, E. T., & Vanderriest, C. 1995, A&A, 298, 343
Chatzichristou, E. T., Vanderriest, C., & Lehnert, M. 1997, A&A, in press

INTEGRAL: An Optical-Fiber System for 2-D Spectroscopy at the 4.2-m William Herschel Telescope

S. Arribas, L. Cavaller, B. García-Lorenzo, A. García-Marín, J. M. Herreros, E. Mediavilla, M. Pi, C. del Burgo, J. Fuentes, J. L. Rasilla, and N. Sosa

Instituto de Astrofísica de Canarias, 38200 La Laguna, Tenerife, Spain

D. Carter[1], and L. Jones

Royal Greenwich Observatory, Madingley Road, CB3 0EZ Cambridge, United Kingdom

R. Edwards, B. Gentles, D. Pollacco, and P. Rees

Isaac Newton Group, Ado. 368, 38700-S/C de la Palma, Spain

Abstract. INTEGRAL is a new optical-fiber system to perform 2-D spectroscopy at the WHT, in combination with the WYFFOS spectrograph. It is situated at the GHRIL Nasmyth platform, mounted together with newly built acquisition, guiding, and calibration units. INTEGRAL currently contains three science-oriented bundles, which have different spatial configurations on the focal plane (their fiber core diameters being 0.45, 0.9, and 2.7 arcsec, respectively). As they are mounted in a wheel they can be easily rotated into the focal plane (with an overhead of a few seconds). Hence, depending on the prevailing seeing conditions the instrument can be optimized for the scientific program. Here we will present a brief description of the instrument, and the first results.

1. Introduction

INTEGRAL is a new optical fiber system for the WHT which uses WYFFOS (see Bridges, these proceedings). This spectrograph was originally conceived to allow science exploitation of light carried by fibers from the AUTOFIB2 prime-focus robotic positioner (e.g. Bridges, these proceedings). However, it quickly became apparent that other fiber-fed instruments could take advantage of the spectrograph's optimized geometry. INTEGRAL extends the power of WYFFOS allowing integral-field spectroscopy of extended objects.

Previous systems of this type installed at the WHT were HEXAFLEX (Arribas et al. 1991), and 2D-FIS (García et al. 1994). The main advantages of INTEGRAL with respect to these previous systems are: i) it has been de-

---

[1] Also at: Astrophysics Research Institute, Liverpool John Moores University, Liverpool, L3 3AF, United Kingdom

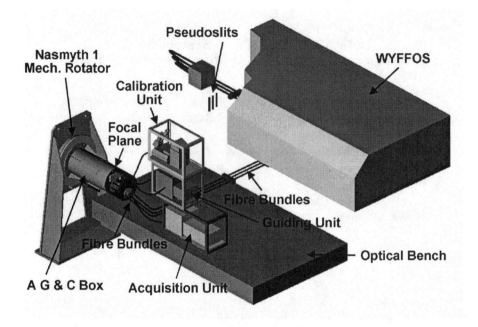

Figure 1. Outline of the INTEGRAL system.

signed as a common-user instrument, open to the whole community, ii) it allows more flexibility to interchange bundles with different spatial configurations at the telescope focal plane, iii) in most of cases "seeing limited" observations are possible, and iv) the use of WYFFOS makes this system more efficient, permitting the use of bundles with a larger number of fibers. INTEGRAL also gives all the infrastructure needed to implement new bundles and it can be used in combination with new spectrographs. Another project for the WHT based on this technique is TEIFU (Haynes, these proceedings), which will extend and complement INTEGRAL's capabilities.

2. Description of INTEGRAL

2.1. Acquiring, Guiding, and Calibration

INTEGRAL is mounted in the Nasmyth No. 1 platform (GHRIL) of the WHT. The general layout of the system is shown in Fig. 1, and a more detailed description of the focal plane area is shown in Figure 2.

A cylindrical structure (the acquisition, guiding, and calibration box structure) is bolted to the Nasmyth mechanical rotator. As this structure rotates synchronically with the focal plane when the telescope tracks, the sky position angle is fixed during observation, with no need for pre-optics.

The swing plate (SP) is located in the telescope focal plane area, and this is the connection point for the acquisition and scientific fiber bundles. By rotating the SP any installed bundle can be placed at the center of the focal plane, while

INTEGRAL: Optical-Fiber 2-D Spectroscopy at the WHT

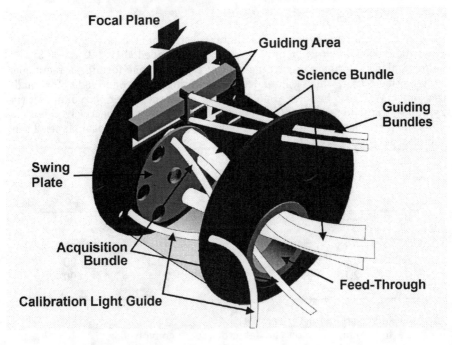

Figure 2. Focal-plane area.

the other bundles are light blocked. Hence, only one bundle at any one time can feed the spectrograph. This *multi-bundle arrangement* allows high flexibility to modify the relevant observational parameters, without implying extra losses of light. Note that optical systems which allow the the telescope scale to be modified imply losses of light.

For acquisition a coherent fiber bundle transmits the light from the focal plane (via the SP) to a standard TV camera.

Guide stars may be acquired by moving an XY slide containing two coherent fiber bundles. This unit moves in an off-axis region and the bundles feed a TV based autoguider.

A liquid fiber is used to feed a calibration source (currently copper-argon, copper-neon, or tungsten lamps) to the focal plane.

Therefore, the standard observational procedure is as follows. First, moving the SP the acquisition bundle is located at the center of the focal plane. The object is then observed in the TV acquisition screen, and centered on the desired position (even the sky position angle may be modified to orient the bundle along a particular direction). Secondly, a guide star is selected, and the XY slide unit is moved to the required position. Then the guide star is observed on the TV guiding screen, and the autoguider connected. Thirdly, the SP is moved to locate the selected scientific bundle at the center of the focal plane. Then we can start an exposure. Finally, for calibrating, the SP is moved again so that the selected bundle can reach the region illuminated by the calibration lamps.

2.2. The Bundles

Up to six scientific bundles can be simultaneously mounted in the SP, although in the standard configuration only three are used (called sb1, sb2, and sb3). At the focal plane the fibers are arranged in two groups, one forming a rectangle, and the other a ring, which is intended for collecting background light (for small-sized objects). These bundles are 5.5 m long, while their fiber core diameter (in sky units) are: 0.45 (sb1), 0.9 (sb2), and 2.7 (sb3) arcsec, respectively. Figure 3 shows the main characteristics of each bundle at the telescope focal plane. Their main characteristics are listed in Table 1.

Table 1. Characteristics of INTEGRAL fiber bundles

Bundle	1	2	3	4	5	6
sb1	0.45	0.7	205 (175+30)	7.80×6.40	variable	90
sb2	0.90	1.4	219 (189+30)	16.0×12.3	4.6	90
sb3	2.70	4.0	135 (115+20)	33.6×29.4	7.4	90

1 – fiber core diameter (arcsec)
2 – fiber image size on the detector (to be convolved
    with a 1.4–2.6 pixel PSF)
3 – total number of fibers (rectangle + ring)
4 – spatial coverage of the central rectangle (arcsec × arcsec)
5 – distance between adjacent fiber/spectra at the CCD
    (in pixels)
6 – diameter of the outer ring (arcsec)

As commented above, the bundles are simultaneously connected at the swing plate and at the entrance (*pseudo-slit*) of the WYFFOS spectrograph. Therefore, they can be interchanged very easily, with an overhead of a few seconds. Hence, depending on the prevailing seeing conditions the instrument can be easily optimized for the scientific program.

These bundles were manufactured at the IAC, using Ceramoptec fibers. The bundles sb1 and sb2 use "ultra-dry special" hydrogen-treated fibers. Bundle sb3 uses Optram "UV" fibres. Each is 5.5 m long. The relative positions of the fibers at the telescope's focal plane are known very accurately, as they were measured in the metrology laboratory with a precision of about 10 μm (equivalent to 0.05 " at the telescope focal plane). The packing fraction was about 65% for the three bundles. The fibers were arranged linearly at the slit. In addition, the slit was curved (with a radius of 1190 mm) to fit the collimator focal plane. Some space was intentionally left between fibers. This guarantees that the spectra are well separated on the detector, minimizing the extra optical cross-talk due to the finite size of the PSF of the spectrograph. The fibers were ordered by "rings", following the procedure proposed by García et al. (1994).

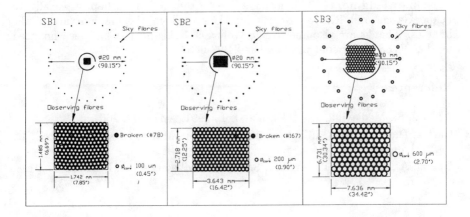

Figure 3. Configuration of the fibers of the standard bundles at the focal-plane end.

This procedure guarantees that two adjacent fibers at the slit: i) correspond to two adjacent fibers at the focal plane, and ii) are at the same distance from the center of the bundle. Some of the outer ring fibers are used in the transition between different "rings". This ordering minimizes optical cross-talk, especially in those objects which have a sharp intensity profile (e.g. galaxies).

Taking into account that the fibers are illuminated with an $f/11$ input beam, we expected to fit the $f/8$ collimator of the spectrograph well thanks to the focal-ratio degradation produced by the fibers. However, the actual bundles are producing a wider output beam, which implies extra losses of light. This can also produce a small extra level of scattered light inside the spectrograph. This point is irrelevant in our case because the spectra are well separated on the detector, which allows a good knowledg of the scattered light to be subtracted.

3. First Results

INTEGRAL was commissioned during two periods in 1997 July and November. A first estimate of the throughput gives a mean value of 1 count s^{-1} Å^{-1} for a star of AB magnitude 17 at 5000 Å. This value has some scatter depending on the relative efficiency of the fibers (of course, the different fiber efficiencies can be corrected to a high degree by flat-fielding), and it is not far from figures obtained with standard long-slit spectrographs at the WHT (e.g. ISIS; Carter et al. 1993), indicating that the advantages of the use of fibers are now compatible with reasonably good efficiencies.

During these periods several astronomical objects were observed, and preliminary analysis are presented elsewere (Serra-Ricart et al., this conference: del Burgo et al., this conference).

Acknowledgments. Thanks are due to the IAC workshop, especially to Felipe García and Ricardo Negrín for their work in the manufacture of the bundles. Thanks are also due to Sue Worswick. This project has been financed by the DGES under contract number PB93-0658.

References

Arribas, S., Mediavilla, E., & Rasilla, J. L. 1991, ApJ, 369, 260

Carter, D. et al. 1993, ISIS users manual, Users Manual No. XXIV, ING, La Palma.

Garcia, A., Rasilla, J. L., Arribas, S., & Mediavilla, E. 1994, Proc. SPIE, 2198, 75

2D Spectroscopy of the Gravitational-Lens System Q2237+0305 with INTEGRAL

M. Serra-Ricart, E. Mediavilla, S. Arribas, C. del Burgo, A. Oscoz, D. Alcalde, B. García-Lorenzo and J. Buitrago

Instituto de Astrofísica de Canarias, E-38200 La Laguna, Tenerife, Spain

L. J. Goicoechea

Departamento de Física Moderna, Universidad de Cantabria, E-39005, Santander, Cantabria, Spain

Abstract. We present two-dimensional spectroscopy of the gravitational-lens system Q2237+0305 (the Einstein Cross) obtained with INTEGRAL in sub-arcsecond (FWHM $\sim 0''.7$) seeing conditions. The four components of the system are clearly separated in the continuum intensity maps. However, the intensity map of the [C III] $\lambda 1909$ line exhibit an arc of extended emission connecting the A, D, and B components. This can be explained if the continuum arises from a tiny source in the nucleus of the lensed object while the line emission comes from a large region.

1. Introduction

The quadruple system of images Q2237+0305 is one of the most interesting gravitational lens systems because of its proximity and high degree of symmetry, for which it is also named the Einstein Cross. Discovered by Huchra et al. (1985), it has been a habitual target of many observers who provided imaging, long-slit spectroscopy, and photometric monitoring (see e.g. Yee 1988; Racine 1991; Rix et al. 1992; Corrigan et al. 1991). All this information is recorded with the aim of both the modeling of the gravitational lens and the study of microlensing. On the one hand, a good knowledge of macrolens parameters, the constraints imposed by the accurate location of the compact images and of the lens nucleus, and the insensitivity of the analysis to uncertainties in the cosmological parameters, have furthered the lens modeling in this system. On the other hand, Q2237+0305 is an ideal laboratory for studying microlensing because the predicted time delays between the four images are very short, of the order of a day, allowing a distinction to be made more easily than in other systems between intrinsic variability and microlensing effects.

For most of the scientific aims, 2-D information is crucial. For some purposes (such as the determination of the location of the four compact images) this can be achieved by means of filter imaging. In other cases (such as in the study of microlensing; see Fitte & Adam 1994), 2-D spectroscopic information is indispensable and due to the dimensions of the system ($\sim 2'' \times 2''$) this is possible only in good seeing conditions and is very difficult to carry out with conventional techniques, such as long-slit spectroscopy, which are strongly affected by the changes in the atmospheric conditions and by differential atmospheric refraction. These problems can be avoided using an integral-field spectrograph able to obtain simultaneously a 2-D distribution of spectra. The only spectral mapping of this kind available until now for Q2237+0305 has been provided by TIGER, a spectrograph based in the use of microlenses (Fitte & Adam 1994).

2. Observations and Data Reduction

The spectra of Q2237+0305 were collected on 1997 June 2 with the 4.2-m William Herschel Telescope (WHT) on the island of La Palma. We used INTEGRAL, an optical-fiber based system to perform 2-D spectroscopy. INTEGRAL links the Nasmyth focus of the WHT with the WYFFOS spectrograph. A detailed technical description of INTEGRAL is provided by Arribas et al. (these proceedings). The core of this system consists of three optical-fiber bundles with different spatial sampling and coverage which are simultaneously mounted at the focal plane (in a swing plate) and at the entrance of the spectrograph. Thanks to this configuration, the observer can match on line the spatial resolution and coverage to the atmospheric conditions and the scientific objectives. To observe Q2237+0305 we used the bundle with better spatial sampling, which basically consists of an array of 175 fibers $0''.45$ in diameter covering a rectangle of $7''.9 \times 6''.7$ on the sky, and an additional ring $90".2$ in diameter formed by 30 fibers. The relative positions of the fibers at the telescope's focal plane are known very accurately, the distance between two adjacent fibers (spatial sampling) being $0''.5$. At the entrance of the spectrograph the fibers are aligned to form a pseudo-slit. We took an exposure of 1800 s. The seeing, measured from the guide star, was about $0''.7$. The linear dispersion was 1.46 Å pixel^{-1}, and the spectral coverage 4330–5830 Å. The data were reduced in the standard way using the IRAF data-reduction package. The operations performed included bias subtraction, scattered-light subtraction, fiber-spectrum extraction, throughput correction, wavelength calibration, sky subtraction, and cosmic-ray rejection. For additional details on fiber instrumentation and 2-D spectroscopy data analysis see Arribas, Mediavilla & Rasilla (1991). In Fig. 1 we present a sample of the 2-D distribution of the observed spectra. In order to obtain the two-dimensional maps presented in this paper, we created files with the X and Y positions of the fibers in the focal plane of the telescope and the selected spectral feature (continuum, line intensity). With the help of the NAG routine E01SAF an interpolating two-dimensional surface $F(x, y)$ is generated. This routine guarantees that the constructed surface is continuous and has continuous first derivatives. The interpolant $F(x, y)$ is then evaluated regularly on a rectangular grid of pixels to create the maps presented here.

3. Results

3.1. 2-D Distribution of Spectra

In Fig. 1 we show the 2-D distribution of the 175 observed spectra. The represented spectral region includes the [C III] $\lambda1909$ emission line. The spectra showing appreciable [C III] $\lambda1909$ emission are (i) spectra 132 and 147, close to the C component and, (ii) spectra 106, 111, 110, 119, 118, 129, 144, 145, 161, and 162, which connect components A, D, and B. Some traces of emission also appear in spectra 130 and 120 but at a lower level of intensity.

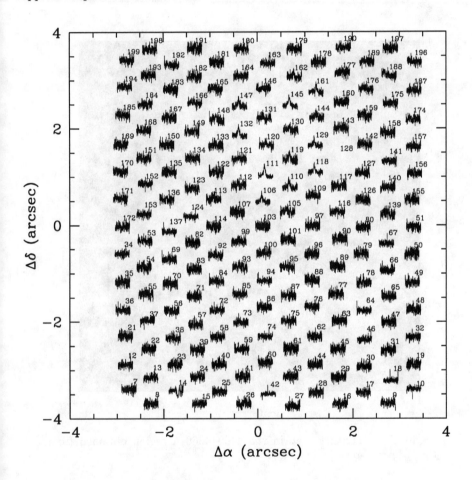

Figure 1.   Two-dimensional distribution of spectra.

3.2. Continuum Maps

In Fig. 2 we present a continuum image obtained by integrating the spectra in the 4400–4600-Å range. The four components appear clearly separated. The averaged displacements of the centroids determined from the present data with respect to the locations given by Ostensen et al. (1996) are $0''.07$ in RA and $0''.02$ in δ. This result is an indirect test of our photometric accuracy and show the usefulness of INTEGRAL and other future fiber-based 2-D spectrographs for filter imaging. We have also obtained other continua (not shown here) in different spectral ranges deriving very similar results.

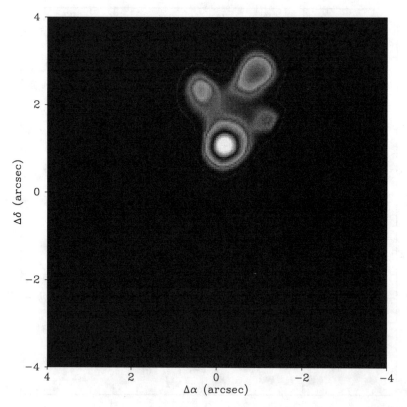

Figure 2. Intensity map in the 4400–4660-Å region obtained from the spectra shown in Figure 1.

3.3. [C III] $\lambda1909$ Emission-Line Map

To compute the [C III] $\lambda1909$ emission we have subtracted the underlying continuum after fitting the points on the blue and red sides of the emission line in each spectrum. After that operation, the emission was integrated. In Fig. 3 we present the corresponding intensity map in the region around Q2237+0305. The main feature of this map is the presence of an arc of extended emission

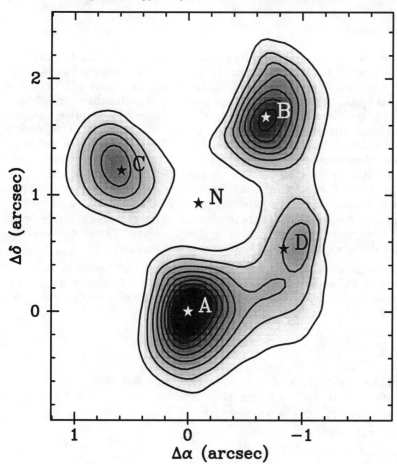

Figure 3. Intensity map of the emission in the [CIII]λ1909 line obtained from the spectra in the region around Q2237+0305.

connecting images A, D, and B. A comparison of this map with the continuum map (Fig. 2) shows no traces of the arc. Note that both maps are obtained from the same spectra and are, for this reason, strictly simultaneous, which implies that the presence of this extended emission can hardly be due to a seeing effect. We are aware that a change in the relative intensity peaks of the components can change the aspect of the map. However, the relative B/A, D/A, and C/A quotients among the intensity peaks are not so different between the 4400–4600-Å continuum (0.5,0.3,0.4) and the [C III] λ1909 emission (0.7,0.4,0.5) maps as to explain the presence of the extended arc as a seeing effect.

The agreement with Ostensen et al. (1996) in the location of the components is even better in this case (see Fig. 3) than in that of the continuum. The averaged differences are, now, $0''.04$ in RA and $0''.02$ in δ. This is also a test of the consistency of our data and analysis.

4. Conclusions

We have observed the gravitational lens system Q2237+0305 with INTEGRAL and have obtained 2-D spectroscopy of this object in sub-arcsecond seeing conditions. In the intensity maps derived from these data we resolve and locate the four compact QSO images with an average accuracy of better than $\sim 0''.05$ with respect to the positions determined using CCD photometry by other authors. This result demonstrates the performance of INTEGRAL, and other similar fiber systems, for working with good spatial sampling and resolution and is an indirect test of its photometric capabilities.

We have discovered an arc of extended [C III] $\lambda 1909$ emission which connects the A, D, an C components. This arc is preliminarily interpreted as the gravitational lensed image of an extended region in the QSO source.

The present results stimulate the routine use of integral-field spectrographs, and add a new type of gravitational-lens system (extended narrow emission line of a QSO + deflecting galaxy) to the observational domain.

Acknowledgments. The 4.2-m William Herschel Telescope is operated on the island of La Palma by the Isaac Newton Group at the Spanish Observatorio del Roque de los Muchachos of the Instituto de Astrofísica de Canarias. We thank all the staff at the Observatory for their kind support.

References

Arribas, S., Mediavilla, E., & Rasilla, J. L. 1991, ApJ, 369, 260
Corrigan, R. T., et al. 1991, AJ, 102, 34
Fitte, C., & Adam, G. 1994, A&A, 282, 11
Huchra, J., et al. 1985, AJ, 90, 691
Ostensen, R., et al. 1996, A & A, 309, 59
Racine, R. 1991, AJ, 102, 454
Rix, H.-W., Schneider, D. P., & Bahcall, J. N. 1992, AJ, 104, 959
Yee, H. K. C. 1988, AJ, 95, 1331

Dynamics and Star-Formation Properties of Giant Luminous Arcs

G. Soucail

Observatoire Midi-Pyrénées, 14 Av. E. Belin, 31400 Toulouse, France

Abstract. The study of the internal dynamics and the star-formation properties of distant galaxies is presented through the observations of a few giant and resolved arcs observed with 2-D spectroscopy. Initial data on the giant arcs in Abell 370, Abell 2218, and Abell 2390 are presented. The potentialities of this method for testing the validity of the Tully-Fisher relation at high redshift are examined.

1. Scientific Context

Before presenting our first results, I would like to highlight some of the scientific results obtained this last decade from the observation of the so-called "giant luminous arcs". These arcs occur when a background field galaxy is serendipitously gravitationally lensed by a massive concentrated cluster of galaxies (Soucail et al. 1987; Fort & Mellier 1994, and references therein). Lensing produces, in the case of spatially resolved sources, a highly distorted and magnified image of the source, with a typical spatial magnification of a factor of 10 in the most favorable cases. Note also that gravitational lensing preserves the surface brightness, an important property for the source reconstruction (see below). Up to now, most of the results of the gravitational lensing phenomena in clusters of galaxies concern mass determinations. Indeed, through an accurate lens modeling, it is possible to reconstruct the cluster mass distribution. Over the past years, *HST* images of cluster-lenses have allowed the identification of many multiple-imaged arcs, leading to strong constraints on the mass models (Kneib et al. 1996) and a well defined mass reconstruction of a few selected clusters (see Kneib et al. 1998, for example). In addition, the spectroscopic analysis of arcs is an important observational task for many reasons: first, a redshift determination of one giant luminous arc fixes the scales and properties of the given optical bench and allows an absolute determination of the cluster center mass; econdly, the use of the gravitational magnification has proved to be a powerful tool to study the properties of distant field galaxies, up to a redshift of 2 to 4 (Mellier et al. 1991; Ebbels et al. 1996; Franx et al. 1997). Finally, the last hope, not yet fully explored, is to benefit from the spatial magnification to study the morphology and the size of galaxies at reasonably high redshift ($z > 0.5$ to 0.7). A significant evolution of the morphology of distant field galaxies has been revealed by the *HST*, both from the Medium Deep Survey (Abraham et al. 1996a) and by the HDF (Abraham et al. 1996b). Galaxies show a clear evolution in luminosity (as expected from the modeling of their spectrophotometric properties) and also in size, with a strong decrease with redshift (see Ellis 1998 for a review). This last result seems to favor a merging model of galaxy evolution, with the merging of sub-units to form larger objects, in a sort of hierarchical scenario (Broadhurst et al. 1992). It also implies the existence of bursts of star formation associated with the merger phases, and probably a smoother mass evolution.

The goal of the observations presented in this paper is to understand the dynamical state of some distant galaxies spatially magnified by clusters of galaxies. To do this we propose a combined analysis of the *HST* imaging of some selected giant luminous arcs, leading to the morphology of the sources, with their two-dimensional spectrophotometry aimed at determining the dynamics of the sources.

An initial approach of this question was proposed by Soucail & Fort (1991) in the case of the giant "straight arc" observed in the cluster Abell 2390 ($z = 0.23$). Indeed this cluster displays a peculiar arc with a straight shape, and many additional faint arcs and arclets. The redshift of the main arc was measured by Pelló et al. (1991) at $z_S = 0.913$ and a significant velocity gradient was detected along the structure. With a value of $\Delta\lambda = 10$ Å, it can be translated to a velocity difference of ± 210 km s^{-1} in the source rest-frame and was immediately interpreted as due to the internal rotational velocity of the source galaxy. Soucail & Fort (1991) attempted to use the Tully–Fisher relation to translate this rotational velocity into an absolute magnitude for the source. Compared to the apparent magnitude, corrected from the lensing magnification and some corrections due to the redshift (namely the k-correction and the evolutionary e-correction), an estimate of the distance modulus was proposed, giving in principle a "long distance value" for the Hubble constant. The authors proposed an analysis of the sources of error, pointing to the most important ones as the uncertainties in the size and shape of the source, as well as in the inclination corrections, possibly improved by a better lens modeling. Also, the assumption that no luminosity or mass evolution had occurred except for the standard spectrophotometric e-correction was considered as a weak point in the analysis.

2. Observations and Data Reduction

Thanks to the improvement of the available instrumentation on large telescopes, and particularly at CFHT, we proposed to continue this program. Two nights were allocated in 1996 August with the MOS/ARGUS spectrograph. MOS/ARGUS is a very flexible instrument, in which a compact array of 600 optical fibers is positioned at the Cassegrain focal plane of the CFHT and is organized as the slit entrance of the low-resolution MOS spectrograph instead of the standard slit plate. This spectrograph (Le Fèvre et al. 1994) provides many spectral resolutions and spectral ranges, depending essentially on the choice of the grism. In our case we chose the R300 grism, optimized in the red (with a useful spectral range of 5000–9000 Å) with a resolution $R \sim 700$ ($\Delta\lambda = 10$ Å FWHM). The fibers have an aperture of 0.4" on the sky, which offers a good sampling of the average seeing at CFHT, and gives a hexagonal field of view of $13'' \times 8''$. The CCD in use was the STIS2, a thinned CCD with 21-μm pixels. This unfortunately provided an undersampling of the fibers on the CCD with

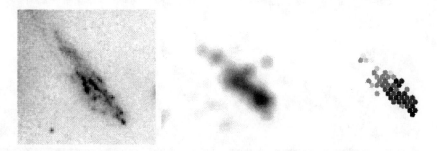

Figure 1. The giant arc #289 in Abell 2218: *HST* F702W image (*left*), reconstructed [O II] image (*center*), and velocity field (*right*).

a scale of 1.65 pixel per fiber. Finally, only two cluster-lenses were observed during our run, partly because of weather conditions.

The data reduction was performed in a rather standard way, although some specific points had to be treated carefully. The data were re-binned in both directions (in X for the wavelength calibration, and in Y to reconstruct one spectrum per line on the 2-D image or one spectrum for each fiber). This step corrects for the optical distorsion and the spectral dispersion, with a resampling of the data. The second step is to flat-field the data, by a pixel-to-pixel correction with a continuum lamp. The sky subtraction was done with a 2-D fit on the sky spectrum after selecting some clean sky regions in the image, away from objects and away from dead fibers. Flux calibration was standard, although we encountered some difficulties above 7500 Å because of some contamination by the second-order spectra. An additional correction for red objects had to be applied for red objects, similar to that published by Crawford & Vanderriest (1997). Finally image reconstruction was performed both with broad-band imaging (used to check the centering of the objects in the ARGUS bundle), and with narrow-band imaging through emission lines, with the line-flux imaging, the velocity field distribution, and possibly the equivalent-width distribution if the underlying continuum had enough signal. All this package, written as IRAF scripts, was kindly provided by Christian Vanderriest and Pascal Teyssandier (see their poster in these proceedings), and adapted in Toulouse for our own purposes.

3. Some Results

3.1. The Arc # 289 in Abell 2218

Abell 2218 is a very rich cluster at $z = 0.17$ with a numerous system of arcs detected by Pelló et al. (1992) and revealed by the spectacular *HST* images (Kneib et al. 1996). The lens modeling was proposed by Kneib et al. (1995, 1996) with a bi-modal potential centered on the two main cluster members. Many arcs and arclets have now been measured spectroscopically thanks to the spectacular work of Pelló et al. (1992) and Ebbels et al. (1998). For the "spectromorphological" study of some selected arcs, object labeled #289 in Pelló et al. (1992) was one of the most favorable candidates. Located at a redshift of

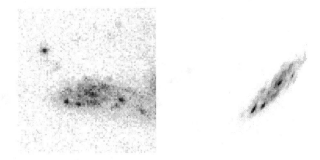

Figure 2. *HST* F675W image of the eastern end of the giant luminous arc in Abell 370 (*left*) and the reconstructed source from the lensing model (*right*).

$z = 1.034$, it is a very blue object and the spectrum displays a strong emission line at 7581 Å ($W_\lambda \simeq 120$ Å) identified with [O II] 3727. This object is located close to the second brightest cluster member and belongs to the secondary system of arcs and arclets around this galaxy. *HST* imaging shows that it corresponds to a single imaged and highly magnified spiral galaxy, with a spatial extension of $7.1'' \times 1.4''$ well matched to the field of view of MOS/ARGUS. Seven exposures totaling an integration time of 4.5 hr were obtained on this object. After the overall data reduction, an S/N of about 1 was obtained on the continuum for a surface brightness of $\mu_R \simeq 23$, and an S/N of 13 on the emission line at the peak location. The velocity-field reconstruction through the [O II] emission line shows a significant velocity gradient which translates to ± 230 km s^{-1} in the rest-frame of the source (Fig. 1). Such a value corresponds to a large Sc or Sd galaxy, seen nearly edge on. The monochromatic image in [O II] can be compared to the more detailed *HST* image, seen in the continuum of the spectrum: a bright spot, associated with a strong H II region is visible in the [O II] image, while the rest of the image is consistent with the global *HST* morphology. Although these results stress the necessity of a source reconstruction to derive the true morphological properties of the source, it already shows the power of the 2-D imaging to select some specific details in these images of high-redshift galaxies.

3.2. The Giant Arc in Abell 370

This rich cluster ($z = 0.37$) is among the first in which giant luminous arcs were discovered (Soucail et al. 1987). It presents one large arc and many arclets. An accurate lensing model was presented by Kneib et al. (1993) and was recently improved, thanks to the identification in *HST* images of several sets of multiple images and the inserting of local components of cluster galaxies as additional deflectors (Bézecourt et al. 1998). Again, the lensing potential is bi-modal, with two clumps of matter located in the two brightest giant galaxies. Concerning the arc, its redshift of $z = 0.725$ was the first to be measured spectroscopically (Soucail et al. 1988). The spatial extent of the arc is $25''$ in length and $2.5''$ in width at maximum in the enlarged eastern part. *HST* imaging has confirmed that this arc is in fact the merging of three images of the same source. The spectrum shows a prominent emission line at 6425 Å identified with [O II], with an equivalent width of about 30 Å. With a surface brightness of $\mu_R = 22.8$, similar to the previous object, it is less favorable than the arc # 289 in A2218 for 2-D spectroscopy and we estimated that 6 hours were required as a minimum to get enough signal for a velocity mapping. Unfortunately, due to poor weather conditions, only 1 hour was available on the enlarged eastern part of the arc. A marginal detection of [O II] was obtained in a few fibers with an S/N slightly smaller than 3, and a possible velocity shift was suspected in the transverse direction (± 60 km s^{-1}). Of course, deeper data are necessary to confirm this detection!

Anyway, thanks to the recent detailed lens model of Bézecourt et al. (1998) we attempted a source reconstruction by simply sending the image pixels in the source plane and building a source image with squared pixels. No optimization of the procedure is done yet, although this may be interesting to obtain on this triple-imaged arc. The results are shown in Figure 2. A little bit more can be said concerning the source size. Indeed, for the reconstructed source, we measured an angular half-light radius of 0.6″, which translates to a linear size of $r_{hl} = 5.7\ h_{50}^{-1}$ kpc ($q_0 = 0$) or $r_{hl} = 4.8\ h_{50}^{-1}$ kpc ($q_0 = 0.5$). These values seem smaller than the local values determined by Mathewson et al. (1992), for which typical spirals have a linear half-light radius of 8.7 h_{50}^{-1} kpc. But their dispersion is also rather large. At larger redshift, Smail et al. (1995) used an empirical model of size evolution to study the statistics of arcs and arclets from HST images, with a half-light radius varying as $r_{hl} \propto 1/(1+z)$. Applied to the redshift of the arc of A370, this gives a typical value of 5 kpc, similar to the one derived from our model. It is at present difficult to draw any conclusions on the size evolution of distant galaxies from this example only, and this analysis has to be performed for many more arcs to test the hypothesis of size evolution of distant galaxies.

4. Conclusions and Future Developments

Soucail & Fort (1991) have initiated the method detailed in this paper. Since that time many improvements have taken place that can reduce the uncertainties in the estimation of the source parameters. Better lens models are now available for clusters in which multiple images offer many strong constraints. They often include the local effects of individual galaxies which can modify the source reconstruction. A safe source reconstruction will allow also to better correct from the magnification and to determine the unlensed magnitude of the source. Il will also allow the analysis of the morphology of high-redshift galaxies with a spatial resolution increased by a factor of up to 10. Finally, 2-D spectroscopy and reconstruction of the velocity field is a more efficient way of recovering the internal rotation field than long-slit spectroscopy, for which geometric corrections to the "slit effect" are necessary (Vogt et al. 1996).

The ultimate goal of this study is to test the Tully–Fisher relation at high redshift and to explore the evolution of distant galaxies, both in luminosity/mass and in size/morphology. This kind of program has also been initiated at Keck, but in long-slit spectroscopy only (Vogt et al. 1996, 1997). All these preliminary observations have confirmed the fact that it is a very difficult observational task, at the limit of feasibility for a 4-m class telescope. Only a few cases are presently accessible, and we need the next generation of VLTs or other 8-m class telescopes equipped with intermediate-resolution spectrographs and IFUs to make significant progress in this very exciting problem.

This work is done in collaboration with J. P. Kneib (Toulouse), J .P. Picat (Toulouse), and R. S. Ellis (IoA, Cambridge).

References

Abraham, R. G., Tanvir, N. R., Santiago, B. X., Ellis, R. S., Glazebrook, K., & Van den Berg, S. 1996a, MNRAS, 279, L47

Abraham, R. G., Van den Berg, S., Glazebrook, K., Ellis, R. S., Santiago, B. X., Surma, P., & Griffiths, R. E. 1996b, ApJS, 107, 1

Bézecourt, J., Kneib, J. P., Soucail, G., & Ebbels, T. M. D. 1998, in preparation

Broadhurst, T. J., Ellis, R. S., & Glazebrook, K. 1992, Nature, 355, 55

Crawford, C., & Vanderriest, C. 1997, MNRAS, 285, 580

Ebbels, T. M. D., Le Borgne, J. F., Pelló, R., Kneib, J.-P., Smail, I. R., & Sanahuja, B. 1996, MNRAS 281, L75

Ebbels, T. M. D., Ellis, R. S., Kneib, J.-P., Le Borgne, J. F., Pelló, R., Smail, I. R., & Sanahuja, B. 1998, MNRAS 295, 75

Ellis, R. S. 1997, ARAA 35, 389

Fort, B., & Mellier, Y. 1994, A&AR, 5, 239

Franx, M., Illingworth, G. D., Kelson, D. D., van Dokkum, P. G., & Tran, K-V. 1997, ApJ, 486, L75

Kneib, J.-P., Mellier, Y., Fort, B., & Mathez, G. 1993, A&A, 273, 367

Kneib, J .P., Mellier, Y., Pelló, R., Miralda-Escudé, J., Le Borgne, J. F., Böhringer, H., & Picat, J.-P. 1995, A&A, 303, 27

Kneib, J. P., Ellis, R. S., Smail, I., Couch, W. J., & Sharples, R. M. 1996, ApJ, 471, 643

Kneib, J. P., Pelló, R., Mellier, Y., Soucail, G., Fort, B., Ellis, R. S., Aragón-Salamanca, A., Smail, I., & Miralda-Escudé, J. 1998, A&A, submitted

Le Fèvre, O., Crampton, D., Felenbok, P., & Monnet, G. 1994, A&A, 282, 325

Mathewson, D. S., Ford, V. L., & Buchhorn, M. 1992, ApJS, 81, 413

Mellier, Y., Fort, B., Soucail, G., Mathez, G., & Cailloux, M. 1991, ApJ, 380, 334

Pelló, R., Sanahuja, B., Le Borgne, J.-F., Soucail, G., & Mellier, Y. 1991, ApJ, 366, 405

Pelló, R., Le Borgne, J. F., Sanahuja, B., Mathez, G., & Fort, B. 1992, A&A, 266, 6

Smail, I., Couch, W. J., Ellis, R. S., & Sharples, R. M. 1995, ApJ, 440, 501

Soucail, G., Fort, B., Mellier, Y., & Picat, J. P. 1987, A&A, 172, L14

Soucail, G., Mellier, Y., Fort, B., Mathez, G., & Cailloux, M. 1988, A&A, 191, L19

Soucail, G., & Fort B. 1991, A&A, 243, 23

Vogt, N. P., Forbes, D. A., Phillips, A. C., Gronwall, C., Faber, S.M., Illingworth, G. D., & Koo, D. C. 1996, ApJ, 465, L15

Vogt, N. P., Phillips, A. C., Faber, S. M., Gallego, J., Gronwall, C., Guzman, R., Illingworth, G. D., Koo, D. C., & Lowenthal, J.D. 1997, ApJ, 479, L121

The PMAS Fiber Spectrograph

M. M. Roth

*Astrophysikalisches Institut Potsdam, An der Sternwarte 16
D-14487 Potsdam, Germany*

U. Laux

Semmelweisstr. 22, D-99425 Weimar, Germany

Abstract. The Astrophysical Institute Potsdam (AIP) is developing a new integral-field spectrograph capable of performing two-dimensional spectrophotometry. The requirements to make this technique work properly are discussed. A novel, very efficient, all-dioptric, immersion-coupled fiber spectrograph is described as the main module of the PMAS instrument.

1. Introduction

Integral-field spectroscopy has been employed now for about a decade, mainly by groups that were involved in the development of their own prototype instruments. It has only been recently, that the usefulness of this technique was discovered by an increasing fraction of the astronomical community, eventually leading to the definition of integral-field units as user instruments for several major new generation telescopes, e.g. ESO-VLT, or GEMINI. Driven by the scientific demand emerging from ongoing research projects at AIP, and going beyond the task of merely taking spatially resolved spectra of extended objects, we have started to develop an instrument that is capable of performing true *2-D spectrophotometry*, i.e. taking flux-calibrated spectra simultaneously for a number of resolution elements of a contiguous two-dimensional field-of-view—in a sense like a CCD with a special spectrograph associated with each pixel. PMAS, the Potsdam Multiaperture Spectrophotometer, is designed to be a traveling instrument which can be mounted on nearly any telescope with only minor modifications. It is scheduled for first light at the Calar Alto Observatory 3.5-m telescope in Spain. The design is based on a fiber bundle, feeding light from a micropupil lens array to a dedicated fiber spectrograph. In what follows, we will concentrate on the peculiar design of the latter, and how it is influenced by the requirements of spectrophotometry.

2. 2-D Spectrophotometry

In order to illustrate our motivation for this project, we start to list a few examples that demonstrate how 2-D spectrophotometry can have a great advantage over conventional observing techniques.

2.1. Extended Emission-Line Objects

The emission-line intensities of planetary nebulae (PNe) and H II regions yield important information about the physical conditions in these gaseous nebulae and their respective excitation sources. In a pioneering paper, Jacoby et al. (1987) introduced the method of narrow-band CCD spectrophotometry, which provides spatially resolved, flux-calibrated line intensities for selected emission lines, each corresponding to a given filter. Instead of taking separate exposures through the whole set of filters of interest, the natural extension of this method is to record simultaneously a spectrum for each pixel and find a way to flux-calibrate the set of spectra. Re-arranging these spectra, the resulting data cube may also be viewed as a stack of monochromatic images. There is an obvious gain in efficiency with the integral-field method if the required number of lines is larger than one. Even more importantly, each monochromatic image has been taken under the same atmospheric conditions, contrary to the filter method which, by definition, must work sequentially and is therefore affected by variations of atmospheric extinction.

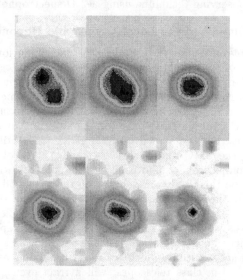

Figure 1. IC 351 monochromatic images (smoothed, 8×16 array, 1" per lens), extracted from a test data cube taken on 1996 November 6 with the MPFS integral-field spectrograph at the 6-m Zhelenchuk Telescope prime focus. *From left to right:* [O III] 5007, H$_\beta$, and He II 4686 (*top row*); [Ar IV] 4711, [Ar IV] 4740, and continuum (*bottom*). The morphology in the different lines agrees well with published narrow-band images. Note the presence of the central star in the continuum.

2.2. Extragalactic PNe and H II Regions

The narrow-band filter CCD method as described above has been extended to the detection and photometry of extragalactic planetary nebulae in the [O III] 5007 line out to distances as far as the Virgo Cluster. Originally developed for PN luminosity function studies in ellipticals, it has also been successfully applied to the detection of PN and H II regions in late-type galaxies (Soffner et al. 1996). Obtaining emission-line intensities for PNe in galaxies beyond the Magellanic Clouds would provide an extremely useful handle on stellar populations, chemical evolution, and other issues. However, due to the intrinsic faintness of the lines of interest, this is not at all a trivial task. Moreover, the fainter lines tend to have poorer contrast with respect to the underlying continuum of the background intensity the closer the PN is located to the center of the host galaxy. A practical problem is pointing accuracy and field acquisition: none of the planetaries would be visible on a TV guider. Very similar to the advantage of crowded field CCD photometry over photoelectric aperture photometry, 2-D spectrophotometry will help to extract even faint emission lines from the noisy and spatially variable background intensity. In addition, one does not have to care about very precise positions and pointing. As long as the target is within the field of view of the instrument, this is no longer of great concern.

Extragalactic H II regions allow us to determine abundances in their host galaxies. However, the error bars on abundance determinations are often huge. For the same reasons as outlined above, also work in this field would benefit largely from an improved observing technique using a 2-D spectrophotometer.

2.3. Stars in Galaxies of the Local Group and Beyond

The first two examples are directly related to the determination of fluxes, hence to the very purpose of spectrophotometry. However even in cases, where fluxes are not really needed in the first place, an integral-field spectrograph with a *calibrated* and *linear* response of each spatial channel allows measurements to be performed that would otherwise not be feasible. Again, let us consider a 2-D array of apertures as opposed to the slit for the situation of taking spectra in crowded stellar fields. For example, Kudritzki et al. (1989) observed luminous early-type stars in the SMC and derived extreme values of up to 113 solar masses from the comparison of photospheric line profiles with NLTE model atmospheres—later on this result was debated on the basis of deconvolved direct images showing that the supermassive star was in fact the superposition of a multiple system (Heydari-Malayeri & Hutsemekers 1991). Clearly, the original spectroscopic analysis would have gained from integral-field spectra to prove or disprove the relevance of contamination from the neighboring stars. The topic of extragalactic stellar spectroscopy, which is already now being exercised for Local Group galaxies with 4-m class telescopes, will attract even more attention as soon as the new 8-m telescopes with excellent seeing conditions become available (Kudritzki 1998).

Certainly there are many more applications that would benefit from the method. In general, we stress that whenever geometrical details of the spectral intensity distribution are of concern (as is the case for the crowded-field situation), the linear behavior of an integral-field spectrograph becomes important.

3. Problems

If we identify a 2-D spectrophotometer with an integral-field spectrograph which is able to ensure linear response and flux calibration for each channel, captures all the flux with no gaps or dead space between spatial elements, and is intrinsically stable enough to maintain calibration over a reasonably long period of time; in reality, we will encounter a number of problems when trying to build such an instrument.

1. *Contiguous Sampling:* This requirement immediately reduces the conceivable types of integral-field designs to the lens-array type of instrument. The conceptually more simple bare fiber bundle will not work. Also, the use of lenses has the advantage of allowing for the presence of an image of the telescope pupil (the micropupil), thus improving stability in the sense of the Fabry lens of the classical aperture photometer. Since the TIGER type of 2-D spectrograph involves limitations in terms of free spectral range and a drawback concerning wavelength calibration (Bacon et al. 1995), the logical choice appears to be to combine a lens array with a fiber bundle, which in turn feeds a fiber spectrograph through a pseudo-slit. On the other hand, the required precision and stability for maintaining the fibers centered on their micropupils behind the lens array is a serious complication.

2. *Linear Response:* At first glance the issue appears to be guaranteed when we are using a linear detector. However, contrary to the simplistic comparison of a 2-D spectrograph with a CCD having attached an individual spectrograph to each pixel, a real instrument involves the process of extracting many spectra that have passed the same optical system and are projected onto one common detector. The overlap of adjacent spectra, stray light which may compromise the inter-spectra background definition, and ghost images will in reality affect the linearity within single spectra. It is therefore essential for a 2-D spectrophotometer which has to rely on linearity to reduce these effects to a minimum. It may be necessary to trade FOV (i.e. number of spectra) for a larger spacing of spectra to ensure linearity.

3. *Dynamic Range*: Based on the same line of reasoning, the dynamic range may be cut at the low-intensity end where faint spectra next to bright ones are adversely affected by extended wings of the latter. Contrary to limitations that are due to e.g. detector noise where it is possible to combine short and long exposures and then tie one to another, here we cannot resort to this strategy: the spatial wings of a standard star profile are always necessary to determine the total flux—but the ratio of corresponding bright and faint spectra on the detector stays the same, irrespective of the total exposure level.

4. *Calibration:* Since our standard stars are point sources and by definition cannot illuminate the entire FOV homogeneously, there is the problem of how to achieve a good flux calibration. Perhaps a combination of several offset standard-star calibrations with sky flats, possibly also an internal integrating sphere will help to alleviate this problem. The appropriate techniques remain to be developed.

5. *Stability:* Mechanical and optical stability is clearly a key issue when considering photometry. Any variation that may affect either the throughput of the system, or modify the spectral image and its position on the detector, will produce errors, either directly or through the extraction process of the spectra. It is therefore necessary to reduce opto-mechanical instabilities and flexure of the spectrograph, and make sure that the fibers operate under identical conditions all the time in order to avoid variable transmission losses.

4. Fiber Spectrograph Optical Design

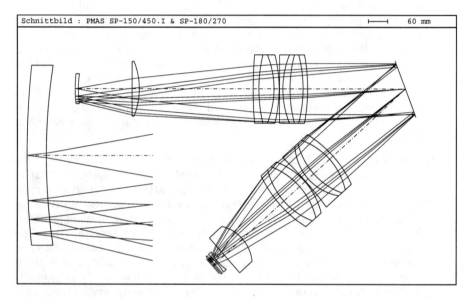

Figure 2. Fiber spectrograph cross-section (design by U. Laux).

The principle of operation involving fibers requires the use of a spectrograph which is optimized with respect to the problems outlined above. After a first design study it became obvious that conventional designs based on reflective collimators and fast Schmidt cameras were inappropriate for meeting all these goals. An alternative, all-dioptric solution was proposed, forming the basis for subsequent improvements and the final design layout as depicted in Figure 2. A fundamental consideration was to introduce a system which was free of any central obscuration. Collimator and camera are required to operate at wide field angles with excellent image quality throughout the optical from 350 to 900 nm. Priority was given to a clean separation in the cross-dispersion direction as a trade-off with respect to the possible number of spectra. The design employs a similar sequence of lenses in both the collimator and the camera arms (for the latter in reverse direction): a field lens, followed by a fused silica singlet, and a characteristic combination of two groups of three and four lenses which are built around a central CaF_2 lens, bracketed by two crown glass lenses (and a flint in the case of the quadruplet). The properties of this basic configuration are described in some more detail in Roth et al. (1997), and Laux (1998). As a consequence of the results of fiber FRD experiments (see Schmoll et al., these proceedings) the basic layout was modified such that the telecentric fiber pseudo-slit couples directly to the first lens in immersion (insert in Fig. 2). This approach improves the optical efficiency without requiring the application of anti-reflective coatings to the fibers, substantially reduces FRD losses due to fiber end face imperfections or dust, and decreases the sensitivity to fiber misalignments. It also allows for the replacement of poorly performing or damaged individual fibers. The system is currently being manufactured at Carl Zeiss Jena GmbH, Germany.

The properties of the spectrograph design can be summarized as follows:

- all-dioptric system with external focus, reflective gratings
- collimated beam diameter 150 mm
- $f/3$ collimator ($f = 450$ mm), 10° full field
- $f/1.5$ camera ($f = 270$ mm), 180-mm pupil, 12° full field
- 78-mm fiber pseudo-slit
- telecentric fiber input in immersion on curved surface
- system fully corrected at 350–900 nm
- 100% spot concentration below 30 μm, $D80 \approx 20$ μm over full field and at all design wavelengths
- nominal fiber diameter 100 μm, projected to 60 μm on CCD
- throughput 58% at 350 nm, $\approx 80\%$ from 550 to 900 nm
- compact lens groups, few glass-air interfaces (stray light, ghosts)

Acknowledgments. The PMAS project is supported by the German Ministery for Science and Education (BMBF) under Grant No. 05 3PA414 1. Support with observations at the SAO 6-m telescope by V. Afanasiev and S. Dodonov, and helpful discussions concerning the optical design with W. Seifert, LSW Heidelberg, are gratefully acknowledged.

References

Bacon, R. et al. 1995, A&AS, 113, 347
Heydari-Malayeri, M., & Hutsemekers, D. 1991, A&A, 243, 401
Jacoby, G. H., Quigley, R. J., & Africano, J. L. 1987, PASP99, 672
Kudritzki, R. P., et al. 1989, A&A, 226, 235
Kudritzki, R. P. 1998, in Stellar Astrophysics for the Local Group, eds. A. Aparicio, A. Herrero & F. Sánchez (Cambridge: Cambridge University Press), 149
Laux, U. 1998, Astrooptik (München: Verlag Sterne und Weltraum)
Roth, M. M., Seydack, M., Bauer, S., & Laux, U. 1997, Proc. SPIE, 2871, 1235
Soffner, T., et al. 1996, A&A, 306, 9

Two-Dimensional Stellar Kinematic and Population Analysis of Galaxies

Reynier Peletier

Department of Physics, University of Durham, South Road, Durham DH1 3LE, UK

Santiago Arribas, Carlos del Burgo, Begoña García-Lorenzo, Evencio Mediavilla, and Carlos Gutiérrez

Instituto de Astrofísica de Canarias, 38200 La Laguna, Tenerife, Spain

Francisco Prada

Instituto de Astronomía, UNAM, Apartado Postal 70-264, México

Alexandre Vazdekis

Dept. of Astronomy, Graduate School of Science, University of Tokyo, Bunkyo-ku, Tokyo 113, Japan

Abstract. Two-dimensional maps of the absorption-line indices of the Lick system have been obtained for a number of bright early-type galaxies using the integral-field system 2D-FIS on the WHT. Here the observations are described, and some preliminary results are given concerning the stellar populations at the center of the Sombrero Galaxy.

1. Introduction

In the recent years technical advances in telescopes and instruments have been such that integral-field spectroscopy (IFS) has started to become a competitive technique. Historically, almost all applications of IFS to galaxies have used emission-line objects. Absorption lines up to now have rarely been studied, because they are so much fainter and more difficult to work with. However, now that instruments with thousands of fibers will soon be appearing, it would be very attractive to be able to study the kinematics and stellar populations of galaxies from their absorption lines. Using the multi-fiber instrument 2D-FIS at the WHT, we have started a pilot study to find out how accurate kinematic and line strength maps are from IFS, and what its limitations are. To do this we have observed a number of "standard" early-type galaxies, for which a significant amount of comparison data is available in the literature, and we describe a summary of the results of a comparison in this paper. More details can be found in Peletier et al. (1998). We conclude that accurate kinematic and stellar-population parameters can be obtained with IFS absorption-line data.

2. Observations and Data Reduction

Observations were carried out on 1997 February 15–17 at the WHT on La Palma using the 2D-FIS system. The heart of this instrument is a bundle of 125 optical fibers distributed in the focal plane of the telescope in a central rectangle, covering an area of about $9.2'' \times 12.2''$ on the sky, and an outer ring of fibers at $38.4''$ from the center, to measure the sky background light (for more details see García et al. 1994). The 95 inner fibers have a core diameter of $0.9''$. These fibers are then fed into the slit of the ISIS double-beam spectrograph so that the gratings, filters, and dichroics of ISIS can be used. Six galaxies were observed simultaneously in the red (resolution of 65 km s^{-1}) and blue (200 km s^{-1}) to measure accurate stellar line-of-sight velocity profiles, and have a large wavelength range to measure simultaneously many absorption lines. The Lick system (Faber et al. 1985) was used, since this has been used by most authors in the field. The data reduction was performed in IRAF, and is mostly straightforward. Sky background was subtracted using the fibers in the outer ring. Finally, we measured for all the wavelength-calibrated, sky-subtracted fiber spectra the equivalent width of about 25 absorption lines, as well as various kinematic parameters, such as the radial velocity, velocity dispersion, and asymmetry of the profile, and made two-dimensional maps of all these quantities.

3. Some First Results

3.1. Comparison with Long-Slit Spectroscopy

It was first established how well line strength could be measured by taking spectra of stars, which were far out of focus. Since in this case the spectrum of each fiber should give the same line strengths, the rms variation between the line indices obtained in the different fibers would be a good measure of the obtainable accuracy. About 20 spectra were obtained, and we found that generally this rms fluctuation was 0.01 mag for Mg$_1$ and Mg$_2$, 0.25 Å for Fe 5270, Fe 5335 and Hβ, and slightly higher for the other indices. These errors are generally systematic, and do not depend much on the signal-to-noise ratio in the data. They are partly due to non-perfect flat-fielding, caused by fringing in the CCD (strong in the red, but also somewhat present in the blue), and partly due to small changes in the continuum depending on varying stresses on the fibers. However, since these rms values are often much smaller than differences between authors in the literature presenting the same galaxy, we conclude that 2D-FIS is perfectly capable of performing this kind of science. As a following check our indices were compared to the long-slit data of Vazdekis et al. (1997), who performed a very detailed study for three of our standard galaxies using most of the lines of the Lick system. Here we show the comparison for NGC 3379 (Fig. 1), but similar comparisons have been made for the Sombrero Galaxy and NGC 4472. Both data sets have been corrected for velocity-dispersion broadening, and have been converted to the Lick system using 20 standard stars of Worthey et al. (1994). Error bars in the upper right corner indicate the error in the conversion to the Lick system. One can see that the comparison is good. Our Ca 4227 line is somewhat higher than in Vazdekis et al. (1997), who found that this line was much lower than predicted by models, but the inferred Ca deficiency is less. The

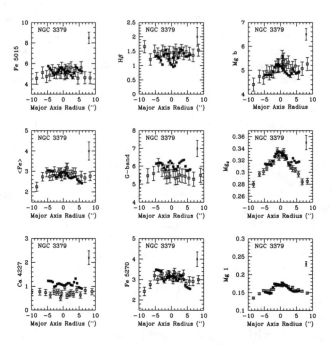

Figure 1. Comparison with the minor-axis long-slit data of Vazdekis et al. (1997) for a large number indices. Open symbols with error bars indicate the long-slit data. Note that the seeing of their observations was about 2.5–3″, as compared to an effective seeing of $\sim 1.2″$ in our case.

hole in Hβ (due to emission near the center) also shows up much better with better seeing.

3.2. Reconstructed Continuum Images Compared with *HST*

Another test we applied was to reconstruct the continuum images in red and blue, fit ellipses to these images, and compare various photometric profiles to recent *HST* archive images. To do this, the spectra were collapsed in the blue between 4236 and 5643 Å, after which ellipses were fitted on the reconstructed images. In Fig. 2 we show the surface-brightness profiles, normalized in the center, of the three galaxies, together with profiles determined on recent *HST*-WFPC2 images in V (F555W) convolved to a seeing of 1.2″ (FWHM). The agreement is striking. The slope of our profiles can be seen to be slightly shallower, which is what one expects from color gradients, since the B-profiles of these galaxies are shallower than their V-band profiles (Peletier et al. 1990). We conclude that the good agreement between our photometry and *HST* shows that scattered light is not a problem in the analysis of our spectra.

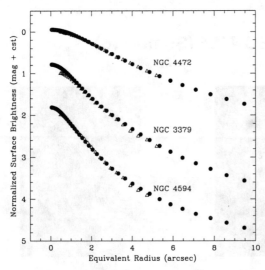

Figure 2. Normalized surface-brightness profiles in B (open symbols), compared to HST profiles in V (555W), convolved to a seeing of 1.2″ FWHM. Plotted on the abscissa is (major × minor axis radius)$^{1/2}$. The surface-brightness profiles are scaled by an arbitrary number.

3.3. The Sombrero Galaxy

In Fig. 3 some of the maps of the Sombrero Galaxy are shown. These can be compared with maps in Emsellem et al. (1996, figs 8, 9, and 18), the only study in the literature where IFS was applied to measure absorption-line strengths. Those observations were done with TIGER, a multi-lens system at the CFHT. The observations of Emsellem et al. have a higher resolution (microlens diameters of 0.39″), but their wavelength range is only 600 Å (5040–5690 Å).

The Sombrero Galaxy has a very large bulge, an outer disk, and a small, fast-rotating inner disk (e.g. Wagner et al. 1989). In fact in Fig. 3 we show the potential of our method to discuss these features. In Fig. 3a the reconstructed relative intensity is shown by averaging all the blue light in the fibers. Figure 3b shows the stellar radial-velocity field. Clearly visible here is the inner disk, responsible for the maximum and minimum velocity on the major axis. It also seems that only this inner disk is rotating rapidly, not the whole bulge, as one would think from just the major-axis rotation curve. The agreement with TIGER is excellent. At the bottom, two line-index maps are shown. They appear noisier, but this is an optical effect, since the noise in the line strength from fiber to fiber is in general smaller than the line-strength gradients across these nuclear areas, which in general are very small. In Fig. 3c we show the Mg_2 index, an often-used indicator of the stellar metallicity, and especially the Mg abundance. Here also the agreement is good. It is clearly evident that the inner disk has a larger Mg abundance than the surrounding bulge. This fact itself is not surprising, but the fact that other absorption lines, like the Ca II IR triplet (shown in Fig. 3d) and also various Fe lines do not show this disk,

178   Peletier et al.

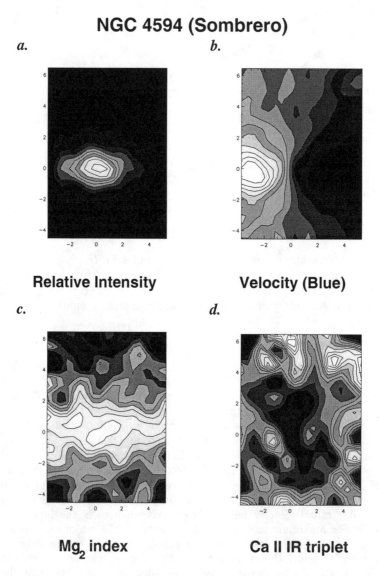

Figure 3.   Some preliminary results for the Sombrero Galaxy. Shown are the reconstructed surface brightness in the blue (a), the radial-velocity field (b), and the absorption line maps of Mg_2 and Ca II IR (c and d). For all plots the highest values are indicated by the lighter colors, and the lowest values by dark colors.

is very interesting. The Ca II IR triplet index is a generally good metallicity indicator, but is also very sensitive to the ratio of the number of dwarfs to giants. It might be that the inner disk is overabundant in [Mg/Fe], and has a larger dwarf-to-giant ratio than the surrounding bulge.

4. Conclusions

This paper shows the great potential of IFS applied to absorption lines of stellar systems. It is clear that these maps can be very useful in studying intrinsic kinematics and stellar populations of early-type galaxies. In addition, integral-field, as compared to long slit, data have the advantage that one integrates in annuli, which increases the signal-to-noise of radial profiles, thereby more than compensating for the loss of light in the fibers. But, most of all, they give us the opportunity to detect two-dimensional features in stellar populations and kinematics, something which is crucial for our understanding of these systems and in particular of their inner components.

References

Emsellem, E., Bacon, R., Monnet, G., & Poulain, P. 1996, A&A, 312, 777
Faber, S. M., Friel, E. D., Burstein, D., & Gaskell, C. M. 1985, ApJS, 57, 711
García, A., Rasilla, J. L., Arribas, S., & Mediavilla, E. 1994, Proc. SPIE, 2198, 75
Peletier, R. F., Davies, R. L., Illingworth, G., Davis, L. E., & Cawson, M. 1990, AJ, 100, 1091
Peletier, R. F., Vazdekis, A., Arribas, S., del Burgo, C., García-Lorenzo, B, Gutiérrez, C., Mediavilla, E., & Prada, F. 1998, in preparation
Vazdekis, A., Peletier, R. F., Beckman, J. E., & Casuso, E. 1997, ApJS, 107, 203
Wagner, S. J., Bender, R., & Dettmar, R.-J. 1989, A&A, 215, 243
Worthey, G., Faber, S. M., González, J. J., & Burstein, D. 1994, ApJS, 94, 687

Two-Dimensional Spectroscopy of Active Galaxies

Evencio Mediavilla, Santiago Arribas, Begoña García-Lorenzo, and Carlos del Burgo

Instituto de Astrofísica de Canarias, E-38200 La Laguna, Tenerife, Spain

Abstract. The kinematics in the central regions of active galaxies is complex and shows strong departures from regular rotation. This implies velocity fields without rotational symmetry that must be studied with 2-D spectroscopy. We present velocity maps for the ionized gas and, in some cases, for the stars in the circumnuclear region of a sample of active galaxies. Two main topics, the connection between the ionized gas and the stellar kinematics, and the location of the rotation centers, are briefly addressed.

1. Introduction

The morphology of the emission and the kinematics of the ionized gas in the central regions of active galaxies are strongly influenced by the presence of the active nucleus. The current models for active galactic nuclei, the so called unified models, suppose that the central engine (an accretion disk) is surrounded by a torus or disk of blocking material which, complemented or not by a larger structure, collimates the emission in two opposite radiation cones. When the observer's line of sight is within the radiation cones the central engine can be directly seen and the object is considered a Seyfert 1 or a QSO; otherwise the object is catalogued as a Seyfert 2 or a narrow emission line radio galaxy. This essentially anisotropic scenario, supported by a considerable amount of observational evidence, requires and justifies the use of 2-D spectroscopy.

In this article we summarize some results derived from 2-D spectroscopy of the galaxies NGC 4151, NGC 5728, NGC 3227, NGC 1068, NGC 7331, NGC 3516, and M 31. The program devoted to the study of the central regions of these and other active galaxies is the scientific back-up project in the development of two-dimensional spectroscopy with optical fibers at the Instituto de Astrofísica de Canarias.

2. Observations

The two-dimensional spectroscopy presented in this communication was performed using different fiber systems. NGC 5728 (Arribas & Mediavilla 1993), and NGC 3227 (Mediavilla & Arribas 1993; Arribas & Mediavilla 1994) were observed with the earlier system, HEXAFLEX (Arribas et al. 1991) at the Nasmyth focus of the William Herschel Telescope. NGC 1068 (Arribas et al. 1996)

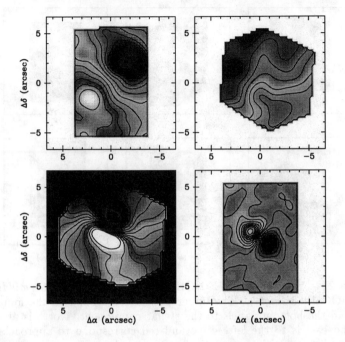

Figure 1. Ionized gas velocity fields of NGC 4151 (*top left*), NGC 3227 (*top right*), and NGC 5728 (*bottom left*). The map in the bottom right panel is the stellar velocity field of M 31. North is at the top, and east is to the left. Blue and red correspond to approaching and receding velocities, respectively.

was observed with a modified version of this system (HEXAFLEX-II; García et al. 1994) at the Nordic Optical Telescope. NGC 4151, NGC 7331 (Mediavilla et al. 1997a), NGC 3516 (Arribas et al. 1997), and NGC 1068 again (García-Lorenzo et al. 1997) were observed with 2D-FIS (García et al. 1994) at the Cassegrain focus of the William Herschel Telescope. Finally, M 31 (del Burgo et al., these proceedings) was observed with the new instrument INTEGRAL at the Nasmyth focus of the William Herschel Telescope.

3. Results

Studies of the narrow emission lines prior to two-dimensional spectroscopy (mainly based on nuclear data) have shown that these lines are very often asymmetrical with wings, shoulders, secondary peaks, or other evidence of substructure. One of the most remarkable results of our 2-D studies is that in all the cases of our sample of galaxies the narrow emission-line profiles exhibit substructure in the circumnuclear region also. This finding implies that several gaseous systems inhomogeneously distributed in space appear projected in the observer's line of sight, hindering the obtaining and interpretation of 2-D maps of features such as velocity, intensity or broadening. In several cases (NGC 5728, Arribas & Medi-

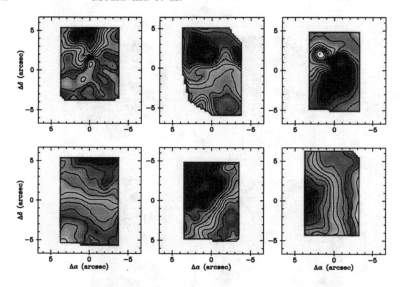

Figure 2. Velocity fields of NGC 7331 (*left*), NGC 3516 (*middle*), and NGC 1068 (*right*). The upper panels correspond to the ionized gas, and the bottom panels to the stellar component. North is at the top and east is to the left. Blue and red correspond to approaching and receding velocities, respectively.

avilla 1993; NGC 3227, Mediavilla & Arribas 1993; NGC 1068, Mediavilla et al. 1997b; NGC 7331, Mediavilla et al. 1997a; NGC 3516, Arribas et al. 1997) we have attempted a multi-component line analysis, which, in NGC 1068 and NGC 3516, has allowed to identify as many as different gaseous components, most of them clearly related to nuclear activity.

In Fig. 1 we present velocity fields corresponding to the ionized gas for NGC 4151, NGC 3227, and NGC 5728, and to the stellar velocity field for M 31. In Fig. 2 we present the velocity fields of both ionized gas and stars for NGC 1068, NGC 7331, and NGC 3516. In spite of the comments in the previous paragraph most of the velocity maps in these Figures have been obtained by cross-correlating the spectra or fitting a single Gaussian function to the lines without performing any multi-component analyses. Almost all the velocity fields in the figures are very distorted, especially those corresponding to the ionized gas. It is clear that essential information about the kinematics was lost in one-dimensional long-slit studies, and that a good understanding of the dynamical processes ruling the ionized-gas movements can be achieved only through 2-D spectroscopy.

4. Discussion

4.1. Kinematics: Decoupling of the Ionized Gas from the Stellar Components

One conclusion that can be derived from comparing the velocity maps of Fig. 2 is the strong decoupling between the ionized gas and the stellar kinematics. In spite of the strong distortion present in the stellar velocity field of NGC 1068, the stars are probably rotating in the three galaxies. This implies that radial movements or a more complex kinematics are ruling the ionized gas movements in NGC 1068, NGC 3516, and NGC 7331. As noted by García-Lorenzo et al. (1997), Arribas et al. (1997) and Mediavilla et al. (1997a), outflow outside the galaxy plane is the most suitable explanation. We lack stellar velocity fields for NGC 4151, NGC 5728, and NGC 3227, and no direct gas–star comparison can be established, but the presence of outflow was claimed in the three galaxies by Mediavilla et al. (1992), Arribas & Mediavilla (1993), and Mediavilla & Arribas (1993). In these articles, it was supposed that a relevant part of the ionized gas could be co-rotating with the stars. However, after the studies of NGC 7331, NGC 3516 and NGC 1068 were made, it was concluded that the hypothesis of global decoupling should be reconsidered.

4.2. Location of the Active Nucleus

An interesting common feature among several of the objects in the sample (NGC 5728, NGC 3227, NGC 3516, NGC 1068, and M 31) is the non-coincidence between the optical nucleus and the kinematic center. In the Seyfert 2 galaxy NGC 5728, this can be explained as the result of obscuration hiding the nucleus located at the kinematic center (Arribas & Mediavilla 1993; Wilson et al. 1993). In the other Seyfert 2 galaxy, NGC1068, the situation is more complex since the kinematic center is not well defined, there being a rotation center for the stars in the inner 5 arcsec area and another at the outer region (García-Lorenzo et al. 1997). In the Seyfert 1 galaxies NGC 3227 and NGC 3516, the Seyfert nucleus coincides with the optical nucleus but appears as offset with respect to the kinematic center. This finding is difficult to reconcile with the standard model for AGNs, in which the off-centering should be explained as the result of a transitory migration or as the result of a merging process. This last hypothesis has been also considered in an attempt to explain the distortion in the stellar velocity field of NGC 1068. In this respect, it is very interesting to mention the case of the center of M 31, where 2-D spectroscopy (Bacon et al. 1994; del Burgo et al., these proceedings) reveals the presence of a compact (and possibly active) nucleus displaced with respect to the galaxy rotation center. Also in this case, the compact nucleus is the brightest point. All this evidence strengthens the case for galaxy merging as the origin of nuclear activity. This idea, however, needs support from the morphological point of view (consistent isophotal off-centering or the presence of two nuclei), something which at the moment has been found only for M 31 and, tentatively (García-Lorenzo et al. 1997), for NGC 1068.

To summarize, 2-D spectroscopy is essential in the study of the strongly anisotropic phenomena present in the circumnuclear regions of AGNs. Here we have addressed four such phenomena: i) the decoupling of the stellar from ionized gas kinematics, ii) the presence of multiple gaseous systems, iii) the dominance of non-rotational motions in the movements of the ionized gas, and iv) off-centering of the active nucleus with respect to the kinematic center.

References

Arribas, S., Mediavilla, E., & Rasilla, J. L. 1991, ApJ, 369, 260

Arribas, S., & Mediavilla, E. 1993, ApJ, 410, 552

Arribas, S., & Mediavilla, E. 1994, ApJ, 437, 149

Arribas, S., Mediavilla, E., & García-Lorenzo, B. 1996, ApJ, 463, 509

Arribas, S., Mediavilla, E., García-Lorenzo, B., & del Burgo, C. 1997, ApJ, 490, 227

Bacon, R., Emsellem, E., Monnet, G., & Nieto, J. L. 1994, A&A, 281, 691

García, A., Rasilla, J. L., Arribas, S., & Mediavilla, E. 1994, Proc. SPIE, 2198, 75

García-Lorenzo, B., Mediavilla, E., Arribas, S., & del Burgo, C. 1997, ApJ, 483, L99

Mediavilla, E., Arribas, S., & Rasilla, J. L. 1992, ApJ, 396, 517

Mediavilla, E., & Arribas, S. 1993, Nature, 365, 420

Mediavilla, E., Arribas, S., García-Lorenzo, B., & del Burgo, C. 1997a, ApJ, 488, 682

Mediavilla, E., Arribas, S., García-Lorenzo, B., & del Burgo, C. 1997b, Ap&SS, 248, 151

Wilson, A. S., Braatz, J. A., Heckman, T. M., Krolik, J. H., & Miley, G. K. 1993, ApJ, L419, 61

Two-Dimensional Spectroscopy with Optical Fibers: the Kinematics of NGC 1068

B. García-Lorenzo, E. Mediavilla, S. Arribas, and C. del Burgo

Instituto de Astrofísica de Canarias, E-38200 La Laguna, Tenerife, Spain

Abstract. We have obtained two-dimensional optical spectroscopy of the central $24'' \times 20''$ of NGC 1068 using an optical-fiber system. These data have been used to discuss the gas and stellar kinematics in the complex environment of this Seyfert 2 galaxy. The velocity field of the ionized gas shows a strong S-distortion with a kinematic center at $\sim 1''.3$ NE of the optical nucleus. The stellar velocity field indicates regular rotation of two distinct stellar systems whose kinematic centers are offset by $\sim 2''.5$.

1. Introduction

Current active galactic nuclei research is focused on the role that the host galaxy plays in the fueling of the central engine and the influence of its own activity on the galactic environment. On the one hand, the activity could be related to morphological structures (bars, rings, etc.) or to interaction phenomena, which generally go with AGNs. On the other hand, nuclear activity involves phenomena of energy ejection which could be perturbing the galactic medium. The kinematics of the circumnuclear region of AGNs allows us to understand both problems but the complex environments of AGNs require the recording of information over a 2-D region.

The lack of two-dimensional stellar and gas kinematic studies in active galaxies, and even in normal galaxies, is mainly due to the observational requirements of this type of study. Traditional techniques (long-slit, Fabry-Perot, etc) require a sequential observational procedure, which carries with it uncertainties due to changes in atmospheric and possible mechanical conditions. New instrumental techniques (optical-fiber systems, multilenses) have been performed in order to overcome this problem (see e.g. Arribas et al. 1991; Bacon et al. 1995).

We present two-dimensional spectroscopy of NGC 1068 using an optical-fiber system (2D-FIS). NGC 1068 is the best studied example of a hybrid starburst/Seyfert galaxy. The galactic environment of NGC 1068 is very complex, with a $16''$ stellar bar, a starburst ring at the end of the bar, and a strong radio jet coming from the nuclear region (Helfer & Blitz 1995; Wilson & Ulvestad 1987). We have obtained 570 spectra of the central $\sim 22'' \times 24''$ of this exceptional galaxy in order to study the ionized gas and stellar kinematics. A possible explanation of the origin of the activity in NGC 1068 has been proposed based on the kinematics obtained from these data.

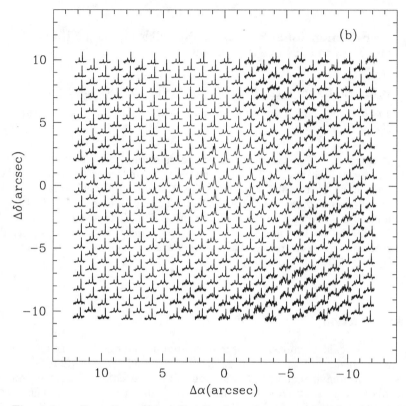

Figure 1. Two-dimensional distribution of spectra from the observed region of NGC 1068 in the range 4916–5082 Å, which includes the [O III]$\lambda\lambda 4959, 5007$ lines.

2. Observations

The data were obtained on 1994 September 16 at the Observatorio del Roque de los Muchachos, on the island of La Palma, using the 2D-FIS System (see García et al. 1994, for a system description). Two exposures, 1800 s each, were taken for six different positions, as a mosaic, of the central region of NGC1068, and the total area observed was $\sim 22'' \times 24''$ (see García-Lorenzo et al. 1997 for more details). The well-separated spectra on the detector were reduced basically in the same way as long-slit spectroscopy. Figure 1 shows the set of reduced data in the spectral range corresponding to [O III]$\lambda\lambda 4959, 5007$ emission lines.

3. Results

The emission-line profiles observed in the circumnuclear region of NGC 1068 are very complex, changing remarkably from one region to another (Fig. 1). It is likely that light coming from several gaseous systems inhomogeneously distributed in space and with different kinematics are integrated along the observer's line of sight, giving rise to multi-component line profiles. In the inner 10″, double-peaked lines are present toward the NE and SW of the optical nucleus. Outside the central 10″, the profiles are narrow and show no sign of multi-components.

Figure 2(a) shows the ionized gas velocity field inferred from cross-correlation using as template one of the observed spectra. The gas kinematics outside the central 10″ is regular and can be interpreted as a disk rotating around the north–south direction (e.g. Kaneko et al. 1992). However, the minor kinematic axis presents a strong S-distortion in the central 5″ × 5″, which can be attributed to radial motion outside the galactic plane (Arribas et al. 1996). The kinematic center of the gas velocity field indicates the origin of this radial motion. The "mean" kinematic center (obtained from the kinematic center of the velocity fields derived from Hβ, [O III]$\lambda\lambda 4959, 5007$, H$\alpha$+[N II]$\lambda\lambda 6548, 6584$, [S II]$\lambda\lambda 6716, 6731$, and [S III]$\lambda 9069$) is placed at $1''.3 \pm 0''.4$ of the optical nucleus in a direction with a position angle of $43°.66$ (García-Lorenzo et al. 1998). The origin of the radial motion must be close to this mean position. The mean amplitude of the gas velocity field in the central 10″ is 422 ± 68 km s^{-1}. It is clear from Fig. 2(a) that the optical nucleus is located asymmetrically with respect to the kinematics axes.

Figure 2(b) shows the stellar kinematics of NGC 1068. The stellar velocity field was obtained by applying the cross-correlation technique to the spectra in the Ca II triplet spectral range using as template the solar spectrum (Kurucz et al. 1991). The kinematics of the stars has been interpreted in terms of two different stellar systems rotating around displaced but parallel axes: (i) the stars of the inner system are rotating around the optical nucleus, and (ii) the outer system is a disk rotating around an axis located $2''.5$ towards the east (see García-Lorenzo et al. 1997). The peculiar stellar velocity field of NGC 1068 suggests the hypothesis that a minor merger event could have occurred in the past history of this active galaxy.

4. Summary

The ionized gas kinematics in NGC 1068 is very complex and seems to be due to several gaseous systems in the circumnuclear region, which are evident from multi-component emission-line profiles. The stellar kinematics reveal two well defined regions (inner and outer) where the stars are rotating around parallel but shifted ($\sim 2''.5$) axes.

The velocity maps obtained make clear the power of two-dimensional spectroscopy with optical fibers for these kinds of studies. With classical techniques we would lose important information about the kinematics in the circumnuclear region of this complex galaxy.

Acknowledgments. The authors acknowledge the help of Adolfo García in developing the instrument and during observations. We also thank all the staff at the Observatory for their kind support. This work has been partially supported by the Spanish Dirección General de Investigación Científica y Técnica (PB93-0658).

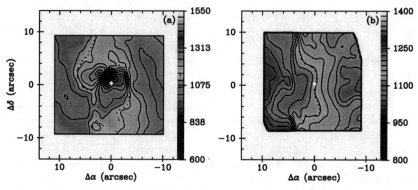

Figure 2. (a) Velocity field of the ionized gas obtained from cross-correlation in the spectral range 4800–5150 Å, including the Hβ and [O III]$\lambda\lambda$4959, 5007 lines. (b) Velocity field of the stars derived from cross-correlation in the range 8444–8882 Å, which includes the Ca II lines. The asterisk marks the position of the optical nucleus.

References

Arribas, S., Mediavilla, E., & Rasilla, J. L. 1991, ApJ, 369, 260
Arribas, S., Mediavilla, E., & García-Lorenzo, B. 1996, ApJ, 463, 509
Bacon, R. et al. 1995, A&AS, 113, 347
García, A., Rasilla, J. L., Arribas, S., & Mediavilla, E. 1994, Proc. SPIE, 2198, 75
García-Lorenzo, B., Mediavilla, E., Arribas, S., & del Burgo, C. 1998, in preparation
García-Lorenzo, B., Mediavilla, E., Arribas, S., & del Burgo, C. 1997, ApJ, 483, L99
Helfer, T. T., & Blitz, L. 1995, ApJ, 450, 90
Kaneko, N. et al. 1992, AJ, 103, 422
Kurucz, R. L., Furenlid, I., Brault, J., & Testerman, L. 1991, Solar Flux Atlas from 269 to 1300 nm. National Solar Observatory Atlas No. 1
Wilson, A. S., & Ulvestad, J. S. 1987, ApJ, 319, 105

Two-Dimensional Spectroscopy of M 31 with INTEGRAL

C. del Burgo, S. Arribas, E. Mediavilla, and B. García-Lorenzo

Instituto de Astrofísica de Canarias, E-38200 La Laguna, Tenerife, Spain

Abstract. We present results of the ionized gas and stellar kinematics in the circumnuclear region (16.42"×12.25") of M 31 obtained using a new optical-fiber system (INTEGRAL) to perform two-dimensional spectroscopy. The velocity field of the ionized gas shows that it is decoupled with respect to the stellar component; whereas the motions of the stars are dominated by a pure rotation, the gas (confined to small clouds) is dominated by radial motions.

1. Introduction

M 31 is the nearest (778 kpc; Stanek & Garnavich 1998) spiral (Sb) galaxy. The velocity field of the central 7.5"×7.5" obtained using TIGER (Bacon et al. 1994) is domained by circular motions and a very fast rotation, indicating the presence of a massive object (10^7 $M_\odot$) at the center of this galaxy.

V and I images obtained with the Planetary Camera of the *HST* show a double nucleus at the center of M 31 (Lauer et al. 1993), already discovered with the Stratoscope II (Light et al. 1974). In the V band, the brightest peak (P1) is offset by 0.49" (1.85 pc) with respect to the lower component (P2), which is almost coincident with the photometric and kinematic centers. In the UV band (at 175 nm), P2 is brighter than P1 (King et al. 1995). This asymmetry in the intensity profile is not produced by a dust line (Lauer et al. 1993; King et al. 1995). The broad-band infrared colors of P1 and P2 are identical, suggesting similar evolutionary histories of P1 and P2 (Davidge et al. 1997). The $V-K$ color of the region between P1 and P2 is the same as that of the surrounding bulge. These results favor models like the bar model of Tremaine (1995), and reject the hypothesis that P1 is an accreted object (Lauer et al. 1993).

2. Observations and Reductions

The present data were obtained using INTEGRAL (see Arribas et al., these proceedings) on 1997 July 23 at the 4.2-m William Herschel Telescope (WHT) at the Observatorio Roque de los Muchachos, on the island of La Palma. The SB1 bundle allows us to obtain the spectra of 175 regions with an aperture of diameter 0.45", covering an area of 7.85"×6.69". The SB2 bundle allows us to study almost 200 regions (with a diameter of 0.9" on the sky), covering a total area of 16.42"×12.25". A 1200-groove mm^{-1} grating provided intermediate resolution (FWHM ~ 2.0 and 2.3 Å for arc lines of the SB1 and SB2 bundles,

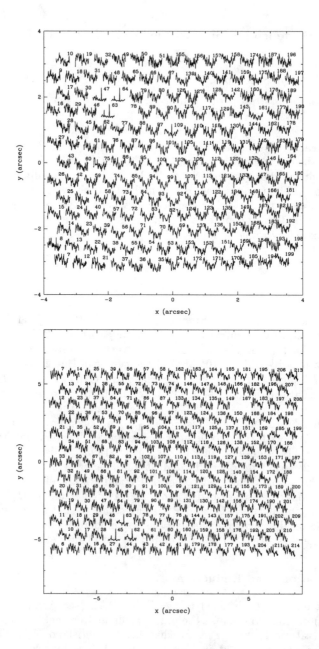

Figure 1. Spectra in the [O III] 4959,5007 spectral range of M 31 obtained using the SB1 (*top*) and SB2 (*bottom*) INTEGRAL bundles. North: 45° counterclockwise; east: 135° counterclockwise.

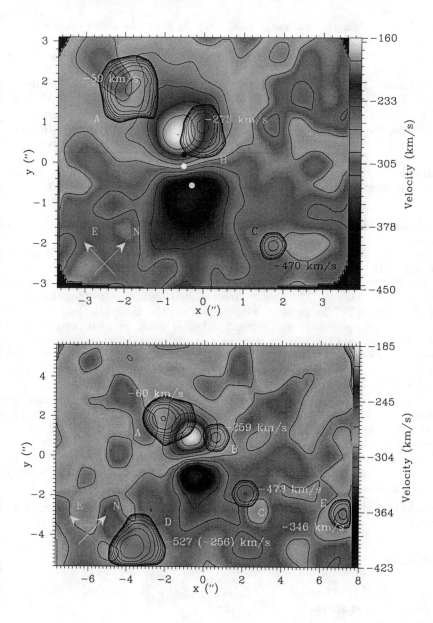

Figure 2. Stellar velocity fields (velocity range in the right column) of M 31 obtained using the SB1 (*top*) and SB2 (*bottom*) INTEGRAL bundles. The clouds of ionized gas (A, B, C, D, E, and F; represented by intensity contours) have been superposed, and the velocities of these clouds are indicated. Note the correspondence between the ionized-gas velocities measured with the SB1 and SB2 bundles. In region D, the ionized gas shows two kinematic components, with a relative velocity of 271 km s^{-1}. North: 45° counterclockwise; east: 135° counterclockwise.

respectively) in the spectral range 4330–5830 Å. With the above configuration, an exposure of 1200 s for each bundle was taken of the central region of M 31. An exposure of a velocity template star was taken using each bundle. The seeing was around 0.7″.

The data reduction was performed with the INTEGRAL package (del Burgo et al. 1998), which includes tasks developed specially for INTEGRAL data reduction and other standard tasks, as included in the SPECRED package. The rms of the wavelength calibration was less than 0.1 pixel in all cases. In order to measure the parameters (center, and FWHM) of the faint emission lines (Hβ, [O III] 5007) of the ionized gas at the center of M 31, we subtract from them a template absorption spectrum obtained from the spectra without emission lines (summed after relative redshift correction).

3. Discussion

In Fig. 1 are shown the spectra in the [O III] 4959,5007 spectral range for the SB1 and SB2 bundles. Note that the emission lines are restricted only to some fibers. Figure 2 shows the stellar velocity fields obtained by cross-correlation between 5075–5300 Å. The stellar kinematics is very regular (dominated by circular motions). The rotation with respect to the kinematic center is very rapid. According to the velocity field obtained using SB1, an amplitude of 284 km s^{-1} between the velocity maxima and minimum, which are separated by 1.58″ (240 km s^{-1} and 1.89″ according to Bacon et al. 1994). The stellar velocity dispersion field is 250 km s^{-1} at the maximum (offset 0.7″ (2.6 pc) SW with respect to the kinematic center), and 150 km s^{-1} at most external distances.

The kinematics of the ionized gas of the center of M 31 is domained by non-circular motions. In the inner region, the small clouds of ionized gas (A, B, and C) are moving radially. The [O III] 4959,5007 emission-line profiles show double components in one of the spectra corresponding to region D. The relative velocity between the components is 271 km s^{-1}. This region extends toward the west and presents a high velocity gradient (del Burgo et al. 1998). The velocity dispersions of all the ionized gas clouds are around 60 km s^{-1}.

Acknowledgments. We thank all the staff of the ING/ORM for their kind support. This work was partially supported by the DGCYT (PB93-0658).

References

Bacon, R. et al. 1994 , A&A, 281, 691
Davidge, T. J. et al. 1977, AJ, 113, 2094
del Burgo, C. et al. 1998, in preparation
King, I. R., Stanford, S. A., & Crane, P. 1995, AJ, 109, 164
Lauer, T. R. et al. 1993, AJ, 106, 1437
Light, E. S., Danielson, R. E., & Schwarzschild, M. 1974, ApJ, 194, 257
Stanek, K. Z., & Garnavich, P. M. 1998, ApJL, accepted
Tremaine, S. D. 1995, AJ, 110, 628

TEIFU: an Integral-Field Unit Optimized for Use with the ELECTRA Adaptive-Optics System at the WHT

Roger Haynes, Robert Content, Jeremy Allington-Smith, and Peter Doel

Astronomical Instrumentation Group, University of Durham, Physics Department, South Road DH1 3LE, UK

Abstract. TEIFU—a Thousand-Element Integral-Field Unit—is being constructed in Durham for use with the ELECTRA adaptive-optics (AO) system and the WYFFOS spectrograph at the Nasmyth focus of the William Herschel Telescope. It has been designed to take advantage of the improved image quality provided by the ELECTRA AO system. The observer is provided with a choice of three spatial scales: 0.125, and 0.25 arcsec with AO correction, and 0.5 arcsec without AO correction. These scales give corresponding object fields of 12, 47, and 191 square arcsec. The system uses microlenses at the input of the fiber bundle to ensure ~100% filling factor, eliminating the dead space associated with bare fiber bundles. It also uses microlenses at the output to ensure efficient coupling to the spectrograph. The unit includes a separate background field to enhance sky subtraction, and the whole device will take full advantage of the wavelength coverage available with ELECTRA and WYFFOS giving good performance from 500–1000 nm. The system is being built in such a way that it will be compatible with the NAOMI common-user AO system for the WHT which is due to come on line in 1999.

1. Introduction

Building on the knowledge gained from the prototype SMIRFS-IFU (Haynes et al., these proceedings), TEIFU is primarily intended to operate with the ELECTRA AO system (Buscher et al. 1995), using the image-correction modes that maximize the energy enclosed within a given diameter. These encircled energy improvements are less significant at shorter wavelengths. Figure 1 shows simulations of the enclosed energy gains achieved using corrected images from the ELECTRA AO system. The uncorrected data correspond to seeing conditions of 0.8 arcsec at 0.5 μm. The corrected data simulates ELECTRA performance using a $V = 10$ mag guide star with full co-phasing of the adaptive mirror and assumes a closed loop time lag of 5 ms. However, further simulations suggest that some gains may be possible using novel correction techniques that are possible with ELECTRA's segmented mirror, e.g. co-phasing sub-areas of the mirror. TEIFU has been optimized to cover the wavelength range 500 nm and 1 μm, this being limited by the ELECTRA system. It is well within the capabilities of the WYFFOS fiber spectrograph. WYFFOS itself has been designed specifically as a fiber-coupled spectrograph and is well suited to the TEIFU requirements.

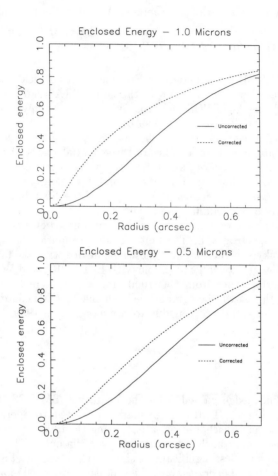

Figure 1.   *Top*: Enclosed energy at a wavelength of 1.0 μm. *Bottom*: Enclosed energy for a point source at a wavelength of 0.5 μm.

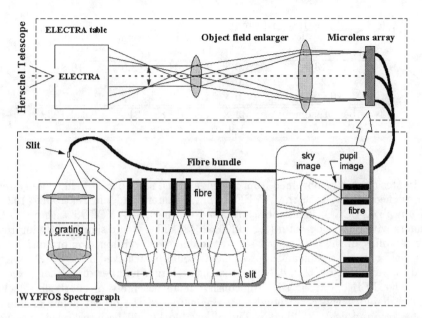

Figure 2.   A Schematic of the TEIFU optical layout.

2. The Design of TEIFU

TEIFU will join other integral-field systems at the WHT including 2D-FIS and INTEGRAL (Arribas et al., these proceedings), which are both bare-fiber systems. However, these are aimed at spatial resolutions of 0.5 arcsec, or coarser, whereas TEIFU has resolutions of 0.13, and 0.25 when used with ELECTRA and 0.5 arcsec without ELECTRA. For example, this fine spatial sampling can provide information at scales matched to the *HST* images on kinematics and emission processes in galaxy nuclei, studies of co-aligned radio and optical structures in distant radio galaxies, and studies of star-forming regions to reveal fine detail of shocks.

A schematic of the TEIFU optical layout is shown in Figure 2. Using an off-axis guide star, ELECTRA provides AO-corrected images. These images are then magnified by the TEIFU enlarger optics and re-imaged onto the input close-packed hexagonal microlens array, which has a pitch of approximately 300 μm. By changing the enlarger optics, the three different spatial sampling scales are achieved. The microlens array forms an image of the telescope pupil on the face of the fibers, which are fixed into a matrix of micro-tubes having the same pitch as the array. The fibers are then re-formatted into a slit and coupled to the WYFFOS spectrograph via the output microlenses.

TEIFU has two spatially separate fields: the main object field and a second field containing ~10% of elements (Table 1) for background subtraction when sky

Table 1. TEIFU: field formats

Image scale (arcsec/element)	Object field (arcsec^2)	Background field (arcsec^2)	Magnification
0.13	4.0 × 2.9	0.9 × 2.6	10.8
0.25	8.0 × 5.6	1.7 × 5.3	5.4
(0.50	16.0 × 11.3	3.4 × 10.6	2.7)
Number of elements	32 × 27	8 × 21	Total 1032
ELECTRA field	80 arcsec diameter		

estimates cannot be obtained from the edge of the main object field. The system is designed for use with the WYFFOS $f/1.2$ camera and the Tektronix (1024^2, 24-μm pixels) CCD. It has been proposed to upgrade WYFFOS with an $f/2.9$ camera with a large detector (e.g. 4096×2048). This would significantly improve the sampling of the IFU output slit at the detector. Without compromising the IFU field of view, significant improvements could be achieved if the $f/1.2$ camera could be upgraded with a detector that has a larger number of smaller pixels.

The TEIFU output slit will consist of an assembly of short slit blocks, and each block will correspond to a single row of the object field (32 fibers) or background field (21 fibers). The object blocks will be interleaved with the background blocks at regular intervals to form a single slit 1032 elements long. Each block will have a linear microlens array attached to couple the fiber output beams to the spectrograph. The output microlenses are rectangular in shape to maximize the amount of light gathered from the fibers. The TEIFU slit assembly will be positioned inside WYFFOS in place of the AUTOFIB2 fiber-bundle slit.

3. Future Plans

The current schedule plans to have TEIFU delivered to the WHT for 1998 August. This fits in well with the ELECTRA schedule, giving the AO team time to fully commission ELECTRA in closed loop with full co-phasing of the adaptive mirror.

Acknowledgments. We would like to thank the staff at RGO and the ING, particularly Sue Worswick. We also thank the staff in Durham, especially, George Dodsworth, David Robinson, David Robertson, and John Webster.

References

Buscher, D. F. et al. 1995, in Topical Meeting on Adaptive Optics Proceedings, ed. M. Cullum (ESO: Garching), 64

Two-Dimensional Fiber-Bundle Manufacture and FRD Characterization

José Luis Rasilla, Ana Belén Fragoso-López, and Adolfo García-Marín

Instituto de Astrofísica de Canarias, E-38200 La Laguna, Tenerife, Spain

Abstract. Several optical-fiber bundles have been designed, manufactured and characterized at the Instituto de Astrofísica de Canarias (IAC) for the INTEGRAL system. Here we briefly describe some details of these processes.

1. Introduction

The INTEGRAL system (Arribas et al., these proceedings) is a new facility that performs two-dimensional spectroscopy with fibers on the William Herschel Telescope. This system links the Nasmyth focus with the WYFFOS spectrograph (Bingham et al. 1994). In this paper we describe the design, manufacture, and testing of the three INTEGRAL fiber bundles. The following sections show the adhesive and fiber selection, the definition of the main parameters of the bundles, the manufacturing process, and the FRD testing. In another paper of these proceedings (García Marín et al.) some studies on bundle design are also presented.

2. Adhesive and Fiber Selection

We have tested the following adhesives: Araldit AY303 + HY951, Araldit AY303 + HY956, Epotek-353D, Trabond F113SC, Trabond F141, Loctite 322, and Loctite 350. We have also tested the fibers: FHP200240270, FHP200220240, FHP320385415, WF200240315A, and WF200220240P. In order to study the FRD several samples of a type of fiber were glued with the different adhesives selected. In addition, samples of the five fiber types were glued with the same adhesive. All the samples were polished simultaneously to reduce possible effects on the results. It was concluded that the WF200220240 fibers (manufactured by Ceram Optec) and the Araldit AY103 + HY951 adhesive show the best preservation of the focal ratio.

3. Definition of Main Parameters of the Bundles

Three fiber bundles (SB1, SB2, and SB3) were designed with fibers of 100, 200, and 600-μm core diameter, respectively.

To determine the maximum number of fibers in the bundles, we needed to study the fiber-to-fiber separation at the pseudoslit. To this end a test bundle built with fibers of different core diameters and a different fiber-to-fiber separation was used in order to:

1) Evaluate the energy enclosed in the fiber image, using a Gaussian fitting (Tables 1 and 2), and

2. Analyze the optical cross-talk between fibers. Figure 1 shows how the cross-talk increases when the fiber-to-fiber separation decreases.

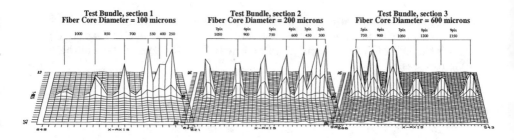

Figure 1.   Visual analysis of the bundles.

Table 1.   Gaussian fitting to the fiber image

Bundle	FWHN radial (pix)	FWHM spatial (pix)	FWHM spectral (pix)
SB1	1.45± 0.20	1.66± 0.47	1.50± 0.19
SB2	1.67± 0.28	1.73± 0.45	1.67± 0.29
SB3	3.37± 0.28	2.95± 0.41	3.06± 0.20

Table 2.   Energy analysis for the bundles

Bundle	Energy (%)	Half-width (pix)	Separation (pix)	Separation (μm)
SB1	99.00	2.58σ=1.65	3.30	491.4
SB2	99.00	2.58σ=1.88	3.77	560.5
SB3	99.00	2.58σ=3.35	6.71	997.1

From these analyses we arrived at the following conclusions for each bundle:

SB1: Following the energy analysis we find that separations > 3.5 pixel are enough to ensure cross-talk less than 1 % between adjacent spectra (when the fibers are equally illuminated). However the visual cross-talk analysis shows that we need a greater separation (4.7 pixel = 700 μm). This disagreement could be due to the underestimation of the FWHM in the Gaussian fitting.

SB2: The energy and visual cross-talk analyses agree. For a 1 % level we need a center-to-center separation > 4 pixel. Finally a separation of 4.57 pixel (680 μm) was used; this permits the inclusion of 219 fibers in the bundle.

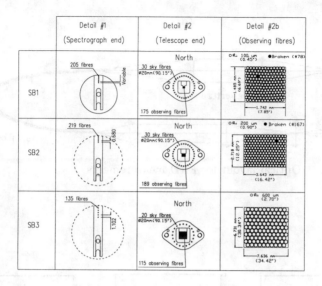

Figure 2. Main parameters of the bundles.

SB3: The energy and visual cross-talk analyses agree. For a 1 % level we need a center-to-center separation > 7 pixel. A center-to-center separation of 7.4 pixel (850 μm) was used; this permits the inclusion 135 fibers in the bundle.

The fiber distribution at the focal plane end of the bundles was defined to be rectangular. A configuration of 14×13 fibers for SB1, 15×14 fibers for SB2, and 11×11 fibers for SB3 was chosen (Fig. 2). For the bundles we used fibers manufactured by Ceram Optec. These are of WF hydrogen type treated. A view of the INTEGRAL standard fiber bundles is shown in Figure 3.

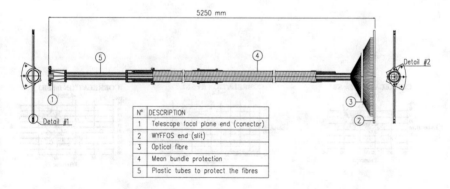

Figure 3. General layout of an INTEGRAL standard bundle.

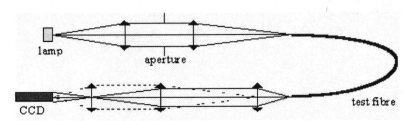

Figure 4. Optical layout for FRD measurements

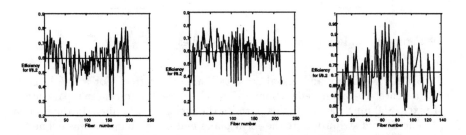

Figure 5. The efficiency for $f/8.2$ vs. fiber number for SB1, SB2, and SB3.

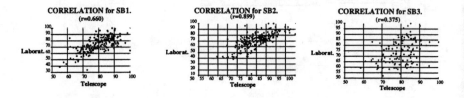

Figure 6. Correlation for SB1 ($r = 0.660$), SB2 ($r = 0.899$), and SB3 ($r = 0.375$).

4. Bundle Manufacture: Main Steps

The INTEGRAL bundles use fibers with two jackets, hard (polyimide) and soft (epoxy), which are used to absorb stress in the fibers. Before the array manufacture the second jacket was removed by immersing the fiber end in acetone.

The rectangular array of the bundles was packed manually. The fibers corresponding to each row at the array were introduced into plastic tubes to minimize stress. On finishing the array, the central fiber position was measured. The connector was manufactured in order to ensure that the center of the array is located at the center of the focal plane of the telescope.

The mechanical parts of the bundle were assembled and the fibers glued to the pseudo-slit. The pseudo-slit was made by drilling a linear array of holes in an aluminum stick. The hole diameters were fitted to the external diameter of the fibers. Both ends of the bundle were polished using diamond abrasives from 25 to 0.25 μm diamond-compound particle size. Finally, in order to fit the fibers to the curved focal plane of the WYFFOS collimator, the pseudo-slit was bent with a radius of curvature of 1190 mm

5. Focal-Ratio Degradation

After testing several set-ups for studying the FRD we decided to use the one shown in Figure 4. The method is reliable and efficient, so long as care is taken with the lens positions.

The three lenses at the bottom of Fig. 4 carry out the following function: the first two lenses control the size of the light distribution on the CCD, while the second and third lenses image the light distribution onto the CCD.

The input focal ratio chosen is $f/11$. The output f/# (for 95% energy) and the percentage of energy contained in f/8.2 are shown in Table 3.

Table 3. FRD measurements

Bundle	SB1	SB2	SB3
$f/\#$ (95% of energy)	4.34	4.96	5.65
Efficiency for $f/8.2$	59.4%	58.8%	71.4%

The results presented are the average of the values obtained for each of the fibers. All the values are presented in the graphs of Figure 5.

In order to test the quality of the measurements taken in the laboratory with this method, the data were compared with the relative signal obtained from a flat-field. The results can be seen in Figure 6. The high degree of correlation between the data obtained in laboratory and at the telescope shows that the method chosen is efficient and reliable.

Acknowledgments. We would like to thank Ricardo Negrín, Felipe García, and the IAC workshop staff for their efforts in the manufacturing process of the fiber bundles.

References

Bingham, R. G., Gellatly, D. W., Jenkins, C. R., & Worswick, S. P. 1994, Proc. SPIE, 2198, 56

Studies on Bundle Design

Adolfo García-Marín, José Luis Rasilla, Santiago Arribas, and Evencio Mediavilla

Instituto de Astrofísica de Canarias, E-38200 La Laguna, Tenerife, Spain

Abstract. Two different bundle designs have been studied with the aim of increasing the number of sampling elements, and consequently the spatial coverage, at the telescope focal plane. The convenience and viability of implementing these designs in INTEGRAL is addressed.

1. Introduction

INTEGRAL (Arribas et al., these proceedings) is a collaboration between the Instituto de Astrofísica de Canarias, the Royal Greenwich Observatory, and the Isaac Newton Group.

The optical fiber bundles used to perform 2-D spectroscopy usually have a small field of view on the sky. This is a consequence of the reduced number of spectra that can be accommodated on the detector. One of the ways of optimizing the number of spectra is to reduce the spacing between fibers at the pseudo-slit to the minimum physical quantity (we will call this Option 1). Another possibility is to use a multi-slit arrangement in order to increase the number of spectra at the cost of reducing the spectral coverage (Option 2). Here we present some studies performed with the aim of implementing these two possibilities in a particular case, the INTEGRAL field unit (Arribas et al., these proceedings) for the WYFFOS spectrograph at the WHT.

2. Option 1

When the fibers are located at the slit with no separation among them some cross-talk can be produced as consequence of the non-zero width of the PSF of the spectrograph. This is obviously a general problem also affecting long-slit observations.

In the case of 2-D spectroscopy, the reconstructed images can be affected by certain artifacts as consequence of this cross-talk. The way in which the PSF of the spectrograph affects the reconstructed images depends on the ordering of the fibers. Here we will consider the ordering proposed by García-Marín et al. (1994) with the aim of minimizing the cross-talk.

To study the effects introduced in the image-reconstruction process by the cross-talk with this design (Option 1), the following assumptions have been made:

- A Gaussian function is used to simulate the seeing, which can vary from 0.5 to 1.5 arcsec

- The fiber parameters are 100/110/125 μm (core/cladding/buffer diameters)

- Other optical parameters are fixed by the WHT (e.g. 1 arcsec=222 μm) and WYFFOS (e.g. magnification=1/6.2; detector pixel size=24 μm)

According to these assumptions we have obtained, (Fig. 1, *left*) the energy collected by a fiber coming from four objects:

- Object 1 centered on the fiber

- Object 2 centered on the adjacent fiber (situated 125 μm from object 1)

- Object 3 situated 250 μm from object 1

- Object 4 situated 375 μm from object 1

Figure 1 (*right*) shows the reconstructed contour map of a star using the information obtained from seven fibers of the bundle, supposing a seeing of 0.5 arcsec. In this figure the star appears strongly distorted by the cross-talk effects. Generally, and depending of the spectrograph's performance, this deformation could be an important limitation in considering Option 1.

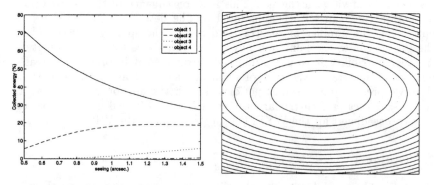

Figure 1. *Left:* energy collected by the extraction of a fiber due to four objects. *Right:* star as viewed for seven fibers of the array after its reconstruction for a seeing of 0.5 arcsec.

3. Option 2

Another possible of increasing the spatial coverage at the focal plane could be to distribute the fibers of the bundle at the spectrograph entrance in several pseudo-slits mounted in parallel.

In the simplest case, two slits separated a distance X along the spectral direction, the spectra will be separated at the detector at a distance $Y = mX$

(where m is the magnification of the system). The spectra in a given order may overlap depending on the spectral range of the dispersing element. So to avoid overlapping spectra the spectral range must be restricted by means of filters.

In the specific case of INTEGRAL + WYFFOS (1200 l mm^{-1} grating, and a Tek CCD of 1K×1k) we have considered configurations involving two, three, and four slits. A summary of the resulting spectral coverages is given in Table 1.

Table 1. Spectral range for four configurations (1200 l mm^{-1} grating)

Configuration	X (mm)	Slits	X (mm)	Y (px)	Y(Å)
1	39	2	39.00	262	384
		3	19.50	131	192
		4	13.00	87	128
2	55	2	55.00	370	541
		3	27.50	185	271
		4	18.33	123	180
3	63	2	63.00	423	620
		3	31.50	212	310
		4	21.00	141	207
4	72	2	72.00	484	709
		3	36.00	242	354
		4	24.00	161	236

Configurations with three or four slits offer very interesting performance, although only a configuration of two slits may be attempted without modifying the slit area of WYFFOS. However, in this last case the increasing in spatial coverage would be only moderate.

In any event, the concept behind this option proves to be very powerful for increasing the spatial coverage of specially designed fiber-linked spectrographs at the cost of losing spectral range: we could have highly dedicated bundles (spectrally) but with greater spatial coverage.

References

García-Marín, A. Rasilla, J. L., Arribas. S., & Mediavilla, E. 1998, Proc. SPIE, 2198, 75

ESPRIT D'Argus: an IRAF-Based Software Package for the Treatment of ARGUS Data

P. Teyssandier[1], and C. Vanderriest

DAEC, Observatoire de Paris-Meudon, 92195 Meudon Cedex, France

Abstract. We describe a software package that combines existing IRAF tasks and new IRAF-style scripts for the treatment of ARGUS data. The present version uses the SPP programming language, which allows calibrated spectra and monochromatic images to be produced almost in real time.

1. Introduction: History and Grammar

The first scientifically useful data obtained with SILFID, in 1985, were processed with the EVE software package (a kind of "MIDAS light" product elaborated at Meudon) running on VMS machines. A few years later, with the invasion of UNIX systems, we decided to base our reduction software on IRAF. In addition to the standard procedures of this package, we thus had to write specialized scripts that are fully compatible with IRAF syntax. Most recently, some of these procedures have been rewritten in SPP (*S*ubset *P*re-*P*rocessor language, an IRAF-designed FORTRAN code generator) in order to speed up the slower tasks.

The resulting package, ESPRIT D'Argus (*E*nsemble *S*imple de *P*rocédures *R*éalisé avec *I*RAF pour le *T*raitement des *D*onnées d'*Argus*) may be used directly at the telescope, providing the observer with calibrated spectra and reconstructed images less than one hour after the data have been obtained. It is also used for the final, more meticulous, data processing.

2. The Chain of Tasks

Our philosophy is to consider ARGUS data in exactly the same way as we would long-slit data during the first part (pre-processing) of the treatment, and then to deal with their two-dimensional nature in an interactive way during the last part (processing). The flow-chart below gives an overview of the chain of tasks to be carried out, and the corresponding procedures.

2.1. Pre-Processing (2-D)

Tasks like wavelength and flux calibrations are done with standard IRAF procedures, without any change. We give a brief comment only for the procedures that were developed in our laboratory (Fig. 1).

---

[1] Software Consultant for Observatoire de Meudon.

ESPRIT D'Argus

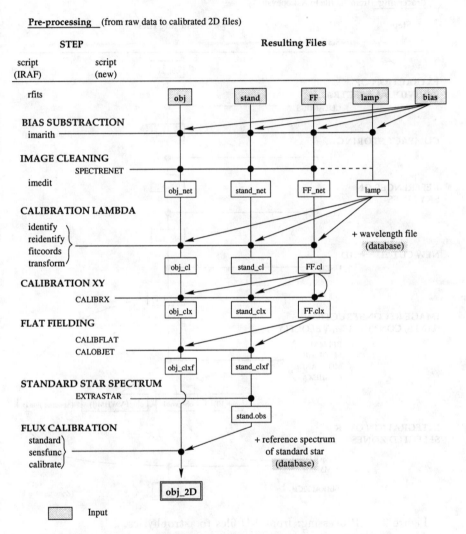

Figure 1. Pre-processing: from raw data to calibrated 2-D files.

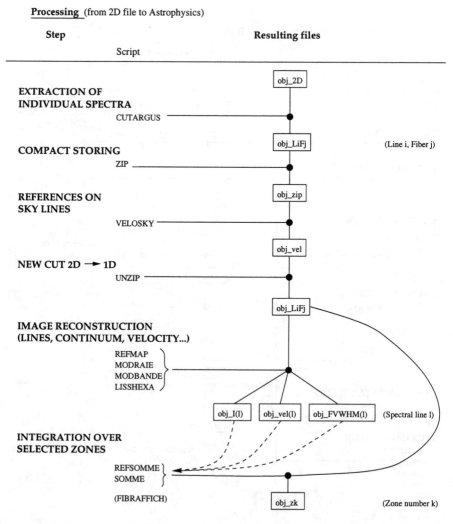

Figure 2. Processing: from 2-D files to astrophysics.

- The removing of cosmic-ray hits on a single frame is done efficiently with our procedure SPECTRENET. However, some stubborn "cosmics" may need a personalized (i.e. manual) treatment with imedit.
- CALIBRX re-samples the data with constant steps along the "slit" by using the separation fiber marks on a flat-field exposure.
- For the flat-fielding, we first generate a normalized flat with CALIBFLAT before correcting with CALOBJET.
- EXTRASTAR automatically extracts the signal from the standard-star file and produces a 1-D spectrum usable for flux calibration.

2.2. Processing (1-D)

This part uses only specially written procedures (Fig. 2).

- First, we cut the fully calibrated 2-D file into a series of 1-D files, each corresponding to the spectrum from one fiber, with CUTARGUS.
- ZIP allows these spectra to be stored in the compact format of a 2-D file (1 line per fiber), while UNZIP does the reverse transformation.
- VELOSKY (an optional procedure) corrects for any remaining shift in the wavelength scale—for instance, because of flexures in the spectrograph between the object and calibration exposures—by using the strong night-sky lines as local velocity references.
- Images are reconstructed with MODRAIE (absorption or emission lines) and MODBANDE (continuum) after the lines or spectral ranges have been chosen with REFMAP. A smoothed image is produced with LISSHEXA.
- Finally, in order to increase the S/B ratio, the spectra from several fibers may be co-added over selected zones. These are chosen on the maps produced at the previous step with the help of FIBRAFFICH and REFSOMME. The resulting 1-D spectra are created by SOMME.

3. Conclusion

The ESPRIT D'Argus package is working well. Together with a short "user's manual", it is available on request (contact: vander@DAEC.obspm.fr). It is now working for the data obtained with SILFID, MOS-ARGUS, and ALBIREO, and could easily be adapted to the data from other integral-field spectrographs. All that is needed is a configuration file describing the geometry of the fiber device.

The image reconstruction is done automatically with a simple Gaussian fitting of the lines. We did not try to implement a more complex routine in the package (e.g. fit with multiple Gaussians or other line profiles, estimate of the continuum, etc.). Such elaborate treatments may be carried out with existing software packages on the spectra of individual fibers.

Acknowledgments. We thank H. Reboul, and M.-C. Angonin-Willaime for their help in the development of the earlier versions of this software, as well as M. Fitz (IRAF specialist at NOAO) for his countless helpful suggestions.

Part 4: Projects for Large Telescopes

Fiber Optics in Astronomy III
ASP Conference Series, Vol. 152, 1998
S. Arribas, E. Mediavilla, and F. Watson, eds.

Integral-Field Spectroscopy with the GEMINI Multi-Object Spectrographs

Jeremy Allington-Smith, Roger Haynes, and Robert Content

Astronomical Instrumentation Group, Physics Department, University of Durham, South Rd, Durham DH1 3LE, United Kingdom

Abstract. The GEMINI multi-object spectrographs (GMOS) will be equipped with an integral-field unit so that they can perform either multi-aperture or integral-field spectroscopy. The changeover between modes will be effected simply by deploying the integral-field unit at the focal plane in the same way as a multi-aperture mask. The design, which uses fibers coupled to lenslet arrays, has ∼2000 elements, giving a sampling of 0.2 arcsec over a field area of ∼60 arcsec^2. A separate, optically-identical field is provided to facilitate accurate background subtraction. The IFU will be installed in the first GMOS, on Hawaii, soon after commissioning in 1999.

1. Introduction

Both GEMINI telescopes will be provided with a GMOS (Davies et al. 1997). The instruments are of identical design, with a 5.5×5.5 arcmin^2 field sampled at 0.07 arcsec per pixel. The detector format is 6144×4608 with 13.5-μm pixels. They are optimized for the wavelength range 0.4–1 μm and provide resolving power up to $R \equiv \lambda/\Delta\lambda = 5000$ with a slit width of 0.5 arcsec (although slit widths of 0.25 arcsec will be possible). The spectrographs have three main modes: *multi-object*, via multi-aperture masks made off-line at the telescope; *imaging*, for target acquisition and mask design; and *integral field* via an integral-field unit (IFU) inserted remotely into the focal plane.

2. The Scientific Case for Integral-Field Spectroscopy

Integral-field spectroscopy provides a spectrum of each spatial element in a two-dimensional field (e.g. Bacon et al. 1995). As well as the advantage for studies of extended objects, work on single unresolved objects or integrated spectroscopy of extended objects also benefits because light from the full extent of the object is collected simultaneously without the need for a wide slit, which would degrade spectral resolution. Target acquisition is easier because it is not necessary to position the object carefully on a narrow slit and because the precise pointing may be determined after the observation by forming a white-light image from the spectral data.

Applications include the following. (a) *Studies of the distribution of star formation in distant galaxies in the field and clusters.* HST reveals their mor-

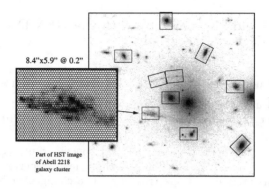

Figure 1. The field of the GMOS integral-field unit compared with objects in an *HST* image of the galaxy cluster A2218. The latest design has 24 % more area.

phologies and colors but spectroscopy is required to study details of the stellar populations. With GEMINI's good images and light-grasp, spatially resolved information can be obtained for distant galaxies. (b) *Kinematical studies of galaxies at intermediate redshift.* This requires both good spectral and spatial resolution and a large wavelength range if results from different spectral features are to be combined. Since nearer galaxies might completely fill the field of the IFU, it is necessary to provide a separate offset field dedicated to background estimation. (c) *Distance estimation at intermediate redshift.* For example, the Tully–Fisher relationship at $z \simeq 0.6$ requires the velocity field to be measured to several scale lengths (1-2 arcsec) via an emission line such as [O II] 3727. This requires good images, high throughput and a full two-dimensional capability to ensure that the *global* velocity field is recovered. (d) *Reconstruction of lensed galaxies.* Gravitational amplification by foreground clusters increases the detectability of faint galaxy populations. By reconstructing the original (x, y, λ) datacube, the above studies can be repeated on even more distant galaxies. Even if the signal/noise in a single element in insufficient, the integrated light may yield the redshift and broad details of the stellar populations. (Fig. 1—see also Soucail, these proceedings).

Especially when extended into the near-infrared (if GMOS was equipped with an IR detector) a number of other applications become possible, including studies of the obscured nuclear regions of active galaxies; the optical–radio co-alignment of distant radio galaxies, and studies of shocks in star-forming regions.

3. The GMOS IFU

The integral-field mode is selected by remote insertion of the IFU at the telescope focus using the mask exchange mechanism. This is possible because the IFU will be packaged into a space not much greater than that of a slit mask (Fig. 2). The main parameters of the IFU for GMOS-North are given in Table 1.

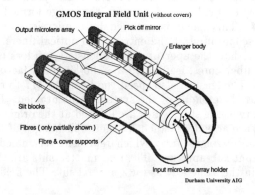

Figure 2. Interim design of the GMOS IFU presented at the GMOS Critical Design Review. The cover and some of the fibers are not shown.

Table 1. Summary of the GMOS baseline Integral Field Unit

Spatial sampling	0.2 arcsec
Object field area	62 arcsec^2
Object field aspect ratio	$\sqrt{2}$
Object:background field area	10:1
Object-background field distance	1 arcmin
Number of reformatted slits	2
Width (FWHM) of spectral-resolution element	0.27 arcsec
Number of spectral-resolution elements	800–1600
Number of spatial-resolution elements	2000

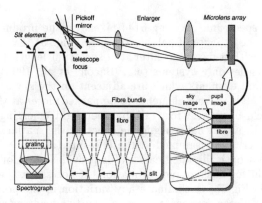

Figure 3. The principle of the GMOS integral-field unit.

216 Allington-Smith et al.

The basic principle is illustrated in Figure 3. The $f/16$ beam from the telescope is sent through an enlarger to magnify the image formed by the telescope by a factor of ~ 4. This image is sampled by a microlens array which forms images of the telescope pupil at the entrance to optical fibers. Since the pupil images are smaller than the aperture of the lenslets, no loss is experienced at the gaps between fiber cores, leading to a ~ 100 % filling factor, in contrast to fiber-only designs. The lenslets also speed the beam to $f/7$ to reduce *focal ratio degradation* so that the etendue of the system is maintained. The fibers are reformatted to form two long slits, which are located at the original telescope focal surface. The slits are actually defined by the pupil images of the fiber outputs formed by a linear microlens array which represent scrambled versions of the sky within each input sub-aperture. The reformatted slits are then dispersed by the spectrograph in the same way as for a normal slit. The field layout is shown in Figure 4.

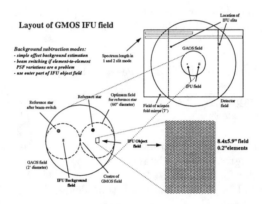

Figure 4. The layout of the field of GMOS, the IFU and the GEMINI adaptive-optics system. The format of the detector and spectra are also indicated.

A full-length spectrum, covering 6144 pixels can be produced by blocking one of the slits at the expense of halving the field (Fig. 5). Even in the two-slit mode, when used with a passband filter, the spectrum length far exceeds that obtainable from lenslet-only systems (e.g. Bacon et al. 1995). The field-slit mapping is such that elements which are adjacent on the sky are also adjacent at the slit (Fig. 6). This allows elements to be packed closer at the slit resulting in an increased field of view. Cross-talk between light emerging from adjacent elements at the IFU output has a negligible effect on the spatial resolution when the input image is critically sampled at the IFU input.

A separate, but optically identical, field is provided at an offset of 1 arcmin from the main field for background estimation. If necessary, the two fields can be beam-switched (Fig. 7) to eliminate any variation in the point spread function produced by different elements. Ideally, the two fields would be the same size, but this is not possible without reducing the overall field of view when beam-switching is not used. Figure 4 also shows how the beam-switching is achieved when GMOS is used with the GEMINI adaptive optics system. Here it is important to ensure that the wavefront sensor can access the reference star during both parts of the beam-switch cycle. This imposes a 1-arcmin limitation on the field offset and requires that the two fields are symmetrically disposed about the field center.

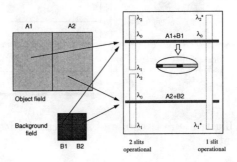

Figure 5. Illustration of the one- and two-slit modes. The one-slit mode more than doubles the available length of spectrum at the expense of halving the field area.

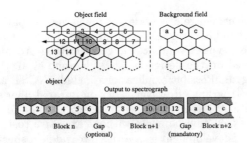

Figure 6. Illustration of the mapping between the field and slit designed to minimize contrast variations between adjacent elements at the slit.

4. Current Progress and Prospects for the Future

The GMOS IFU is expected to be completed in 1999/2000. The design will be finalized in 1998 to allow us to learn as much as possible from the constructions

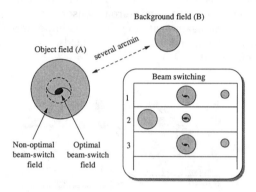

Figure 7. Illustration of the use of a dedicated offset field for background estimation.

of the other IFUs in our program. The interim design presented at the GMOS Critical Design Review is shown in Figure 2.

Our current program of IFU construction comprises the following. (a) The SMIRFS-IFU, a near-IR system retrofitted to the CGS-4 1–5-μm spectrograph on UKIRT. This 72-element device, working in the J and H bands has now been successfully commissioned (Haynes et al. 1998) and used to investigate the distribution of [Fe II]/Paβ in active galaxies. (b) The Thousand Element Integral Field Unit (TEIFU), which will exploit adaptively-corrected images provided by the ELECTRA and NAOMI AO systems at the 4.2-m William Herschel Telescope at scales down to 0.13 arcsec per element (completion 1998). (c) The GMOS-IFU.

To address the challenge of working at much higher spatial resolution (0.05 arcsec per element) and in cryogenic environments (for > 1.8 μm) or in space, we have designed a new type of compact image slicer (the Advanced Image Slicer; Content 1997) based on the design of 3-D (Krabbe et al. 1997). Work on a prototype has begun. Possible applications include NGST and the GEMINI Near-Infrared Spectrograph. However, work on infrared fiber systems continues to investigate whether the lenslet+fiber approach adopted for our current IFUs can be adapted for cold environments without excessive cost.

Another area of development is *multiple* integral field spectroscopy, in which a number of separate IFUs (10–20), each containing a modest number of elements (100–200), are positioned as desired within the field by a pick and place robot to give a combination of multi-object and integral-field spectroscopy. This approach is required for studies of star-formation regions in a number of galaxy-cluster members simultaneously—or in the Hubble Deep Field. A study of such a capability is currently under way with partners in the UK and Australia for the SUBARU 8-m telescope project.

References

Bacon, R. et al. 1995, A&AS, 113, 347
Content, R. 1997. SPIE, 2871, 1295
Davies, R. et al. 1997, SPIE, 2871, 1099
Haynes, R., Allington-Smith, J., Content, R., & Lee, D. 1999, UKIRT Newsletter, in press [see also star-www.dur.ac.uk/~jra/ukirt_ifu.html]
Krabbe, A, Thatte, N., Kroker, H., & Tacconi-Garman, L. 1997, SPIE, 2871, 1179

The Original FUEGOS Project on the VLT

P. Felenbok

Observatoire de Paris-Meudon, 5 Place Jules Janssen, 92195 Meudon, France

Abstract. FUEGOS (Fiber Unit for European General Optical Spectroscopy) is one of the foreseen focal instruments for the ESO VLT (Very Large Telescope). The goal is to provide the European community with a first class instrument allowing high- and low-resolution spectroscopy for stellar and extragalactic observations, in a large field. This large field of view, 26 arcmin, provided by the VLT, is quite unique on 8-m class telescopes and it is fully exploited. This multi-object spectrograph, fiber fed, will be implemented on the Nasmyth platform of one of the 8-m telescopes. It is made of two parts: a fiber positioner attached to the Nasmyth rotator, and a spectrograph sitting on the Nasmyth platform. The two components are linked with optical fibers. For the MEDUSA mode, the positioner is of the robot type, made of two heads traveling simultaneously on an $X-Y$ carriage. It magnetically locks 80 independent optical-fiber bundles at object coordinates, on a metallic focal surface. Two-dimensional spectroscopy is provided through a fiber anamorphoser, called the ARGUS mode, in 6- and 23-arcsec fields, with 0.2- and 0.8-arcsec sampling. An atmospheric dispersion corrector is supplied for this mode. The spectrograph is of the classical type, based on a catadioptric design. It gives access to spectral resolutions ranging between 30000 and 1600. The detector, of the cold-finger type, is made of two 4k × 2k CCD, with 15-μm pixels, located inside the camera optics.

1. Introduction

Since a few years ago, there is a real boom in the use of optical fibers for astronomy, mainly for spectroscopic applications. This is due to the great improvement in the quality of industrial products, which now match the astronomical requirement. In opposition telecommunication needs, which are driving the optical-fiber market, and for which flux transmission is not the paramount constrain, for astronomy, where no enhancement of the light source could be achieved, this is the main challenge. Fibers used in astronomical spectroscopy are of the step-index type with silica cores. New products are covering a large spectral domain, and provided great care is taken in handling and using the fibers, their focal-ratio degradation (FRD) can be minimized. Coupling spectrographs to telescopes with optical fibers has many advantages: physical decoupling giving rise to a new spectrograph concept, the "bench-mounted" one, and, in two-dimensional observations, anamorphosis between the image field and the spectrograph slit is achieved. In many applications, when fibers are packed together to make an area collecting surface, dead space is generated between fiber cores. To overcome this drawback, adjacent microlenses are used to feed unpacked fibers with total field coverage. With the coming of large-size telescopes at the dawn of the third millennium, a new challenge is facing fibers for the spectroscopy of deep-sky investigations where sky subtraction is the main limitation. This niche has been typically devoted to multi-slit spectrographs, but a better knowledge of fiber parameters and the side effects on fiber transmission could change this situation. Fibers in astronomy are used mainly in two modes, the MEDUSA mode, and the integral-field mode.

2. The MEDUSA Mode

This mode, intended for multi-object spectroscopy, uses individual fibers or bundles of fibers to send the light from individual objects spread apart in the field of view of the telescope to the slit of the spectrograph. This practice, started more than ten years ago, has spread all over the world. It started with manual fiber positioning on small telescopes and it is now automatic on 8-m class telescopes such as AUSTRALIS (Taylor, these proceedings), FUEGOS (Felenbok et al. 1994) and Hectospec(Fabricant et al. 1994). The main competitor to this type of instrument, is the multi-object spectrograph using masks in the focal plan of the telescope. This technology was initiated a couple of years ago on small telescopes and is entering general use on 8-m class telescopes with projects such as DEIMOS (James et al. 1998), GMOS (Daviset al. 1997), VIMOS (Le Fèvre et al. 1996). Figure 1 shows the comparison between a mask or slitlet spectrograph and a MEDUSA one.

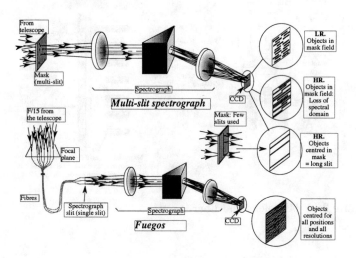

Figure 1.   Multi-object spectrography with masks (*top*) and with fibers (*bottom*).

It might be noticed that in the mask instruments, such as the MOS at the CFHT, the spectra are located on the spectrograph detector at the position where the undispersed source would remain if we were ti take away the dispersing element. So the spectra are spread around the detector and not all of its surface is not used. It is even worse if the spectral resolution is increasing because in this situation, the spectral range is also changing from an object in the center of the field to one at the edge of the field. This is not the case for a MEDUSA spectrograph, for which all the output fibers are aligned along a slit and lead to a detector fully covered with spectra that are of the same spectral range, whatever the spectral resolution. The MEDUSA instrument has the reputation of achieving poor sky subtraction compared to the mask-type spectrograph due to the change in transmission parameters with fiber bending. Photometry at the 1% level is achieved on 4-m class telescopes, and redshift measurements of galaxies up to bj= 22 could be performed (Cuby 1994). As our goal is to use fibers on 8-m class telescopes and to go deep, our main concern will be to improve fiber photometric stability during observation and calibration. This issue is fully investigated in the paper by Baudrand elsewhere in these proceedings. When high spectral resolution on 8-m class telescopes is required, and when the grating is to be kept in a reasonable size, image slicing has to be achieved to overrun the large image scale. In the MEDUSA mode, classical image slicers of the Bowen–Walraven or Richardson type, are impracticable. Image slicing could be made directly with fibers, by replacing a single fiber by a bundle of seven spread over the spectrograph slit. To do this in an efficient way without losing light, micro-lenses are fitted at the top of the fibers. This is shown on Fig. 2 for FUEGOS.

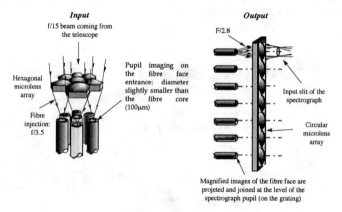

Figure 2.   Microlens and fiber layout.

3. The Integral-Field Mode

The integral-field unit (IFU) or the so-called IFU mode, is a 2-D spectroscopic sampling facility for extended objects recorded with adjacent fibers. The fibers are packed tightly together at the entrance of the bundle and spread out in a line at the output to form a spectrograph slit. This type of instrument was built by Vanderriest (Vanderriest & Lemonnier 1988) and Felenbok and Lemonnier (Felenbok et al. 1994) for the CFHT, Garcia (Garcia et al. 1994) for the WHT and Barden (Barden & Wade 1988) for KPNO. This approach competed with bare-microlens systems such as Tigre (Courtes) at the CFHT, each instrument with its own advantages and drawbacks. Nowadays, the progress in microlens technology and the advent of large telescopes with a large scale on the sky, have rendered this debate obsolete. New IFUs combine microlenses and fibers to supply full field coverage and overcome the FRD of the fiber. This is illustrated in Fig. 3 for the FUEGOS project. In this case, the microlenses feed the fibers with the appropriate beam aperture, projecting the telescope pupil at the fiber entrance and effectively providing the unbaffled VLT through the cladding masking.

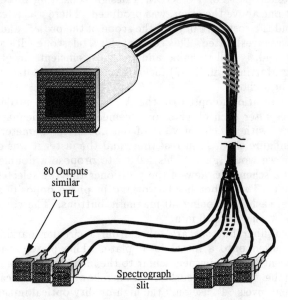

Figure 3. Microlenses and fibers in the integral-field unit.

The main claim to originality of FUEGOS lies in two unique features: the high spectral resolution for a multi-object spectrograph and the integral-field capability for an 8-m class telescope. Since our project was initiated, two other projects, GMOS on Gemini and VIRMOS on the VLT, have decided to implement IFU in their concept. The maximum spectral resolution goal was 30000, and this led to a very large grating, even with the fiber-slicing technique, but it was still feasible with a mosaic design. Contacts with our Russian colleagues have provided an opportunity to produce a 300-mm × 500-mm monolithic grating which could fulfil the requirement. Due to the fact that it was quite impossible to build a spectrograph with a slit able to deal with more than 600 fibers and to use a CCD detector bigger than 4k × 4k, only 80 objects could be taken simultaneously.

3.1. The Positioner

As seen above, we had to design a positioner that would deal with 80 fiber bundles in the MEDUSA mode. The focal surface is spherical and the incoming beam is not "telecentric". Two classical solutions were investigated: arms and a robot.

Our first approach was to see if an arm positioner was feasible. Arms are fast and minimize the time lost on the sky because they are moving simultaneously. They allow real-time position corrections and quick changes of configuration for wavelength calibration. So they are able to compensate for atmospheric refraction variation during an exposure and also for position offset if some errors are spoiling the computed field coordinates. Our laboratory has experience of such systems, having build MEFOS (Felenbok 1997), an arm positioner for the 3.60-m ESO telescope. For FUEGOS, a design consisting of two circles of 40 arms each, set one above the other, was produced. There was a feasible solution, but the size and the cost were beyond the scope of the project and an alternative robot design was investigated. This concept, for a telescope with a corrector that delivers a flat field, is still valuable and the most efficient. In particular, if the GTC (Alvarez Martín et al. 1997) focus is corrected to be flat, the arm design would be the best choice.

The robot solution adopted on the AAT for the 2dF project, exchanging the focal plates after each observation, seemed to us inadequate for the VLT Nasmyth focus, with a field of view of one meter in diameter. So we were doomed to configure the field in real time, and the faster it was done, the more telescope time we would gain. This led us to propose a double-robot design. Figure 4 gives a schematic view of the positioner that we selected as the phase A study output. Two robot heads working in phase opposition travel on an $X-Y$ carriage, each positioning 40 magnetic buttons. The configuration time expected is between 5 and 10 min.

The optical-fiber buttons are taken from a circular parking storage and magnetically locked onto a metal focal plate. This plate, consisting of four spherical adjacent rings, is perpendicular to the incoming beams throughout the field of view (the "telecentric surface") and follows the spherical focal surface via four different levels. To use only the high-quality optical images delivered by the telescope and to avoid vignetting generated in its optical train, we limited ourselves to a 26-arcmin usable field. The focal plate and the positioner are fixed to the Nasmyth rotator, which compensates the field rotation generated by the alt-azimuthal telescope mount. To avoid mechanical flexure and metrological trouble, all configurations are made at the same rotator stop position.

4. The Optical-Fiber Link

There are two types of fiber links connecting the positioner and the spectrograph: the MEDUSA mode for single objects spread out in the field, and the IFU mode for 2-D spectroscopy of extended objects. The fiber links are 7 m long.

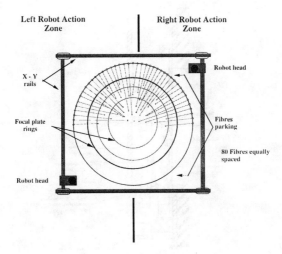

Figure 4.   The double-positioner design.

4.1.   The MEDUSA Fiber Link

In this mode, 80 fiber links are used. The fiber input is locked trough the magnetic button at any place inside the 26-arcmin field. The focal surface, as illustrated in Fig. 5, is approximated by four adjacent flat rings. At the junction of two rings, a step prevents access to the magnetic button. This represents only an 8 % surface loss. The scientific goal that we intend to reach needs a spectral resolution of 30000. This is quite difficult to obtain with an 8-m telescope if we wish to avoid the use of a huge grating. Some slicing technique has to be employed. As we are coupling the focal surface to the spectrograph with optical fibers, we use this opportunity to make an image slicer with a fiber bundle acting as an "anamorphoser". Instead of collecting the light spot at the telescope focus with a single fiber, we use a fiber bundle of seven fibers covering the same front surface. At the input, the seven fibers are packed together in a hexagonal distribution, and at the output they are aligned to form a slit. Microlenses are fitted to the bundle input end to make a full-surface collecting area, without dead spaces. The input microlenses also widen the $f/15$ Nasmyth beam to $f/2.85$, a value that minimizes focal-ratio degradation. They also baffle the telescope stray light, the fiber cladding playing a light-stop function. The fiber output is also fitted with microlenses to change the fiber $f/2.85$ output beam to $f/10$, which is a better choice for the spectrograph collimator.

4.2.   The IFU Link

For FUEGOS, we foresee a 600-fiber IFU, which corresponds to the detector storage capacity. The individual fibers, as for the CFHT, will also be 100-μm core fibers. Due to the large image scale of an 8-m telescope, we hope to couple

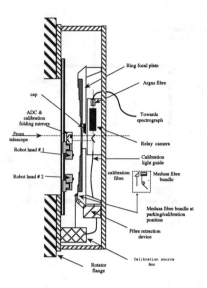

Figure 5. The focal-surface design.

the IFU input to a microlens matrix made of adjacent hexagonal lenses as in the MEDUSA seven-fiber individual bundle. The manufacturing process is the same, only the number of components is larger. In such a design, the image is fully sampled without dark spaces. At the output, as in the MEDUSA mode, microlenses are fitted to the fibers to match the spectrograph collimator aperture. Two or three spatial samplings, between 0.2 and 0.8 arcsec in fields ranging between 5.7 and 23 arcsec, will be provided by dedicated focal-transformation optics. Because the IFU field is small, an ADC (atmospheric dispersion compensator) will be inserted, at a low cost, in the incoming telescope beam.

5. The Spectrograph

The main design difficulty of the spectrograph comes from the high spectral resolution request and from the long slit that has to accept 80 ×7 fibers in the MEDUSA mode or 600 fibers in the IFU mode. The FUEGOS spectrograph is extensively described by Casse (1995). In fact it has the same slit height, 316 mm, in both modes, some spacing being needed between the fibers coming from different bundles to avoid cross-talk from information emanating from independent objects.

Two optical designs were investigated: a "white pupil", and a "classical".

The "White Pupil" Type This leads to a limitation of the camera size down to the monochromatic beam size, which is very important in such a large instrument. It is also near the Littrow mount, with high efficiency, being closer to the blaze angle. The collimator is catadioptric, and four gratings are mounted on a carousel for fast interchange. Three gratings are of standard size, except the highest-resolution one, which is made of a mosaic of two.

Two designs were investigated for the camera, one with catadioptric optics and another with dioptric optics. The catadioptric one has an internal detector which prod;uces a shadow in the incoming beam. The dioptric design does not give higher throughput and requires the use of large exotic optical elements. WE opted for a full catadioptric optical design, with spherical mirrors and spherical correctors

The "Classical" Type   This design was undertaken at the request of ESO to see if its throughput was higher. In fact, after a first feeling that its transmission was better, in the end, it appeared that it was equivalent to the "white pupil" design. We decided, nevertheless, to go ahead with the classical design because it could be used more easily with a pre-monochromator for order selection. The pre-monochromator is of the prism type. The spectrograph design layout is given in Figure 6.

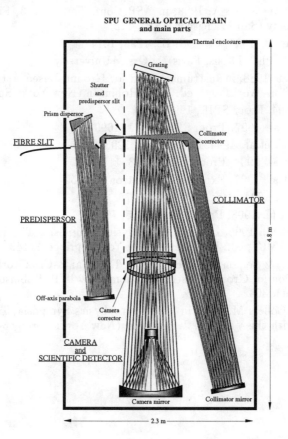

Figure 6.   The "classical" spectrograph design with pre-monochromator.

Four spectral resolutions, 30000, 11000, 4500, and 1600, are supplied through a carousel fitted with four gratings. The spectrograph detector selected is made of two 4096 ×2048 pixels CCD with a 15-μm pixel size. The pixel size samples the fiber-projected image well, and the large geometrical size of this 4k CCD gives a large spectral range. The detector is located inside the catadioptric camera in a vacuum vessel of the cold-finger type. The last camera corrector lens is also the dewar window, which will have to withstand the atmospheric pressure without optical aberrations. If necessary, the window will be calculated to take this strain into account.

References

Allington-Smith, J. et al. 1997, Proc. SPIE, 2871, 1284

Alvarez Martín, P. 1997, Gran Telescopio Canarias Conceptual Design (La Laguna, GRANTECAN, S. A.)

Barden, S. C., Wade, R. A. 1988, in ASP Conf. Ser. vol. 3, Fiber Optics in Astronomy (San Francisco: ASP), 113

Baudrand, J. et al. 1994, Proc. SPIE, 1298, 1071

Casse, M. 1995, PhD Thesis, Paris 11, Orsay University

Courtès G et al. 1988, in Instrumentation for Ground-Based Astronomy (IXth Santa-Cruz Workshop), ed. L. B. Robinson (New York: Springer), 266

Cuby, J. G. 1994, Proc. SPIE, 1298, 98

Davis, R. L. et al. 1997. Proc. SPIE, 2871, 1099

Fabricant, D. G. et al. 1994, Proc. SPIE, 1298, 251

Felenbok, P. et al. 1994, Proc. SPIE, 1298, 115

Felenbok, P. et al. 1997, Experimental Astron., 7, 65

Garcia, A. et al. 1994, Proc. SPIE, 1298, 75

James, E. C. et al. 1998, Proc. SPIE, 3355, 70

Le Fèvre O. et al. 1996, in ESO Astrophysics Symposia, The Early Universe with the VLT, ed. J. Bergeron(New York, Springer), 143

Vanderriest, C., & Lemonnier J. P. 1988, in IXth Santa-Cruz workshop, Instrumentation for Ground-Based Astronomy, ed. L. B. Robinson (New York: Springer), 304

Taylor, K., & Colless M. 1996, in ESO Astrophysics Symposia, The Early Universe with the VLT, ed. J. Bergeron (New York: Springer), 151

A Wide-Field Integral-Spectroscopy Unit for the VLT-VIRMOS

E. Prieto, O. Le Fevre, M. Saisse, and C. Voet

Laboratoire d'Astronomie Spatiale, Marseille, France

L. Hill

Observatoire de Haute Provence, St Michel de l'Observatoire, France

D. Mancini

Osservatorio Astronomico di Capodimonte, Naples, Italy

D. Maccagni

Istituto di Fisica Cosmica e Tecnologie Relative, Milan, Italy

J. P. Picat

Observatoire Midi-Pyrenees, Toulouse, France

G. Vettolani

Istituto RadioAstronomia, Bolognia, Italy

Abstract. An integral-field unit (IFU) is being developed for the VIRMOS instrument to be placed at one of the 8-m VLT Nasmyth foci. This IFU, to be placed at the edge of the multi-slit field of the instrument, will acquire a field of 1×1 arcmin^2 at low spectral resolution (200–270) and 24×24 arcsec at medium resolution (2000–2700). Two spatial resolutions of 0.7 and 0.3 arcsec will be available (with a focal elongator). The solution investigated includes the coupling of a microlens array with $\sim$6000 lenses, each coupled to a fiber. The output fibers are rearranged on a mock slit 112-arcmin long. Several options for the microlens array, including hexagonal epoxy or silica lenses, and cylindrical lenses, are being studied.

1. Introduction

VIRMOS is an instrumentation development for the ESO-VLT which will produce two imaging spectrographs: VIMOS for the 370nm–1μm domain, and NIRMOS for the 0.8–1.8μm domain. VIMOS will have its first light in early 2000 and NIRMOS in 2001. Two resolutions will be available on the instrument: 200–270 for low resolution, and 2000–2700 for medium resolution. The VIMOS

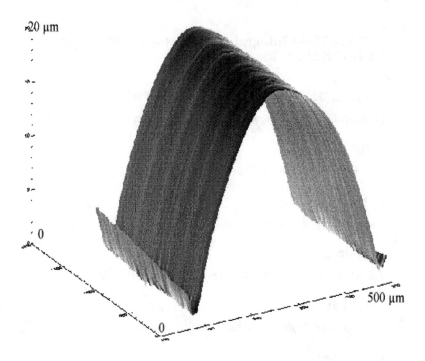

Figure 1.   Microlens surface measurement.

multiplex gain will be 840 simultaneous slits in low-resolution mode. The field of view is separated into four channels of 8 × 7 arcmin^2 each. Both instruments will include a wide-field integral-spectroscopy unit. This unit, situated 10 arcmin from the center field, will have a field of view of 1 arcmin square. The primary spatial resolution is ∼0.7 arcsec. While 0.3-arcsec resolution can be obtained with a focal elongator. The light is collected by a microlens array and guided into 6400 fibers. The output fibers are rearranged in 16 lines, each simulating a slit 7-arcmin long at the entrance of the spectrograph. We are are producing a prototype of the IFU in order to validate our technological choices.

2.   Entrance Microlenses

The first part of the IFU is the fiber entrance bundle. After a study of different technical solutions, we gave silica cylindrical lenses array preference over other solutions for the hexagonal array (silica or epoxy). This choice was guided by a comparison of prices, image quality, filling factor, surface quality, and regularity.

The advantage of the cylindrical LIMO[1] microlens array is that the shape of the surface is aspherical. Their process allow very good surface quality to

---

[1]LIMO GmbH, Hauart 7, 44227 Dortmund, Germany.

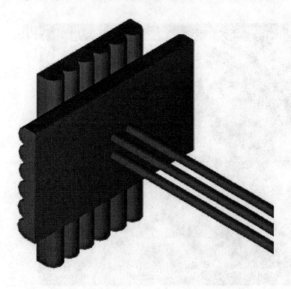

Figure 2. Crossed cylindrical microlenses array.

be attained. A measurement made by ourselves on a silica standard sampler (pitch 400 μm, focal length 2.2 mm, 7 μlenses) gave a very good result for the roughness measurement. We found a 5-nm rms roughness, which is an excellent result for this kind of optics.

The entrance μlens array increases the filling factor, and couples the telescope beam ($f/15$) with the acceptance fiber cone ($\sim f/3$).

In our system, we will use two lens arrays crossed face to face. This superposition of two arrays gives images of the pupil on the last face. We glue fibers here in order to collect the starlight. The equivalent lens obtained has the following characteristics:

Size: 387μm × 387μm
Focal length: 1.3 mm

Fibers are placed by a ferrule with holes.

After optimization of the two surface profiles, we obtain excellent theoretical image quality of the telescope pupil on the fiber core (100-μm diameter). The goal is to have maximum light at the fiber entrance. As we see in the Fig. 3, the theoretical image is completely enclosed within the fiber core dimension.

For a center-field object, 99 % of the encircled energy is in a disk of 94 μm. This dimension is smaller than the fiber core, and allows some tolerance in the fiber positioning and optical aberrations. We expect to have a fiber core positioning error better than ± 2–3μm.

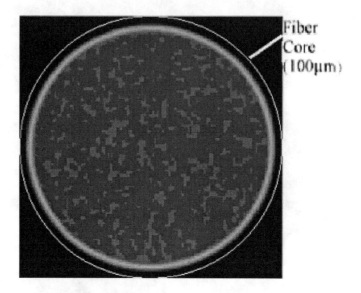

Figure 3.   Pupil image on the fiber entrance.

3. Fiber Output Coupling

The 6400 fibers guide the light at the spectrometer entrance where they are arranged in four lines of 400 fibers in each of the four quadrants. The $f/3$ output beam is adapted to the $f/15$ entrance aperture of the spectrometer by μlens arrays.

These arrays are made following the same philosophy as for the entrance array. They are the superposition of two arrays crossed and face to face. The beam is folded by a prism in order to pass through the mask and enter the instrument. A field lens adapts the pupil to the instrument one just after the mask.

Microlenses image a 0.86-arcsec pseudo-slit at the entrance of the spectrometer. Focal-ratio degradation blurs the pseudo-slit. In order to minimize this phenomenon, we will adapt the fiber aperture to the output μlens speed. The fiber numerical aperture requested to fit our evaluation is 0.16.

4. Fiber Bundle

The choice of fiber is critical. As the fiber length is $\sim$ 3 m, the system is not very sensitive to small differences in fiber transmission. The present choice is:

- Ceramoptec Optran WF
    Core: 100 μm
    Clad: 170 μm
    Acrylic: 250 μm
    Ultra-dry with hydrogen coating

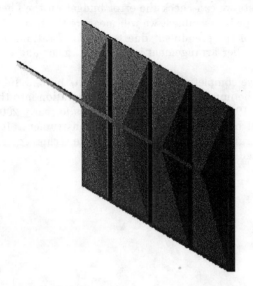

Figure 4. Entrance Fiber positioning mask in VIMOS.

        Screen test 100KPSI (for better mechanical strength)
- Heraeus STU with the same mechanical dimensions

As we are working in the visible, the core/cladding ratio is not so important for the transmission problem. However we need to have a large clad for good stiffness, which will minimize fiber breakage in the ferrules during the integration phase. In addition, a large clad gives the fiber less sensitivity to micro-bending, and the addition of an epoxy coating will increase the protection again this.

5. Work in Progress

We are in a prototyping phase. This prototype will be made with a 10 × 10-μlens array, 100 fibers, and 1 × 100 output lens array. From this prototype, we expect to have more information on:

- μmlens surface quality: we will measure the roughness of the arrays in order to quantify the flux lost by diffusion

- μmlens image quality: we will first measure the profiles and simulate the image quality expected. Secondly, we will measure the image quality over the array directly and cross check with the simulation.

- Hole regularity on the ferrules

- Hole quality on the ferrules

- Focal-ratio degradation with the 0.16 numerical aperture fibers

From the previous tests we can check the error budget on the fiber positioning.

Once the prototype is assembled, we will measure the transmission over the field and the quality of the pseudo-slit due to the focal-ratio degradation. We will also work on the fiber arrangement in the instrument and check the effect on the fiber behavior.

We expect to have completed the prototype tests by mid-July. The definitive version will then be immediately produced. Integration into the instrument is scheduled for early 1999. First light on the VLT is for early 2000.

Another IFU will be produced for the second instrument. It will be optimized for the near infrared. We plan to use the same technology as for the first one. First light is scheduled for early 2001.

Fiber-Optic Instrumentation and the Hobby–Eberly Telescope

Lawrence W. Ramsey

Department of Astronomy and Astrophysics, The Pennsylvania State University, University Park, PA 16802, USA

Abstract. The Hobby–Eberly telescope (HET) design marks a fundamental departure from the usual paradigm for building large optical telescopes. Central to the HET approach is specialization: the HET is tailored for spectroscopy, and in particular, fiber-coupled spectroscopy. By limiting observational flexibility, extremely cost-effective technical solutions are possible and these have been implemented in the HET's design. At the same time the HET places high demands on fiber performance. I will describe the major features of the HET design, particularly those relevant to the fiber-coupled instrumentation. In addition, I will outline the capabilities of the two fiber-coupled instruments currently under construction.

1. Introduction

The Hobby–Eberly Telescope (HET) is an international collaboration and involves The Pennsylvania State University, The University of Texas at Austin and Stanford University in the United States, and Ludwig-Maxmilians Universität München, and Georg-August-Universität Göttingen in Germany. It is located at the University of Texas's McDonald Observatory near Ft. Davis Texas at an altitude of 2000 m. This southwestern site is exceptional among developed US mainland sites for its dark sky. The HET design has been presented previously by Ramsey et al. (1994, 1998), Sebring et al. (1994), and Sebring & Ramsey (1997). Several key subsystems are also described in detail by Booth et al. (1998), and Krabbendam et al. (1998).

The Hobby–Eberly Telescope had first light on 1996 December 11 and is currently in the commissioning phase. We anticipated limited science operations to begin in fall 1998. Initially there will be three facility instruments on the HET. The first of these is a prime-focus low-resolution spectrograph (LRS), currently scheduled for installation during fall 1998. This instrument has multi-object and long-slit capability as well as an imaging mode (Hill et al. 1998). The two fiber-coupled instruments, a medium-resolution spectrograph (MRS) and a high-resolution spectrograph (HRS) will be described in Sections 5 and 6.

2. Hobby–Eberly Telescope Science Drivers

The HET concept results from the fact that a large collecting area will produce a spectrum of a given astronomical source rapidly, thus enabling many objects to be observed in a short period of time. This enables spectroscopic surveys that are fundamental to understanding how the Universe and its components work. The unique design of the HET derives from considering that telescopes are largely used for exposure times of an hour or less at modest zenith distances (Benn & Martin 1987). The technical approach adopted in the HET was to make considered trade-offs between cost and performance. This leads to a powerful yet limited telescope, a fact that is clear when one looks at the ~$15 million cost for this 9-m class telescope. The limitations inherent in the HET design primarily affect scheduling and operational flexibility. Given this, the HET will be especially competitive when used with the following criteria in mind:

- Target classes are uniformly distributed on the sky
- Target objects have sky surface densities of a few per square degree or a few per square arcminute
- Time-critical observations with time scales of days and longer are of interest
- Spectroscopy in the visible and near infrared yield the required astrophysics

Figure 1. The Hobby–Eberly telescope facility looking north.

Queue-scheduling is an integral part of the HET concept. This allows the HET to be especially useful for time-domain astrophysics within the time scale limitations mentioned above. We anticipate that HET will be especially competitive in planetary searches using radial-velocity variations, monitoring of active galactic nuclei and quasar emission-line strength and shape, and studies of activity and structure on stars and in accretion disks using Doppler imaging. For the latter program, the ability of the HET to acquire spectra rapidly with minimal phase smearing is essential. Survey programs such as optical identification of flux-limited X-ray and EUV samples from space missions, investigations of the intergalactic medium by absorption of light in clouds on quasar lines of sight and precision abundance determinations in support of cosmology, stellar population studies and stellar evolution are also well suited to the HET.

3. Telescope Facility

Figures 1 and 2 show the major features of the HET facility. The telescope dome (see Fig. 1) is a geodesic aluminum space-frame structure about 86 feet in diameter. The inside dome area where the telescope is located (Fig. 2) is thermally conditioned in the daytime to hold the interior at temperatures near the expected nighttime conditions. When the dome shutter is opened, six downdraft fans pull air through the opening at rates of up to 20 dome volumes per hour. This system minimizes the effects of the facility on seeing.

Figure 2.   HET inside dome.

As can be seen in Fig. 2, the HET has a segmented primary mirror. (Note that some distortion apparent in Fig. 2 is due to the fisheye lens used to obtain it.) The central (vertex) axis of this spherical primary mirror is tilted at a fixed angle making the HET a tilted optical Arecibo telescope with zero elevation freedom. However it has full azimuthal freedom allowing it to access different declination zones via this azimuth rotation. A motion of an object across the sky is followed by the tracker which carries the spherical-aberration corrector and instrument package. The tracker is apparent in Fig. 2 as the bridge-like structure at the top of the telescope. This design makes HET a nearly fixed zenith distance telescope within a declination zone delimited by the focal-surface tracker ±6° field of view (FoV).

All fiber-coupled instruments are located in a cylindrical instrument room under the telescope within the wall of the concrete pier that makes up the azimuth bearing. Optical fibers have a 33-m path from the prime-focus fiber instrument feed (Horner et al. 1998) though a hole in the pintle bearing. This room is on bedrock and its floor is isolated from the rest of the HET Facility. Additionally, the wall of the instrument room is mostly underground and backfilled leading to a thermally stable environment.

A visually striking element of the HET facility is the 90-foot tall tower to the north-east of the telescope (see Fig. 1). An advantage of a spherical primary mirror is that a point at the center of curvature (CoC) is re-imaged at that point. This provides a simple mechanism for alignment of the segments without having to acquire and track a star and to do so in the daytime if desired. However, this is not normally done as the best results are achieved if the mirror is aligned at a temperature very close to the nighttime use temperature.

A control and service building is located adjacent to the dome (right of dome in Fig. 1) in the prevailing downwind direction. The control area, which is the only part of the facility routinely heated, is encapsulated inside this building.

3.1. Optical System

The HET primary mirror is spherical and has a radius of curvature of 26.165 m. It is an array of 91 identical hexagonally-shaped spherical and unphased segments. The mirror substrates are of Schott Zerodur and all the segments were figured by Kodak in Rochester, New York. Figure 2 also illustrates the primary-mirror array geometry. The packing fraction loss due to gaps and bevels is 3.44%. The total reflecting area of the primary mirror is 77.6m^2. Each segment is coated with Denton FS-99 protected silver. Krabbendam et al. (1998) describe the primary-mirror system in detail.

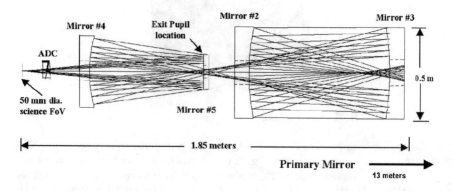

Figure 3. The HET spherical-aberration corrector (SAC).

An all-reflecting four-element spherical-aberration corrector (Fig. 3) removes the formidable amount of spherical aberration present at the prime focus of the primary. Mirrors 2, 3 and 4 are conics, whereas mirror 5 is a low-order asphere. The exit pupil of the system is near the surface of mirror #5 and projects to a 9.2 diameter entrance pupil on the primary-mirror array. This is the maximum effective aperture of the HET. An atmospheric-dispersion corrector (ADC) is part of the HET design. Since the telescope is tilted at a fixed zenith distance and the excursion of the tracker is nominally ±6°, a simple two-element ADC design is possible.

The corrector must move across the focal surface, which is a spherical surface 13.08 m from the primary-mirror CoC, all the time keeping its optical axis aligned with the CoC. This is done by the tracker, whose capabilities are described in detail by Booth et. al. (1998). The measured absolute blind pointing of the HET, including the tracker system, is better than 15″. While this meets our design requirements, we expect future mount modeling to improve on this.

An important effect that must be considered in the HET is a changing pupil illumination as an object is tracked across the sky (see Booth et al. 1998 for examples). The corrector defines a 9.2-m diameter entrance-pupil size, which can "see" off the primary mirror when the tracker moves significantly off center. The vignetting resulting when the tracker is significantly off center makes the average aperture of the HET a function of tracking time. For example, the effective aperture remains about 9 m for short 10-min tracks near the center of the tracker field, but diminishes to about 7.2 m in the worst case off-axis 40-min track. This aspect of the HET demands careful consideration of baffling.

The science focal plane of the HET is 50 mm in diameter. With an image scale of 0.205 mm arcsec^{-1}, this yields a science FoV of 4′. This focal surface is flat to about 10 μm to accommodate the focal-plane low-resolution spectrograph. The fastest f/ratio of the HET is f/4.68, which occurs when the tracker is near center. This $f/\#$ is well suited to fibers and represents a balance between the scrambling, which favors slower f-ratios (Barden et al. 1993) and the focal-ratio degradation properties of fibers which favor faster f-ratios (Ramsey 1988). The flatness, however, leads to the fibers seeing a ±1.6° radial variation in the telecentric angle. This must be addressed or excessive losses will result (Wynne 1993).

At the latitude of McDonald Observatory (30° 40′), the primary-mirror vertex axis points at a declination $\delta = -4° 20′$ at an azimuth position of 180° (due south). When the HET rotates in azimuth, different declinations will be on the primary-vertex axis. The HET has access to declinations from $-10° 20′ < \delta < 71° 40′$ when the 6° tracker FoV is included. Sky coverage is limited to about 70% of what a general-purpose telescope at the same site would normally achieve.

4. The Fiber Instrument Feed

The Fiber Instrument Feed (FIF) is described more completely by Horner et al. elsewhere in this volume and we will not repeat that information here. We will, however, outline its functionality. The FIF provides the link between the HET prime focus and all fiber-coupled instrumentation on the HET. That not only includes the MRS and HRS described below but any visitor fiber experiments that are used with the HET.

The FIF has twelve fiber probes, two of which can translate in one dimension and ten which can translate in two dimensions. The first one-dimensional probe supports an integral-field unit and three fiber synthetic long slits. The second one-dimensional probe supports the fibers for single-object MRS observations, the HRS fibers and fiber feeds to auxiliary user instruments. Both the one-dimensional probes are configured to place the selected fiber(s) at the center of the FoV. The ten two-dimensional probes support the multi-object spectroscopy fibers for the MRS. These probes are distributed around the FoV with each MOS probe covering about 17% of the FoV. There are two types of MOS probes to accommodate the telecentric angle variation; one is optimized for the field center region (Type 0) and the other for the outer regions (Type 1). One of the Type 1 MOS probes also caries the IFU sky fibers as the MOS and IFU are never simultaneously used. The positioning requirement of all the fibers probes is 0.05″ (10 μm).

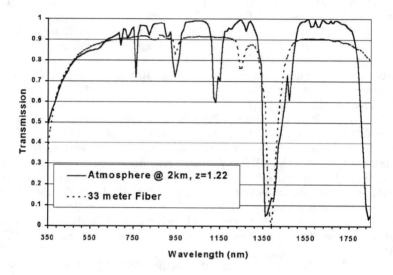

Figure 4. Fiber and atmospheric transmission.

The broad spectral coverage of the MRS places stringent demands on the optical fibers. We must utilize fibers that have good transmission from blue to the NIR. Previously, one selected high-OH fibers to optimize transmission in the blue and low-OH fiber for the red and NIR. Recently, Heraeus Amersil (Schötz et al., these proceedings) has made available new fiber preforms that have an excellent balance of visible and NIR properties. In Fig. 4 we show the transmission, including Fresnel reflection losses at both ends, for one of these fibers in the 33-m length required for the MRS. We have included the nominal atmospheric transmission at the HET site for a 35° zenith angle. With the exception of some of excess absorption in the J band, fibers made from this material are appropriate for the MRS.

There is one other important consideration that the MRS places on the fiber specification. To minimize NIR transmission losses, the cladding thickness must be of the order of 10 times the maximum guided wavelength (Schötz et al., these proceedings). The NIR beam is specified to work out to 1.8 μm, which means that the minimum cladding thickness must be 18 μm. This implies that all 200- and 300-μm fibers must utilize the standard 1.2 cladding/core ratio. The 400-μm and 600-μm fibers can be standard 1.1 cladding-to-core ratios.

The HET fiber focal plane is designed to be utilized with an ADC. This is vital to achieving good throughput for all the fiber-fed instruments. The fraction of light from a point source is very sensitive to both the centering of the image on the fiber as well as the direction of ADC secondary dispersion. Figure 5 illustrates the modeled throughput for three different MRS/HRS fibers. We assume perfect centering and $1.0''$ seeing folded with the dispersion characteristics of the atmosphere and ADC. If the object is decentered, the wavelength dependence can vary greatly depending on the direction of the secondary dispersion relative to the decenter direction.

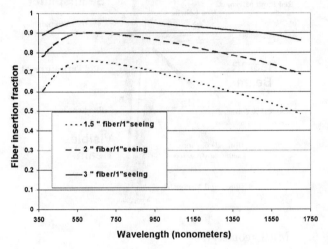

Figure 5.   HET fiber insertion losses as a function of wavelength.

5. Medium-Resolution Spectrograph

The Medium-Resolution Spectrograph (MRS) is a versatile, fiber-fed echelle spectrograph for the HET. This instrument is designed for a wide range of scientific investigations; it includes single-fiber inputs for the study of point-like sources, synthetic slits of fibers for long-slit spectroscopy, multi-fiber inputs for multi-object spectroscopy, and an optical-fiber integral-field unit. The design concepts for the MRS are described by Ramsey (1995) and more recently, Horner et al. (1998). The MRS is a dual-beam system where visible and NIR spectrographs are mounted on an optical bench in an environmentally controlled light-tight room under the telescope. The basic cross-beam geometry of the two white-pupil spectrographs is illustrated in Figure 6. A common collimator and slit system is used to allow spectra to be obtained in both beams simultaneously. A beamsplitter is mounted in the collimated beam. The visible beam is reflected at a $\sim 45°$ angle by a beamsplitter that transmits at $\lambda > 950$ nm. Both the visible and NIR spectrographs employ echelle gratings and grating cross-dispersers.

As the MRS is a fiber-feed instrument, detailed information on the telescope pupil is lost by scrambling and focal-ratio degradation (FRD). This is seen as an advantage for the HET, as the pupil shape is highly variable due to the basic nature of the HET—as the tracker moves, the telescope aperture changes size and shape (see Booth et al. 1998; Hill et al. 1998). The maximum native $f/\#$ of the HET is $f/4.68$. To eliminate losses due to FRD and the variable telecentric angle in the focal plane, the collimator must have an effective focal ratio for each fiber of $f/4.2$.

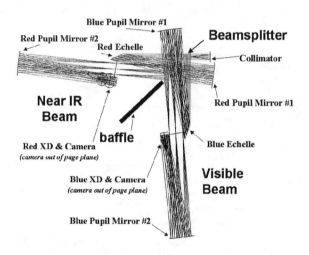

Figure 6. MRS geometry.

The MRS fiber input and slit system is complex. It must selectively place in the focal plane of the collimator one of nine options. There are three fiber sizes each for the long slit and MOS probes yielding a total of six options. A fiber slicer, single object fibers and an integrated field unit bring the total to nine. For the long slit and MOS options, each with three fiber size selections, the input slit system must have a 400-, 300-, 200- or $100 - \mu$m slit placed within 10 μm of fiber the output. This design for the slit system has each fiber input, for example the 200-mm MOS, input centered on narrow (6 mm wide) but deep fiber holder. Two slit jaws positioned by precision piezo-electric actuators are about 1.5 mm thick for stiffness. This system is currently in the concept stage but the baseline design is a rotary mechanism to move the selected fibers sequentially behind the slit. The slit system allows for the following resolution options:

- $R = (\lambda/\Delta\lambda) = 3388$ with 3″ fiber (resel = 13.38 pixels un–binned)

- $R = 5081$ with 2″ fiber or 3″ fiber with 400-μm slit (resel = 8.92 pixels un–binned)

- $R = 6352$ with 1.5″ fiber or 2 or 3″ fiber with 300-μm slit (resel = 7.6 pixels un-binned)

- $R = 10163$ with 2 or 1.5″ fiber with 200-μm slit (resel = 4.46 pixels un-binned)

- $R = 20325$ with 1.5″ fiber with fiber slicer or 100-μm (resel = 2.23 pixels un-binned)

Resel is a resolution element defined by the fiber/slit image on the CCD. Clearly, for all but the highest resolutions, on-chip binning will be advantageous.

The collimator optics is shared by both the visible and NIR beams and begins with a bi-concave fused-silica field lens 3 mm from the fiber surface. This lens places a pupil on the grating and is required, as all the input fibers are parallel. The parallel fibers have the same effect as placing the telescope entrance pupil at infinity. A minimum distance of 3 mm between the fibers and the field lens is needed to accommodate the slit mechanism. To minimize obstruction this lens will consist of a 10-mm wide central slice of a 15-mm diameter circular lens. The baseline collimator mirror is a ~170-mm diameter asphere with a conic constant of -0.4703. This two-element, basically achromatic, system places a pupil on the grating 1911 mm in front of the reflecting element.

The visible beam will be the first to be implemented and its design is well advanced. It is configured to yield a single-exposure full spectral coverage in the range from 450 to 900 nm. The system is also designed to allow extended coverage down to 390 nm and up to 950 nm with different cross-disperser settings. The visible beam begins after the beamsplitter, where the collimator places a pupil on a 79 l mm^{-1} R2 ($\Theta_{\text{blaze}} = 63.5°$) 102 × 280 mm echelle grating. This gives us a 100-mm diameter beam with no vignetting at the grating.

Central to the MRS design are the two parabolic pupil transfer mirrors. These mirrors re-image the pupil that is on the echelle grating onto the cross-disperser (XD). This is the so-called white-pupil design. This could also be accomplished with a single large spherical mirror that has the both the echelle and XD grating at opposite sides of the radius of curvature, but the current system is more compact and has no spherical or comatic aberrations. Both pupil mirrors are identical 2000-mm focal-length parabolas and are aligned axially so their foci coincide midway between them. The center of the echelle is displaced 235 mm from this center line on one side and the cross-disperser is displaced the identical amount on the other side. The second pupil transfer mirror is identical and they both can and should come from the same parent parabola. Centered at the focus between the pupil mirrors is a baffle with a height of 130 mm and a width of 336 mm. This will control scattered light. The second pupil mirror completes the task of imaging the echelle on the XD grating. The baseline XD is a 316 l mm^{-1} grating with a blaze of 6.8°. It is used in plane ($\gamma = 0$) with a $\theta = 20°$ so the camera clears the incident beam.

Dr. Harland Epps designed the visible camera which is a ten-element all-refracting system based on the design developed for the Keck ESI spectrograph (Epps 1998). It has a 320-mm focal length and a 197-mm entrance aperture with a 78-mm field of view. This $f/1.63$ system produces an average rms image diameter of 13.1 ± 4.6 m averaged over all field angles within the 390- to 1000-nm design passband without refocus.

The CCDs, the visible camera, and the 316 l mm^{-1} 6.8° blaze cross-disperser grating combine to provides full spectral coverage from 450 to 900 nm when used with any of the single fibers or fiber slicer. The spectral coverage using a mosaic of two SITe 4096 × 2048 CCDs is illustrated in Figure 7. The horizontal line in this figure illustrates the chip butting region. By rotating the CCD so the central order is roughly parallel with the CCD, it should be possible to have only one order disrupted by the butting region. In a single-fiber mode it is possible to have no losses due to the CCD mosaic. To allow clean order separation when the long slits, MOS probes or IFU are used, a higher-dispersion cross-disperser must be employed.

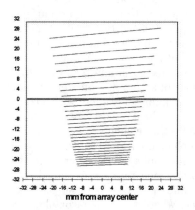
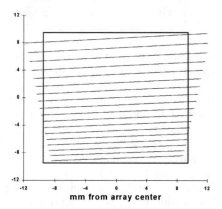

Figure 7. *Left panel:* Format of visible beam spectrum on CCDs. Only one free spectral range is shown for each order. *Right panel:* Spectrum from 950–1350 nm on 1024^2 Rockwell array

The NIR system is system is in an earlier stage of development but will be similar in layout to the visible beam. However, the two beams do differ in detail. We will utilize a 31.6 l mm^{-1} R2 ($\Theta_{\text{blaze}} = 63.5°$) echelle grating with a ruled area of 102 × 280 mm. The geometry of the pupil mirrors is identical. The NIR system utilizes a Rockwell 1024^2 HgCdTe array with a format that will allow near complete spectral coverage from 900 to 1300 nm in a single exposure when used with a 300 mm fl camera and a 150 l mm^{-1} 5.4° blaze cross-disperser grating. Figure 7 illustrates the coverage on the detector. As with the visible beam, this only allows adequate order separation with single fibers and the fiber slicer. A higher-dispersion cross-dispersing grating has yet to be selected for the long-slit and MOS fibers.

The NIR system will operate out to 1.8 μm but it will not have full spectral coverage utilizing the 1024^2 array. The camera, however, will be designed for 40-mm field of view to accommodate an upgrade to a 2048^2 NIR array when they become available and funding can be identified. It should be noted, however, the NIR beam will be operated at ambient temperature, which will lead to a high thermal background through most of the *H* band.

6. High-Resolution Spectrograph

The high-resolution spectrograph (HRS) for the HET has been designed by Dr. Robert Tull at the University of Texas at Austin. This instrument is described in more detail by Tull (1994, 1998). It is basically a single-channel adaptation of the ESO UVES spectrograph for the VLT (Dekker & D'Odorico 1992; Delabre 1993). The design is optimized for the 420- to 1100-nm spectral region at resolving powers $30000 < R < 120000$. It is also designed with the requirement of having a radial-velocity stability of 10 m s^{-1} or better over an observing season.

The HRS is a white-pupil design like the MRS. Indeed the MRS design was derived directly from the UVES/HRS concept. The HRS is fed by the FIF. The $f/4.6$ input beam from the FIF is converted via a doublet transfer lens to $f/10$ at the spectrograph end with the image of the fiber end illuminating a spectrograph slit. The HRS uses an R-4 echelle mosaic with a ruled area of 210 x 836 mm. This yields a basic resolving power of 34000 with a 1-arcsec slit width. From the echelle the dispersed beam returns to the collimator mirror, which now serves as the first pupil transfer mirror. After an intermediate focus the second pupil transfer mirror images the echelle onto the cross-disperser grating. There are two cross-dispersing grating choices. One provides for the maximum spectral coverage and the other provides increased order separation for sky-background spectra. The camera is an all-refracting design and is described by Epps (1998). It has a 280-mm entrance aperture 330 mm from the pupil at the cross-disperser grating. It provides for coverage from 390 to 1100 nm without refocus with outstanding image quality over an 88-mm field. The detector is a similar mosaic of two 4096 $times$ 2048 CCDs.

The HRS has a variety of fiber resolution options. It will utilize two fiber sizes in the focal plane: 1.5 and 3.0″. For each size, there will be two star and four sky fibers. In each size, one star fiber would be used with and the other without an image slicer. In all cases the fiber output will be re-imaged onto the entrance slit of the HRS. The typical resolution options that result are as follows:

- $R = (\lambda/\Delta\lambda) = 30000$ with a 1.12″ slit width (resel = 8.3 pixels unbinned)
- $R = 60000$ with a 0.56″ slit width (resel = 4.2 pixels unbinned)
- $R = 120000$ with a 0.28″ slit width (resel = 2.1 pixels unbinned)

The throughput with the smaller slit widths with be maintained by using a fiber slicer (Tull 1988).

Acknowledgments. We thank Robert E. Eberly of Pennsylvania and Lt. Governor William P. Hobby of Texas and other contributors that have made the Hobby–Eberly Telescope possible. Construction of the HET Medium-Resolution Spectrograph is supported by NSF grant AST9420645 and Pennsylvania State University matching funds. The HRS is supported by NSF grant AST9531674, National Aeronautics and Space Administration, and the University of Texas.

References

Barden, S. C., Armandroff, T., Massey, P., Groves, L., Rudeen, A. C., Vaughnn, D., & Muller, G. 1993, in ASP Conf. Ser. 37, Fiber Optics in Astronomy II, ed. P.M. Gray, (San Francisco: Astronomical Society of the Pacific), 185

Benn, C. R. and Martin, R. 1987 QJRAS, 28, 481

Booth, J. A., Ray,F. B., & Porter, D. S. 1998, Proc. SPIE, 3351, in press

Dekker, H., & D'Odorico, S. 1992, ESO Messenger, 70, 13

Delabre, B. 1993, ESO-Very large Telescope: UVES Preliminary Optical Design Report (ESO Doc. No. VLT-TRE-ESO-13200-0272)

Epps, H. W. 1998, Proc. SPIE, 3355, in press

Hill, G. J., Nicklas, H., MacQueen, P. J., Tejada , C., Cobos, F.J., & Mitsch, W. 1998, Proc. SPIE, 3355, in press

Horner, S. D., Engel, L.G. & Ramsey, L. W. 1998, Proc. SPIE, 3355, in press

Krabbendam, V. L., Sebring, T. A., Ray, F. B., & Fowler, J. R. 1998, SPIE, 3352, in press

Ramsey, L. W. 1988, in ASP Conf. Ser. 3, Fiber Optics in Astronomy, ed. S. Barden, (San Francisco: Astronomical Society of the Pacific), 26

Ramsey, L. W., 1995, Proc. SPIE, 2476, 20

Ramsey, L. W.,Sebring, T. A., & Sneden, C. 1994, Proc. SPIE, 2199, 31

Ramsey, L. W. et al. 1998, Proc. SPIE, 3352, in press

Sebring, T. A., Booth, J. A., Good, J. M., Krabbendam, V. L., & Ray, F. B. 1994, Proc. SPIE, 2199, 565

Sebring, T. A., & Ramsey, L. W. 1997, Proc. SPIE, 2871, 32

Tull, R. G. 1994, Proc. SPIE, 2198, 674

Tull, R. G , 1998, Proc. SPIE, 3352, in press

Wynne, C. G. 1993, MNRAS, 260, 307

The Hobby–Eberly Telescope Fiber Instrument Feed

Scott D. Horner, Leland G. Engel, and Lawrence W. Ramsey

Department of Astronomy and Astrophysics, The Pennsylvania State University, University Park, 16803 Pennsylvania, USA

Abstract. The Fiber Instrument Feed (FIF) for the Hobby–Eberly Telescope is located at the focal plane of the telescope and positions optical fibers to feed the off-telescope instruments. The FIF has a straightforward and economical design which employs commercial precision translation stages and actuator control systems. This simple design allows for a wide range of fiber feeds to the off-telescope instruments—consisting of the Medium Resolution Spectrograph and the High Resolution Spectrograph—facilitating a wide range of scientific investigations. The FIF includes single-fiber inputs for the study of point-like sources, synthetic slits of fibers for long-slit spectroscopy, multi-fiber inputs for multi-object spectroscopy, and integral-field units for synthetic imaging spectroscopy.

1. Introduction

The Hobby–Eberly Telescope (HET) marks a fundamental departure from standard telescope design and is described by Ramsey (these proceedings). The unique design of the HET, with its fixed elevation, necessitates that it be operated in a queue-scheduled mode, requiring rapid transition between observational configurations and instrumentation. The HET design also requires that large instruments must be located off-telescope and thus must be coupled to the telescope with optical fibers. The HET subsystem that performs this coupling is referred to as the Fiber Instrument Feed (FIF).

The FIF mounts to the HET at the corrected prime focus. It is mounted to the tracker, which moves the FIF along the curved focal surface of the telescope to follow the motion of target objects across the sky. While the telescope can rotate in azimuth, it does not do so to track objects as Earth rotates; the telescope azimuth rotation is only used in pointing and acquisition of target objects. The tracker thus tracks the target objects across the sky—moving the FIF across the focal surface—and rotates the FIF to remove field rotation. The position of target objects thus remain fixed in the field of view (FoV) of the FIF. The tracker can follow objects across up to 12° on the sky, which means that it can track objects for a maximum of 0.75 to 2.5 hours depending on the declination of the object.

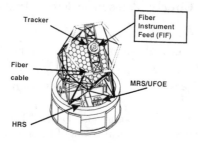

Figure 1. The HET showing the FIF, mounted on the tracker, which translates and tilts the FIF to follow target objects across the curved focal surface of the primary as the Earth rotates. The Medium and High Resolution Spectrographs (MRS & HRS) are located in a room below the telescope and are fed by cables of optical fibers.

2. Basic Design

The FIF is responsible for positioning the required optical fibers in the 4-arcmin diameter science FoV of the telescope. This FoV is flat and has been corrected for (primarily) spherical aberrations by a four-element corrector also mounted on the tracker. The FIF must be able to feed the current planned instrumentation suite and be expandable for future instruments. Because it is located on a difficult-to-access moving platform in the central obstruction of the telescope, the FIF must be fully automated, reliable, lightweight, and have a low profile.

The requirements on the FIF are:

- Position the optical fibers in the HET FoV for all off-telescope instruments:
  - Upgraded Fiber Optic Echelle (UFOE)
  - Medium Resolution Spectrograph (MRS)
  - High Resolution Spectrograph (HRS)
  - Auxiliary and future instruments
- Position fibers to < 10 μm
- Have full spectral coverage from 350 to 1800 nm
- Capability for excellent sky subtraction (control FRD variation)
- Lightweight (< 50 kg)
- Withstand the environmental conditions at the top of the telescope

Due to the queue-scheduled operation of the HET, the FIF needs to be able to rapidly change operating modes. The design of the HET—which limits the region of the sky accessible at a given time—means that the HET will carry out a larger number of separate observing programs during each night than other

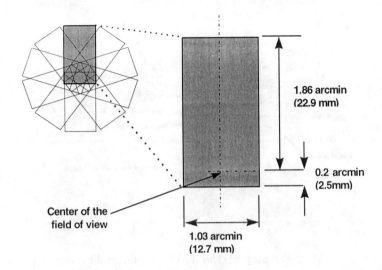

Figure 2. Each of the FIF MOS probes will cover 17 % of the HET science FoV, with all of the probes covering part of the central region of the FoV.

telescopes, including other queue-scheduled telescopes. Therefore, the FIF needs to be able to rapidly reconfigure for new observing programs. This, combined with the small physical size of the science FoV (50 mm in diameter), resulted in our designing the FIF using the one-probe, one-robot design (also known as the "fishermen around the pond" design) used by Hill et al. (1982), as opposed to the one-robot, many-probe design used in Hydra (Barden, these proceedings) and many of the other multi-object systems (see Perry, these proceedings).

The FIF has a total of twelve probes, two of which can only translate in one dimension, and ten of which can translate in two dimensions. The two one-dimensional probes carry fibers that are intended to be used at the center of the FoV, thus requiring only one dimension of motion to position the proper fibers at the center of the FoV or to move them out of the FoV to make room for other fiber probes. The fiber inputs positioned on these one-dimensional probes are the synthetic slits, integral-field unit, and single-object fibers (see Table 1).

The ten two-dimensional probes are used for multi-object spectroscopy (MOS). Each of the ten probes covers a rectangular area of the FoV, with each probe able to cover 17 % of the FoV. These probe coverage areas are distributed around the circular FoV like flower petals; each probe can cover the central region of the FoV and the coverage areas are distributed at 36° angles to each other, as shown in Figure 2.

The MOS probes are further divided into two groups: Type 0, which are optimized for use in the central region of the FoV, and Type 1, which are optimized for use at the outer region of the FoV. These optimizations will be discussed further in Section 3 of this paper.

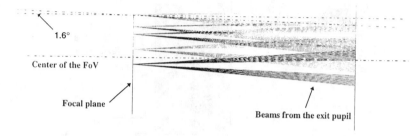

Figure 3. At the center of the FoV the image beams from the corrector-telescope exit pupil are normal to the focal plane. At the edge of the science FoV, the image beams are at a 1.6° angle to the norm of the focal plane. The HET focal plane is essentially flat across its 50-mm diameter science FoV.

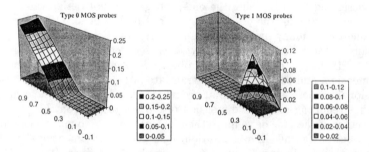

Figure 4. The loss due to the telecentric angle for (a) Type 0 MOS probes as a function of position (*left*), where 0 indicates the center of the FoV (these fibers are normal to the focal plane and thus are optimized to be used in the central core of the FoV), and (b) the Type 1 MOS probes (*right*), which are mounted at a 1° angle from the norm of the focal plane, and are optimized for use outside the central core of the FoV.

The fiber probes are mounted on single (for the one-dimensional probes) or two orthogonal (for the two-dimensional probes) precision translation stages. The stages are standard Newport Corporation low-profile crossed roller bearing translation stages, actuated by DC encoder-micrometers from DynaOptic Motion, and controlled using Programmable Multi-Axis Controllers (PMACs) from Delta Tau Data Systems. The use of these off-the-shelf parts will make the FIF easy to maintain and will help control costs. The PMACs are used throughout the HET control systems, thus the operations staff is familiar with their operation and software development. The PMACs have on-board DSPs for which command macros can be written. They will be commanded from a UNIX workstation, running graphical user-interface software written in Tcl/Tk and C, via an RS-232 serial interface.

3. Telecentric Angle

A significant design challenge for the FIF MOS system is the telecentric angle variation of 1.6° between the center of the HET FoV and the edge. The telecentric angle is illustrated in Figure 3. If a fiber is mounted normal to the focal plane, the telecentric angle at the edge of the FoV would result in a faster output beam than the input beam; the azimuthal scrambling of the fiber will produce an output beam that is a combination of the input beam $f/\#$ plus the telecentric angle. To resolve this issue, we have increased the speed of the MRS collimator to $f/4.2$ and have designated two types of MOS probes. Type 0 MOS probes, of which there will be two, will have fibers mounted normal to the HET focal plane; these probes will be the preferred fibers to use for objects near the center of the FoV. Type 1 MOS probes, of which there will be eight, will have fibers mounted 1° normal of the focal plane; these probes will be the preferred fibers to use outside the central region of the FoV.

We should point out that all of the MOS probes can be used over their entire range shown in Fig. 2; it is just that Type 0 probes will be more efficient (less light will be lost overfilling the collimator) in the central region of the FoV than Type 1 probes, and vice versa (see Fig. 4). The FIF is on a platform that rotates 230°, so the probe allocation can be determined by the observer to best optimize the efficiency of the system. The synthetic slits, integral-field unit, and single-object fibers are mounted normal to the focal plane since they will primarily be used at the center of the FoV.

4. Fibers

The HET instruments cover a wide spectral range. The Medium Resolution Spectrograph will have both a visible (350–950 nm) and a near-infrared beam (950–1800 nm) beam, which can be used simultaneously. The fibers used with the FIF must therefore cover a wide spectral range. Optical fibers made from the new preforms available from Heraeus Amersil (Schötz et al., these proceedings) will likely be used since the fiber transmission profiles are a good match to the silver coatings of the HET optics.

Table 1 lists the probes in the FIF and the fibers attached to those probes. The MOS fibers are organized into slits at the Medium Resolution Spectrograph by size; only one size of MOS fibers is used at a time. Each MOS probe has two fibers of each size so one fiber can be used on the object and the other on the sky for sky-subtraction. Probe 12 includes fibers to be used for sky-subtraction with the Integral-Field Unit (Bershady et al., these proceedings).

Table 1. A list of the FIF probes and the optical fibers on those probes. The "MRS single fibers" include sliced and double–fiber scrambled fibers. All the fibers feed into the MRS except for those indicated as HRS or auxiliary.

Probe	Probe description	Fiber description	# fibers	Fiber size (μm)
1	1-D translation probe	1.5 arc-second slit	17	300
		2.0 arc-second slit	13	400
		3.0 arc-second slit	9	600
		Integral Field Unit	45	200
2	1-D translation probe	HRS Fibers	13	200–600
		MRS single fibers	4	200–600
		Auxiliary fibers	6	200–600
3 & 4	Type 0 MOS probes	1.5 arcsec MOS	2	300
		2.0 arcsec MOS	2	400
		3.0 arcsec MOS	2	600
5–11	Type 1 MOS probes	1.5 arcsec MOS	2	300
		2.0 arcsec MOS	2	400
		3.0 arcsec MOS	2	600
12	Type 1 MOS probe	1.5 arcsec MOS	2	300
		2.0 arcsec MOS	2	400
		3.0 arcsec MOS	2	600
		IFU sky fibers	6	200
	Total		173	

5. Conclusion

Integration of the FIF onto the Prime Focus Instrument Platform will occur in Spring 1998. The FIF will become fully operational as the associated subsystems become operational in Fall 1998.

Acknowledgments. Construction of the HET Fiber Instrument Feed and Medium Resolution Spectrograph is supported by NSF grant AST9420645 and matching funds from The Pennsylvania State University. We also thank Robert E. Eberly of Pennsylvania and Lt. Governor William P. Hobby of Texas and other contributors that have made the Hobby–Eberly Telescope possible.

References

Hill, J. M., Angel, J. R. P., Scott, J. S., Lindley, D., & Hintzen, P. 1982, Proc. SPIE, 331, 279

Galaxy Kinematics with Integral-Field Spectroscopy and the Hobby–Eberly Telescope

Matthew A. Bershady[1]

Department of Astronomy, University of Wisconsin, 475 N. Charter Street, Madison, WI 53706, USA

David Andersen, Lawrence Ramsey, and Scott Horner

Pennsylvania State University, University Park, PA 16803, USA

Abstract. We describe two fiber-optical arrays under construction for the 9m Hobby–Eberly Telescope's (HET) Medium Resolution Spectrograph (MRS). These arrays optimize integral-field spectroscopy for kinematic studies of individual nearby and moderately distant galaxies. The arrays are *formatted* to deliver simultaneous rotation-curve and disk-velocity dispersion measurements for galaxies over a range of look-back times. From these kinematic measurements the evolution of galaxy mass-to-light ratios (M/L) can be explored on spatially resolved scales. One array spans ~30 arcsec using 2-arcsec (400-μm) fibers to densely sample slits at four position angles at spectral resolutions from 5500 to 11000. The second array spans ~15 arcsec using 1-arcsec fibers to densely sample a core area of ~60 arcsec^2 at spectral resolutions from 11000 to 22000. The first array will be installed during the commissioning phase of the MRS in early 1999. This two-beam, fiber-fed echelle spectrograph covers from 0.5 to 0.95 μm in a red beam, and 0.8–1.6 μm in a near-infrared beam. Below we describe the scientific motivation, design rationale, and expected performance of these two integral-field units (IFUs) for the HET's MRS. The arrays are the first formatted IFUs designed for astronomical application.

1. Niche Science Mission

The primary purpose of the fiber arrays (or IFUs) under construction for the HET is to study galaxy disk kinematics in an extremely cost-effective and efficient way (Bershady 1997). Rotation-curve and disk velocity-dispersion measurements yield independent estimates of galaxies' total and disk masses. From these kinematic measurements, halo masses can be directly inferred without further assumptions (e.g. the assumption of maximal disks). This will provide a sensitive probe of the nature of dark matter in external galaxies, complementary to, e.g., micro-lensing experiments which probe within the Milky Way. A critical limitation in gathering the essential kinematic data has been

---

[1] Also at Pennsylvania State University.

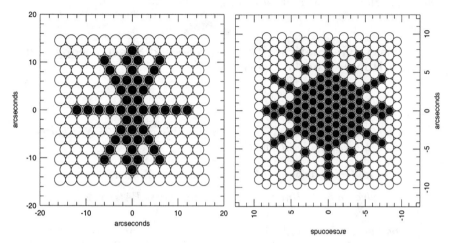

Figure 1. *Left panel:* HET IFU #1, a 400-μm (2-arcsec) fiber array with 10:1:1 ratios of core to clad to buffer. *Right panel:* HET IFU #2, a 200-μm (1-arcsec) fiber array with 10:1:1 ratios of core to clad to buffer. *Dark fibers* (45 and 95 in number for IFU #1 and #2, respectively) are the science fibers that go through to the spectrograph; *open fibers* are short fibers for mechanical packing. The interface yoke is not shown. Seven and nine sky fibers (IFU #1 and #2, respectively) on an independently positionable stage are also not shown.

the inefficiency of long-slit measurements made on aging spectrographs on 4-m class telescopes. Integral-field spectroscopy on 10-m class telescopes using high-throughput, echelle spectrographs promises to open a new window on the mass and mass distributions of galaxies over a range of look-back times.

For example, a key survey with the larger HET IFU (#1, Figure 1) will be to measure disk M/L for samples of nearby spirals. Rotation curves and non-axisymmetric motions can be probed via line emission to intermediate redshifts using the smaller IFU (#2, Fig. 1). Such surveys will map the evolution of M/L to epochs at half the current age of the Universe. On 4-m class telescopes such instrumentation is impractical for low surface brightness spectroscopy at medium spectral resolution; fiber arrays planned for other 10-m class telescopes are generally optimized for smaller areas and higher angular resolution (see these proceedings). However, spiral galaxy disks are predominantly at low surface brightness; at high redshift their surface brightness fades further still. The MRS fiber arrays fill a powerful niche for low surface brightness spectroscopy at medium spectral resolution over relatively large areas.

2. Design and Fabrication

<u>Sampling geometry</u>: To maximize science returns given a finite number of fibers, two patterns are adopted (Fig. 1), sampling four position angles over large areas ($\sim$700 arcsec^2 and 300 arcsec^2, respectively, for IFUs #1 and #2). For both IFUs, there is a densely sampled inner region. The remaining fibers are for

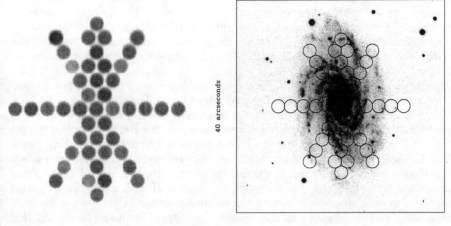

Figure 2. *Left panel:* CCD image of a test version of HET IFU #1, a 400-μm (2-arcsec) fiber array with 10:1:1 ratios of core to clad to buffer. *Dark fibers* (45 in number) are the science fibers that go through to the spectrograph. Packing fibers are not visible because only the long (science) fibers are illuminated. This type of image is used to check the fiber array's astrometric alignment and polishing. The uneven illumination of some fibers is due to the fact that their back ends are not polished. *Right panel:* HET IFU #1 superimposed on a typical nearby galaxy. Shown are only the science fibers. From this spatial sampling it will be possible to determine kinematics position angles, non-axisymmetric motions, and six independent measures of the disk velocity dispersion at each radius for three unique projections of the velocity ellipsoid—all in *one* exposure. Note that the sampling in galaxy azimuth is, to first order, independent of galaxy scale.

mechanical packing, are short, and of little cost. Figure 2 shows the larger IFU (#1) artificially mapped onto a nearby galaxy. Many trade-offs were considered when allocating "science" fibers; chosen patterns reflect compromises between linear size and dense sampling while avoiding preferred angular scales.

Packing fraction: To minimize cost and maximize construction efficiency, the fiber arrays do not employ lenslets, but simply rely on dense packing (e.g. Barden & Wade 1988). The coverage on sky, including 5:1 ratio of fiber core to fiber clad+buffer, is 62 %. This can be increased by 10 % if it is possible to remove the fiber buffer. For the initial arrays, the buffer will remain to avoid making the fibers too brittle. The covering factor has little consequence on the science performance of the IFUs given the formatted nature of their design.

Spectral resolution: For the 2-arcsec fiber array (#1), the ambient resolution is 5500. A slit mechanism at MRS focus will be used to achieve resolutions up to 11000 by stopping the slit down to a 1-arcsec width and passing 61 % of the light. Because of the greater fiber diameter and favorable geometry (unblocked in the spatial dimension) this configuration provides *2.4 times higher throughput for an extended source* than an array of 1-arcsec fibers. The 1-arcsec fiber array is designed for regimes where spatial sampling is critical, e.g. for distant galaxies where image sizes are small. Likewise, the 1-arcsec array can be stopped down to achieve resolutions approaching 22000 with comparable gains over yet smaller fibers.

Fabrication: Fiber arrays are epoxied together using Epotek 354 in a pressing jig consisting of a U-shaped clamp adjusted by a micrometer stage. To date, five test arrays have been made from 200-μm and 400-μm fibers in a variety of configurations. To meet mechanical constraints in the HET Fiber Instrument Feed (FIF), the final 14×15 400-μm fiber array uses 2.5-cm short, packing fibers. The FIF interface consists of an aluminum yoke connected to a one-dimensional translation stage for removal from, and insertion into, the field of view. Fiber arrays are glued to one side of the yoke, and held rigid with a soft rubber gasket on the opposing side. Sky fibers are mounted together in a mini-bundle on a separately positional probe on x, y-translation stages. Both sky fibers and IFU will be telecentrically corrected to within 0.2 deg.

3. Performance Summary

The expected IFU performance should be comparable to that of individual fiber feeds for the MRS (Ramsey 1995, and these proceedings). In particular, the IFUs are not expected to have substantial additional focal ratio degradation; a pupil simulator has been constructed to confirm this for a 1-m test array of 400-μm fibers. Taking the design goal of 15 % total throughput for the MRS (atmosphere through to the detector), the following Table gives the expected performance for IFUs in terms of *limiting surface-brightness* in the V band for a 1-hour exposure yielding a signal-to-noise of 10 per resolution element. The astrometric performance should also be excellent; fiber rows and columns are placed within 7 μm (rms) of their optimal positions. In summary, the integral-field units for the Hobby–Eberly Telescope's Medium Resolution Spectrograph will deliver high-throughput spectroscopy for medium-resolution studies of extended sources at low surface brightness.

HET IFU characteristics		
	Array # 1	Array #2
Fiber size	2 arcsec	1 arcsec
	(400-μm fibers)	(200-μm fibers)
Science fibers	52 (7 sky)	104 (9 sky)
Linear active area	27×26.4 arcsec	18×17.6 arcsec
Filling factor	34 %	39 %
Resolution ($\lambda/\Delta\lambda$)	5500–11000	11000–22000
$\lambda/\Delta\lambda$	Limiting V-band surface brightness	
5500	23.5 mag arcsec^{-2}	22.6 mag arcsec^{-2}
11000	22.8	22.0
22000	21.7	21.1

Acknowledgments. This instrument is funded through NST AST-9618849. We are indebted to Sam Barden for consultation and advice on fabrication.

References

Barden, S. C., & Wade, R. A. 1988, in ASP Conf. Ser, vol. 3, Fiber Optics in Astronomy, ed. S. C. Barden (San Francisco: ASP), 113
Bershady, M. A. 1997, in ASP Conf. Ser, vol. 117, Dark and Visible Matter in Galaxies, ed. M. Persic & P. Salucci (San Francisco: ASP), 547
Ramsey, L. W. 1995, Proc. SPIE, 2476, 20

Part 5: Multi-Object and 2-D Infrared Fiber Spectroscopy

AUSTRALIS: a Multi-Fiber Near-IR Spectrograph for the VLT

Keith Taylor

Anglo-Australian Observatory, PO Box 296, Epping, NSW 2121, Australia

Abstract. An Australian consortium of astronomers and engineers (based at AAO, MSSSO, and UNSW) were contracted by the European Southern Observatory to carry out a one-year concept design study for a near-infrared multi-object spectrograph for the VLT. The underlying instrumental philosophy was to supply a significant object multiplex at a high enough spectral resolution to resolve the internal kinematics of galaxies. A full contiguous-wavelength coverage from 0.9 to 1.8 μm is achieved through the use of multiple HgCdTe-based spectrograph cameras. A preliminary optical design for the spectrographs has been achieved, as has a detailed concept design for the 400-fiber positioner and multiple integral-field units. With these capabilities, the proposed instrument is highly effective both for statistical studies of large numbers of objects and detailed studies of individual objects.

1. Introduction

The AUSTRALIS Concept Design Study was initiated in response to an ESO *Announcement of Opportunity* for the study of optimal approaches to near infrared multi-object spectroscopy on the VLT in parallel with a similar study by a consortium of European astronomers, which later became known as the VIRMOS group. At the conclusions of the two studies the AUSTRALIS consortium (Taylor et al. 1996) had opted for a fiber-based facility in contrast to the multi-slit approach of the VIRMOS group. The VIRMOS concept (two separate instruments: VIMOS for the optical and NIRMOS for the near-IR) was accepted by ESO in preference to the AUSTRALIS design; nevertheless the arguments for a fiber-based system were regarded as sufficiently strong to encourage ESO to explore ways of incorporating such a facility for the VLT in their longer-term plans. This eventually led to the Wide-Field Fiber Facility proposal (WF^3) for one of the Nasmyth platforms of the Unit 2 telescope (Avila et al. 1998) which is planned to be commissioned in mid-2001.

The main science drivers for a near-infrared multi-object spectrograph on the VLT concentrated on studies of galaxies at redshifts $z \leq 1$. However, an instrument optimized for such work inevitably offers potential for a large variety of statistical and direct studies. The AUSTRALIS team considered it vital to enhance, as far as practical, the versatility of such a device in order to ensure its relevance through changing scientific imperatives. This, of course, implied designing for maximum efficiency and object multiplex but, more critically, opening up the resolving power and wavelength range as far as possible. AUSTRALIS should not be viewed as a simple redshift engine.

The following sections consider the implications of this and the other scientific desiderata for the design of the instrument.

2. Basic Design Strategy: Fibers vs. Slits

We elucidate here the key issues determining the basic instrument concept: why we elected a multi-fiber, rather than multi-slit, based scheme.

2.1. Field of View (FoV)

Since no information re-formatting is possible, a multi-slit spectrograph has a FoV limited in the spatial direction by its direct-imaging characteristics and in the spectral direction by the fraction of the spectral axis not occupied by the wavelength range of interest. On the other hand, the FoV of a fiber system is limited only by that of the telescope focal plane. Now the comparison of FoV for multi-object spectrographs is only of relevance in the low surface density regime where the number of targets is significantly less than the potential object multiplex. However this will always be a significant regime whenever selection criteria become refined.

2.2. Object Multiplex

In contrast to the FoV arguments, object multiplex is of relevance only in the high target surface density regime. Since the multi-slit technique does not permit information re-formatting the tendency is to devote all non-object detector space to sky to facilitate sky subtraction. However, the multi-fiber technique leads to a minimization of sky fibers compatible with the sky-subtraction requirements of the observations. Typical sky-to-object fractions here might be ~ 3 for multi-slits and ~ 0.1 for multi-fibers; however, such numbers vary over a large range.

On the surface this argument strongly favors fibers, however practical constraints entirely dominate. While it would seem feasible to contemplate an object multiplex approaching 1000 (at low dispersion) for a multi-slit system based on a 4k-by-4k detector, multi-fiber systems based on robotic positioners are heavily constrained by cost and complexity issues. As a result we allowed object multiplex to be constrained by ancillary issues to do with resolving power, detector space, and positioner complexity to a value of 400.

2.3. Integral-Field Units

The concept of a fiber-based multi-object spectrograph naturally leads to the consideration of fiber-based integral-field units (IFUs). There are two broad choices here. One is to produce a massive single IFU using as many fibers as can be accommodated; the second is to have several smaller IFUs deployable on the field-plate so that IFU observations can be performed on a number of objects simultaneously. The design of our focal-plane changer and fiber-optic switchyard (detailed in Section 3.8) permits both types.

2.4. Sky Subtraction

It is generally believed that sky-subtraction with a multi-slit system is more accurate than with a multi-fiber system and so to do spectroscopy at the faintest possible levels one should avoid using fibers because they do not work as well. This view is simply incorrect. The limiting signal-to-noise ratio that can be achieved for both fibers and slits is fundamentally set by Poisson statistics. This has been demonstrated in practice on a number of occasions (eg. Elston & Barden 1989; Cuby & Mignoli 1994).

The fact is that much greater care is required when making the fiber observations and doing the data reduction in order to eliminate systematic errors. For a multi-slit system the sky and the object-plus-sky spectra are adjacent both on the sky and on the detector and hence the systematic errors are minimized. However, for a fiber system the two spectra to be subtracted from one another are typically neither adjacent on the detector nor on the sky and so the systematic errors associated with each often do not cancel completely if a simplistic data-reduction algorithm is used. The most dominant systematic errors are due to a lack of adjacency at the detector and include wavelength sampling, scattered light and spectrograph vignetting. With care and attention these systematic effects can be successfully dealt with to give results approaching those of a multi-slit system.

For AUSTRALIS several features have been designed in to help in making the sky-subtraction process accurate. AUSTRALIS is a fixed-format spectrograph so many calibration procedures to determine wavelength calibration, scattered-light contributions and spectrograph vignetting can be repeated and the properties of the system can be accurately measured and understood and stored as a database. This, coupled with a sophisticated, dedicated data-reduction package, will make sky subtraction both accurate and painless for the observer. The white-pupil design also helps because fibers at the ends of the slit have similar spectrograph vignetting to those at the center, and so the color of the sky spectrum to be subtracted off any individual fiber is invariant. For redshift work which mainly requires the identification of features and the determination of their positions, the sky to be subtracted off is relatively featureless because of the digital OH suppression and the relatively high spectral resolution. Finally, the spectrographs do not move with respect to the gravity vector giving greater stability.

Most fundamentally, however, the fact that AUSTRALIS relies on digital OH suppression gives it an in-built calibration that is immune to any fiber transmission variations. The OH lines recorded in each object spectrum automatically monitor the on-source, time-averaged, throughput of each fiber, as required for Poisson-limited sky subtraction. The process of digital OH subtraction itself maps out the relatively on-source throughput, not only as a function of fiber number but also across the full wavelength range. In other words, provided the time-averaged OH sky spectrum is invariant over the field, changes in fiber throughput are self-calibrated by the recorded object spectrum itself even if those variations have *slow* wavelength terms. By normalizing the average sky spectrum obtained from the sky fibers to the individual object fiber transmission as obtained by their own OH line fluxes, Poisson-limited sky subtraction should be a realizable goal.

3. Instrument Concept Design

3.1. Nasmyth Corrector

In order to avoid the inherent disadvantages (stepped focal-surface zones, fiber de-focusing, non-telecentricity, pupil motion) of feeding fibers at the raw $f/15$ Nasmyth focus, we choose instead to specify a corrector which would give a mildly curved focal surface whose radius of curvature (4 m) matched that of the re-formed telescope pupil. Thus, wherever the fiber is placed in the corrected focal surface, it sees the same angle to the incoming principal ray equivalent, therefore, to a telecentric fiber feed. Such a design is also readily compatible with an 2dF-like fiber robotic positioner using magnetic buttons on a field-plate (Taylor et al. 1997; Smith & Lankshear 1998).

3.2. Optimum Aperture Size

Determining an optimal fiber aperture size for multi-object work is a key design issue which drives many of the instrumental constraints. For this reason we took great care to quantify the effects of aperture size on predicted S/N performance as a function of target image size, seeing, and instrumental performance.

In order to quantify this for typical faint galaxies the predicted S/N was obtained as a function of aperture size and seeing for a sample of suitably faint galaxies from the Hubble Deep Field (HDF) convolved with Gaussians to simulate a range of seeing conditions. The I-band HDF magnitudes were converted to J using the mean colou $I-J \sim 2$ expected for galaxies at $z \leq 1$. The S/N was then computed using this J magnitude along with the suitable values for J-band sky continuum levels, dark current, and readout noise.

While an optimum aperture of $\sim 1.5''$ is obtained it is clear from the S/N curves that some significant leeway is available should ancillary arguments, such as spectral resolution, push the design to smaller values ($\rightarrow 1''$).

3.3. Requirement for 1-to-7 Hex Relay

The camera f-ratio required to image a $1.5''$ aperture onto 2.35 (18.5-μm) pixels of a 2k-by-2k Rockwell array on the VLT is $f/0.75$. Furthermore, this camera has to have a linear field of 53.6 mm equivalent to an angular field (at $f/0.75$) of $\sim \pm 10°$ assuming a beam size of 200 mm. This is clearly impractical; the most tractable solution is to adopt some form of fiber image slicing. Inevitably, by re-formatting aperture information along the input slit of the spectrograph, such a solution automatically compromises object multiplex, however, this is seen to be an acceptable compromise.

Instead of segmenting the incoming $f/15$ Nasmyth image, we place a $1.5''$ aperture stop at the focal plane and image the telescope pupil onto a single fiber thus avoiding the geometrical losses inherent in such a non-telecentric design. When fed at $f/3$, the resulting single fiber has a diameter of 175 μm whose output is then segmented downstream in order to supply a narrower slit for the spectrograph. The segmentation is provided by relay optics (located in the fiber-optic switchyard (as detailed in ??) which take as their input the $f/3$ output of the single fiber, re-imaging this onto the a close-packed hexagonal lenslet array. These lenslets then form individual $f/5$ pupil images onto the input faces of seven smaller, 130-μm, fibers which are themselves re-formatted to form individual object slitlets at the spectrograph input slit.

This technique possesses a number of significant advantages.

- The fiber input feed is performed by collimation optics which preserve the $\mathcal{A}\Omega$ product; image segmentation by microlenses located at the Nasmyth focal plane require the deployment of significantly thicker fibers to accommodate all rays into the fiber, thus reducing the potential $\mathcal{R}$ by a similar amount.

- The fiber button supports only one thick/robust, rather than seven thin/fragile, fibers;

- It is this single *thick* fiber which is re-configured by the robotic positioner on the field-plate and constrained over pulleys as it is routed to the stationary spectrograph. Being relatively thick, the single fiber will suffer less focal-ratio degradation (FRD).

- The thin, post-relay, fibers are fixed in space from the fiber optic switchyard to the spectrograph slit and as will be detailed in ??, the switchyard provides a great deal of functionality and versatility to the observational and calibration processes.

3.4. Focal-Plane Changer

The demand to facilitate the positioning of a large number of fibers, both single object and IFUs, is a familiar problem which has received a great deal of attention through the years that astronomical fiber systems have evolved. AUSTRALIS uses magnetic buttons on a magnetic field-plate with a double plate changer permitting field configuration on the second plate while observing with the first. This exactly follows the 2dF design philosophy; we see no reason to make any fundamental change for the VLT.

The corrected VLT Nasmyth platform does, however, make fiber positioning substantially easier and a design is proposed which allows for the fiber positioning to be performed on a stable, orientation-invariant, mount. The basic 2dF fiber-button design is retained as is the fiber-retractor scheme which keeps each fiber mildly tensioned so that its locus is well determined. In contrast to the 2dF, however, the AUSTRALIS positioner is an $r - \theta$ robot whose radial arm is curved to match the curvature of the corrected focal plane.

3.5. NIR Spectrograph Concept Design

The basic requirement is to disperse the light from the fiber input slitlets and feed the resulting spectra at an f-ratio which matches the projected slit width onto $\sim$2 detector pixels. In order to capture all the light from the 130-μm, $f/5$ input fibers we have chosen a faster, $f/4.5$, collimator to accommodate the expected fiber FRD. Given the final camera f-ratio then needs to be faster than $f/1.3$, delivering a very wide angle of view, a white-pupil design has been employed.

The system is an off-axis Schmidt, near-Littrow, spectrograph (205-mm beam size) giving a curved intermediate spectrum. In the case of full 900- to

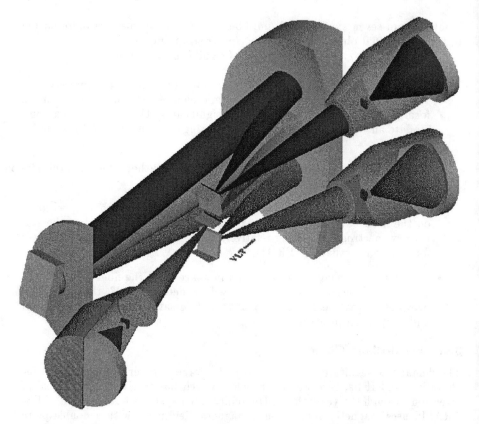

Figure 1. A rendered view looking onto the slit side of one of the IR spectrographs. On the top right is the large half spectrograph mirror and the "red" and "blue" cameras. On the lower left is the "green" camera, the 452-l mm^{-1} diffraction grating and, behind it, the half-Schmidt corrector plate. The slits and twin mirrors that dissect the spectra are in the center of the image.

1800-nm wavelength coverage, the input slit is located in the $J-H$ absorption band to avoid spectral vignetting. The 452 line mm^{-1} grating (1.14-μm blaze) delivers a $\sim$430-mm long spectrum at $\mathcal{R}$ $\sim$4000 which is split into three sections (the two ends by reflection facets; the central region undeviated) which are individually re-imaged by $f/1.23$ finite conjugate Schmidt cameras onto their respective 2k-by-2k detectors. Such a configuration implies that each seven-fiber object slitlet projects onto $\sim$1.9 pixels in the dispersion direction. The length of each slitlet, defined by close packing fibers with a core/clad ratio of 1:1.4, is $\sim$18 pixels. With two pixels between each object spectrum, 100 critically sampled spectra can thus be recorded over the full wavelength range.

Since only half of a large aspheric corrector plate is employed it is natural to build two such spectrographs implying 6 2k-by-2k HgCdTe detectors servicing 200 objects in all. Furthermore, by deploying slits at the ends of the intermediate spectrum as well as at its center, options for recording just half the spectral range (*semi*-spectra: j=900–1350 nm or h=1350–1800 nm) but with twice the number of fibers become available. Thus a 400-fiber system is proposed.

3.6. Thermal Background for the NIR Spectrograph

The number of photons per pixel detected from the sky background is very low so it is vital that AUSTRALIS contributes significantly less than this in terms of the background due to thermal emission from within the spectrograph. Since the detectors have a dark current of $\leq$0.1 electron s^{-1} pixel^{-1} and this is below the background from the sky a useful goal is to ensure that the thermal emission is less than or equal to this. Calculations demonstrate that it is therefore necessary that the AUSTRALIS NIR spectrographs are completely enclosed in a light-tight actively refrigerated cool-room to a temperature of $\sim$220 K.

3.7. Integral-Field Units

The IFUs are designed to provide the same number of fiber spatial samples as there are *spectroscopic* fibers (i.e. seven times the object multiplex, hence the *on-sky* IFU fibers are fed, through the switchyard, into the same set of invariant spectrograph fibers thus requiring a 7-to-7 hex fiber relay in the switchyard rather than the 1-to-7 hex relay for the multi-object mode.

Large-IFU Here we propose a contiguous field of 2800 lenslets feeding 130-μm fibers through a variable focal plane re-imager to effect scale changes at its telecentric input. Scale changes over a wide range can be envisaged but we would suggest sampling options of perhaps 0.2″, 0.4″ and a maximum of 0.7″, corresponding to fields of 11″, 21″ and 36″ on a side.

Small-IFU Given that the deployable IFUs are positioned on the field-plate and hence mounted on fiber buttons, to effect image segmentation they have to receive the raw Nasmyth f/15 beam. In this case the $\mathcal{A}\Omega$ product of the telescope is not easily retained at the f/5 input of the fiber and hence the 130-μ fiber cores will subtend a maximum 0.45″ implying lenslet sizes up to 0.23 mm (center to center).

The number and size of these deployable IFUs is an astronomical specification which is completely uncritical to the design; however, the fiber positioner has two plates to allow for rapid field changes and hence two choices can be made. As an example, it is suggested that one field-plate has eight deployable IFUs, each with 350 lenslets in a square array with a 4.7″ FoV sampled at 0.25″, while the other has 28 deployable IFUs, each with 100 lenslets in a 5-by-20 array with a 2.2-by-8.6″ FoV sampled at 0.43″.

3.8. Fiber-Optic Switchyard

AUSTRALIS uses a fiber-optic switchyard to connect fibers from the active focal plane to the cooled-grating spectrographs housed in the cool-room while enabling back-illumination of the configuring set for accurate positioning. A single fiber runs $\sim$6 m from each button on the telescope focal plane to the switchyard mounted on top of the NIR spectrograph cool-room. At the switchyard a microlens collimates the f/3 beam emerging from each *thick* fiber and

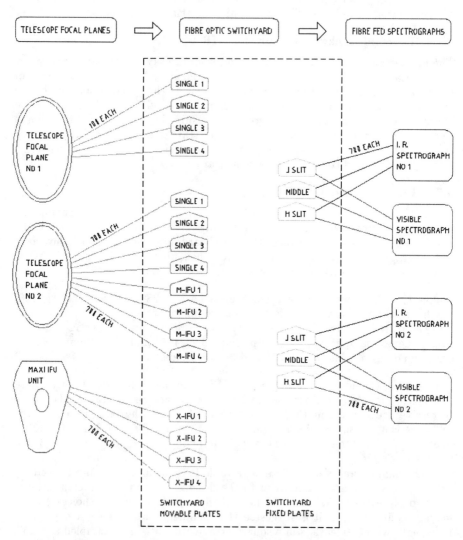

Figure 2. A schematic view of possible interconnections across the AUSTRALIS switchyard. Working from left to right across the figure, signal from one of the two telescope focal planes or the maxi integral-field device is fed to movable plates in the fiber-optic switchyard. Two or four of these movable plates at a time are placed on top of the fixed plates in the switchyard. This directs signal to one or two of the slits in the IR and visible spectrographs.

forms a new pupil ∼2 mm in diameter onto a group of seven hex-packed microlenses. These hex-packed microlenses segment the new pupil and feed a group of seven *thin*, 130-μm fibers at $f/5$. These *thin* fibers are relatively short and lead from the switchyard through the cool-room roof and down to the slits in the spectrographs. The slits and the last 1.5 m of the fibers are cooled along with the spectrographs to minimize background signal at longer wavelengths. There are no back-illumination lamps inside the cool-room, this function is performed at the switchyard and prevents signal contamination inside the spectrograph.

In addition, by use of the 7-to-7 hex relay all the IFU modes detailed in ?? have access to all the spectrograph modes. The switchyard offers an impressive variety of options including:

- simultaneous observing with the active field-plate while back-illuminating the configuring field-plate;
- immediate swapping between field-plates on completion of observation/configuration cycle;
- selection of wavelength coverage (full- [jh] or *semi-* [jj or hh] spectra
- selection of the L-IFU—either in full, recording *semi*-spectra or an inner spatial zone, recording complete 900–1800-nm spectra;
- selection of one or other of the deployable IFUs optionally mixed with single-object fibers;
- flux and wavelength calibration of spectrographs.

This is all achieved without any motion of the spectrograph fibers.

Most important of all of the switchyard functions, however, is the ability to place a dichroic between input and output feeds of the relay to feed a separate optical spectrograph; all of the above options then become available simultaneously between 450 and 1800 nm.

4. Conclusions

The basic concept for the AUSTRALIS positioner has now been incorporated into ESO's Wide-Field Fiber Facility (WF^3). The double-buffered positioner facility (OzPoz) has a 560-fiber (or deployable IFU) capacity (a 1320 capacity in total) and will initially supply eight fibers to UVES (the VLT UV Echelle Spectrograph) and 150 fibers to GIRAFFE (a new optical-only intermediate-dispersion fiber spectrograph dedicated to WF^3. The long-term goal is to acquire an AUSTRALIS-like, near-IR spectrograph facility targeted for higher dispersions than NIRMOS can supply: i.e. $\mathcal{R} \geq 5,000$ to give adequate digital OH suppression while being complementary to NIRMOS.

Acknowledgments. The author would like to thank Ian Parry (IoA), Peter Conroy (MSSSO), and Damien Jones (Prime Optics, Qld) for their assistance during the technical phases of this program. The AUSTRALIS Science Team was headed by Matthew Colless (MSSSO) and compromised Michael Ashley,

Warrick Couch, and John Storey from the University of New South Wales; Paul Francis from Melbourne University; Heath Jones, Charlene Heisler, Peter McGregor, Jeremy Mould, Bruce Peterson, and Peter Wood from Mount Stromlo Observatory; and Joss Bland-Hawthorn, Karl Glazebrook, and Fred Watson from the Anglo–Australian Observatory. More recently, Peter Gillingham and Stan Miziarski put together the costed proposal for OzPoz.

References

Avila, G. et al. 1998, Proc. SPIE, 3355, In press
Cuby, J. G., & Mignoli, M. 1994, Proc. SPIE, 2198, 98
Elston, R., & Barden, S. 1989, NOAO Newsletter, 19, 21
Smith, C. A., & Lankshear, A. F. 1998, Proc. SPIE, 3355, In press
Taylor, K. et al. 1996, "AUSTRALIS Concept Study Report" in Internal report document (AAO/MSSSO), September 1996
Taylor, K., R. D. Cannon, R. D., & and Q. A. Parker, Q. A. 1997, in IAU Symp. 179, Ne3w Horizons from Multi-Wavelength Sky Surveys, eds. B. J. McLean, D. A. Golombek, J. J. E. Hayes, & H. E. Payne (Dordrecht: Reidel), 135

SINFONI: a High-Resolution Near-Infrared Imaging Spectrometer for the VLT

Matthias Tecza

Max-Planck-Institut für extraterrestrische Physik, Postfach 1603, D-85740 Garching, Germany

Niranjan Thatte

Max-Planck-Institut für extraterrestrische Physik, Postfach 1603, D-85740 Garching, Germany

Abstract. The SINFONI[1] project combines the MPE cryogenic near-infrared imaging spectrometer SPIFFI[2] with an ESO adaptive-optics system on the ESO-VLT to perform high spatial and spectral resolution studies of compact objects. This paper describes the optical design of SPIFFI and the novel techniques used in building its integral-field unit.

The image slicer comprises of a bundle of 1024 silica/silica fibers, where each fiber tip is flared to increase the core diameter by a factor of 15. The tapered end is polished to form a spherical microlens with a hexagonal cross-section to couple light into the optical fiber. This not only yields a high light-coupling efficiency and a high geometrical filling factor but also allows us to use the fiber bundle at a working temperature of 77 K without losing positioning accuracy.

1. Introduction

One of the key goals of the VLT is the detailed astrophysical study of single objects, including faint, distant (proto-) galaxies. Building an efficient instrumental capability in this domain requires merging two recent developments, both with major scientific uses, integral-field spectroscopy and adaptive optics.

The HST Deep Field showed that faint (distant) galaxies exhibit considerable structure on 0.1-arcsec scales. In addition, searching for massive black holes in the center of galaxies (Eckart & Genzel 1997), checking the unified scheme for galactic nuclear activity or probing star formation processes (Böker & Förster-Schreiber 1997) calls for 0.1-arcsec resolution. Adaptive-optics systems offer such spatial resolution on ground-based telescopes, especially in the near infrared.

Integral-field spectroscopy is designed to obtain a full three-dimensional (x, y, λ) data set on a two-dimensional detector in one exposure. An "integral-

---

[1] *SI*Ngle *F*aint *O*bject *N*ear-infrared *I*nvestigation.

[2] *SP*ectrometer for *I*nfrared *F*iber-fed *F*ield *I*maging.

field unit" rearranges the two-dimensional field of view on the sky into a one-dimensional format which represents the entrance slit of a long-slit spectrometer. Integral-field spectroscopy not only increases the observing efficiency over conventional scanning techniques, like Fabry–Perot imaging and long-slit spectroscopy, it also reduces the effect of changing observing conditions like varying atmospheric transmission and seeing. Especially in the near infrared, atmospheric conditions vary on short time-scales and require a very stable and well calibrated instrument.

SINFONI is being developed through a collaboration of the European Southern Observatory (ESO) and the Max-Planck-Institut für extraterrestrische Physik (MPE) in which ESO will build the adaptive-optics system and MPE will build the spectrometer, including the integral-field unit.

The following describes the new techniques used in building the integral-field unit and the design of the spectrometer.

2. SPIFFI Design Criteria

The success of the MPE 3D imaging spectrometer (Weitzel et al. 1996) triggered a new effort at MPE to build a successor instrument (SPIFFI) using the larger-size infrared detectors now available, independently of plans for the SINFONI project.

3D uses a NICMOS III detector with 256^2 pixels yielding a 16 × 16-pixel field of view and a spectral resolving power of $\sim$ 1000 in either the H or K band. SPIFFI will be equipped with a Rockwell HAWAII focal-plane array with 1024^2 pixels. This gives a basis for the following design criteria for SPIFFI.

- A Rockwell HAWAII focal-plane array with 1024^2 pixels is used as the detector. A low readout noise of less than 8 e^- per read and dark current of less than 0.1 e^- per second means sky background limited instrument performance for all of the foreseen operating modes. The operating temperature of the detector allows convenient liquid-nitrogen cooling of the instrument.

- The instantaneous wavelength coverage can be chosen to cover any part of the the H and K atmospheric windows, at a spectral resolving power from $R \sim 2000$ to $R \sim 4500$. This wavelength range specifically includes the thermal part of the K band, which requires cooling of the entire instrument.

- A maximum detector utilization should be achieved with 1024 spectra each with 1024 spectral-resolution elements.

- Since the small number of spatial pixels limits the small field of view of integral-field spectrometers, the chosen pixel scale is the largest practically feasible value of 0.5 arcsec per pixel at a 4-m telescope. This pixel scale is also a good sampling width for seeing-limited observations. A scale-changing mechanism for smaller pixel scales allows optimal instrument performance under all seeing conditions and for AO-assisted operation.

- High throughput is essential to all spectrometers but even more so in conjunction with adaptive optics. Further, a high dynamic range requires minimum cross-talk between pixels and excellent stray-light baffling.

- The operation mode foreseen for SPIFFI was that of a traveling instrument, imposing tight constraints on size and weight of the instrument. In addition, for operation with adaptive optics, a small light-weight instrument reduces mechanical flexure which could degrade the performance of adaptive optics. The entire instrument design should therefore be very compact, yet fully cryogenic.

The last two points especially forced us to deviate from the successful "3-D image slicer" concept to a new design using a fiber bundle with microlenses for the integral-field unit. The positioning tolerances of fibers and microlenses had to be met not only at room temperature but also at a working temperature of 77 K.

3. The Integral-Field Unit

Driven by the design criteria listed in Section 2 we chose a fiber bundle with microlens coupling as the integral-field unit. The fiber core diameter of 50 μm yields a compact integral-field unit with a fiber slit length of ~ 60 mm. Because the integral-field unit in SPIFFI is cooled down to cryogenic temperatures a novel technique was used for the microlens fiber coupling involving flared fibers.

3.1. A Fiber Bundle for Image Slicing

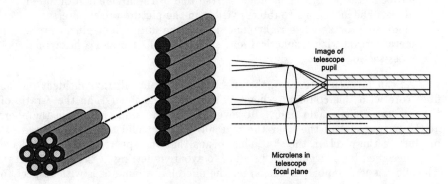

Figure 1. A fiber bundle rearranges a two-dimensional geometry into a one-dimensional format to feed a long-slit spectrograph. Using the principle of light coupling into the fiber, a microlens images the telescope pupil onto the fiber core. With an array of microlenses it is possible to slice a contigous two-dimensional field of view into individual pixels and rearrange them into a long-slit format.

Using the light-guiding properties of an optical fiber, it is relatively easy to build an integral-field unit. A bundle of fibers is configured to have a two-dimensional geometry on one end while the other end is arranged in a one-dimensional pattern. A closed-packed arrangement of fibers with no cladding in a two-dimensional geometry (see Fig. 1) yields a maximum geometrical filling factor of 91 % but it decreases with increasing cladding-to-core diameter ratio to an unacceptable number for the fibers we use (see Section 3.5.).

3.2. Efficient Light Coupling with Microlenses

The poor efficiency due to the geometrical filling factor of the naked fiber bundle can be overcome by using a microlens array with a high geometrical filling factor. Each microlens in the array images the telescope pupil onto the fiber core. The advantage of this coupling scheme is not only the high efficiency, which is now limited by the geometrical filling factor of the microlens array, but also a change in beam focal ratio at the fiber core thus reducing focal-ratio degradation problems.

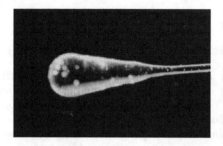 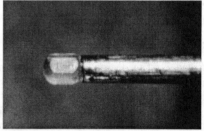

Figure 2. On the left panel is the flared tip of a fiber before it is glued into a metal ferrule. This fiber is from one of the first batch of flared fibers. The inclusions (bubbles) visible in the picture are suppressed in modified process. The right panel shows the flared fiber glued into a metal ferrule with a spherical lens surface on the tip and a hexagonal cross-section.

As can be seen in Fig. 5, the effective f-ratio for all rays incident on the fiber core with this coupling scheme is numerically smaller than the f-ratio of the rays for a single field point. However, due to the mode mixing in the fiber, at the output end of the fiber the rays fill a larger solid angle than the $A \cdot \Omega$ product of the pixel on the sky. This means that the spectrometer optics have to be designed for a larger $A \cdot \Omega$ product to avoid vignetting in the spectrometer. The effect is less important the larger the microlens diameter is with respect to the fiber core diameter.

To achieve a high coupling efficiency it is essential to position the fiber exactly centered on the optical axis of the microlens. For a core diameter of 50 μm an inaccuracy in positioning of 5 μm leads to a light loss of 6 of fiber core and image of telescope pupil. For a cryogenic fiber bundle this positioning constraint is very difficult to fulfil with "conventional" methods, using a microlens array attached to a fiber bundle. Although modern CNC machines achieve positioning tolerances of a few microns, the predictability of the effects of thermal contraction is not accurate enough. A solution to this problem is a monolithic sytem combining a microlens and fiber into one unit. The fiber bundle is then built from these integrated units.

In a process akin to the reverse of drawing a fiber, the tip of a fiber is flared over the required length to the right diameter. In this process core and cladding are flared by almost the same factor. Since the microlens should only consist of core material, the fiber is flared to twice the final microlens diameter. The next step in the process is glueing the fiber into a metal ferrule with an exact inner and outer diameter. The ferrule is used in the subsequent steps of the process as a mount and as a reference for the fiber axis. The third step is polishing the tapered end to a spherical surface with the required radius of curvature and correct taper length. As a last step the microlens is ground to a hexagonal cross-section. At this time all of the cladding material surrounding the core is removed and a microlens of pure core material is left over.

If cooled to low temperatures the position of the fiber with respect to the microlens is preserved and the coupling between microlens and fiber is not affected. The influence of low temperature on fiber properties is discussed in Section 3.5.

3.3. The Microlens Array

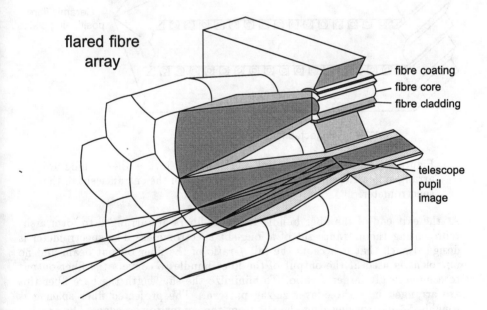

Figure 3. 1024 microlens/fiber units are close-packed to form the entrance end of the fiber bundle.

Each microlens has a hexagonal cross-section of width 0.6 mm. The outer diameter of the metal ferrule is chosen to be only a few microns larger than the width of the hexagonal microlens, so that close-packing the metal ferrules yields the same regular honeycomb pattern as close-packing of the hexagonal microlenses would. The microlens/fiber array is put together by holding the ferrules rather than the microlenses/fibers. This has the advantage that no mechanical stress is directly applied to the microlens or fiber, which could break the microlenses or introduce additional focal-ratio degradation.

1027 of the microlens/fiber units are close-packed to form a hexagonal field of view with 18 closed rings around a central microlens. Three of the bundle of 1027 fibers are dummies while the other 1024 fibers are actually used for the integral-field unit. The accurate outer diameter of the metal ferrules guarantees a regular honeycomb pattern of the microlenses. The accuracy of the hexagonal microlens cross-section yields a geometrical filling factor larger than 96 %.

3.4. The Fiber-Slit Geometry

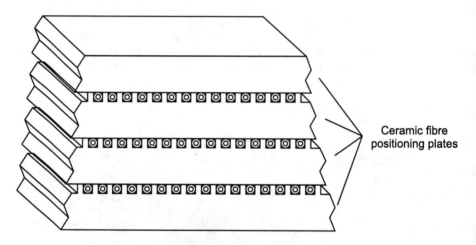

Figure 4. 1024 bare fiber tips at the output of the fiber bundle are arranged in a three-layer zigzag pattern to form the entrance slit of the spectrometer.

At the exit end of the fiber bundle the 1024 fibers are arranged to form a slit representing the entrance slit of a long-slit spectrometer. The spectrometer is designed such that it accepts the full f-ratio of the fiber output beam and no microlens is used at the output of the fiber bundle to convert the fiber output to a numerically larger f-ratio. To minimize the slit length the bare fiber tips are arranged in a three-layer zigzag pattern. The projected fiber spacing is roughly 55 μm varying over the slit from the center to the edge. The spacing of the fibers along the slit is such that image distortion in the spectrometer is compensated, and the fiber images are equidistant and centered on the columns of the detector. Because the mean fiber spacing is 10 % larger than the fiber core diameter (50 μm), the size of the fiber core image on the detector is 90 % of the pixel width. The resulting under-illumination of the detector pixels reduces cross-talk between adjacent spectra on the detector.

3.5. The Choice of Fiber

Table 1. Fiber bundle characteristics

Fiber bundle		Throughput
Fibers		
pure silica core	50-μm diameter	
doped silica cladding	94-μm diameter	
polyimide buffer	100-μm diameter	
fiber length	160-mm	
fiber transmission	> 99 % for λ < 2.0 μm	
	90 % for λ = 2.45 μm	> 90 %
Microlens array and fiber slit		
microlens size	0.6 mm	
focal length	3.2 mm	
taper length	4.7 mm	
fiber slit length	56.3 mm	
Fresnel losses	7 %	93 %
coupling efficiency	95 %	95 %
array filling factor	> 96 %	> 96 %
		> 77 %

The design criteria listed in Section 2 and the novel technique used in the integral-field unit determined the choice of fiber.

For SPIFFI, we selected OH-free silica/silica fibers with polyimide coating. They are robust compared to zirconium fluoride fibers and still have adequate, high infrared transmission for the short length of fiber used in the fiber bundle. Because the core material (pure fused silica) and the cladding material (fluorine-doped fused silica) have similar linear thermal expansion coefficients and similar changes of refractive index with temperature, they also operate at a temperature of 77 K.

4. The Spectrometer

The integral-field unit described in Section 3 is the "heart" of SPIFFI. The rest of the instrument can be functionally divided into two parts, the fore-optics, which includes the image scale changer and the filter wheel, and the spectrometer section, which actually disperses and detects the light.

The compact design achieved with the integral-field unit makes it possible to design compact fore-optics and spectrometer sections with a minimum number of surfaces and moderately-sized optical elements.

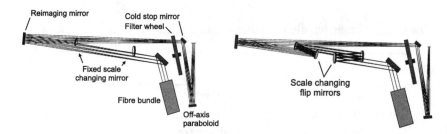

Figure 5. Layout of the fore-optics with filter wheel, cold-stop mirror and scale-changing optics. The figure on the left shows the coarsest image scale, while the one on the right shows the scale-changing optics flipped into the beam to magnify the image by a factor of 5/3.

4.1. The Fore-Optics

In the thermal part of the near infrared, the dominant background is blackbody radiation from the warm parts of the telescope or instrument. To block the radiation from telescope parts other than the mirrors, a cold stop is placed at the location of the image of the telescope pupil. Since, due to mode mixing in the fiber bundle, the information about the telescope pupil size and position is lost, the cold stop in the instrument has to be placed before the fiber bundle. This led to an optical layout where a set of cold fore-optics, including a cold pupil stop, is incorporated into the cryostat.

SPIFFI should be versatile and allow rapid change of image scales to adapt to varying seeing conditions or switch from seeing-limited image scales to diffraction-limited image scales when operating in adaptive-optics mode. This image-scale changing must be done before the integral-field unit slices the image into individual pixels.

A scale changing-mechanism is built into the fore-optics for this purpose. It allows choosing between three image scales during observing. The elegance of the optical design lies in placing the scale-changing optics after the cold stop, thus removing any requirement to select a different cold stop for each pixel scale.

In the collimated beam of the fore-optics the filter wheel for wavelength band selection is placed. The collimated beam diameter in the fore-optics is $\sim$ 10 mm, which makes it possible to use a standard-size band-selection filter.

4.2. The Spectrograph

"Conventional" long-slit spectrometers normally have slow input f-ratios. Integral-field units using fibers built for existing instruments incorporate microlenses at the fiber output end to convert the fiber output beam to the correct f-ratio of the spectrometer. In SPIFFI the collimator directly accepts the fiber output beam at a moderate f-ratio of $f/4.3$. The focal length of the collimator is 535 mm and the collimated beam diameter 125 mm. The optical design is a three-lens achromat covering the wavelength range from 1.1 μm to 2.5 μ.

The f-ratio of a spectrometer camera is given by the slit width in arcseconds on the sky, the diameter of the telescope and the size of a detector pixel. In the case of SPIFFI, the slit width is given by the pixel size in arcseconds. For

Table 2. SPIFFI instrument characteristics

Instrument section	Instrument details				Throughput/ efficiency
Fore-Optics	image scales at 4-m telescope in arcsec				
	Tip-tilt	Adaptive optics			
	0.5, 0.3	0.5, 0.07, 0.04			96 %
	order-selection filter				
	H, K, $H\&K$				75 %
Fiber bundle	1024 fibers with microlenses				> 80 %
Collimator	three-lens achromat				
	focal length	550 mm			
	focal ratio	$f/4.3$			95 %
Gratings	band	groove mm^{-1}	blaze angle	order	
	H	128.57	13.32°	2	70%
	K	102.86	14.23°	2	70%
	$H\&K$	100.00	6.07°	1	77%
Camera	folded Schmidt				
	focal length	185 mm			
	focal ratio	$f/1.2$			85 %
Detector	HAWAII focal-plane array				
	1024^2 pixels	18.5 μm			
	readout noise	< 8 e^- per read			
	dark current	< 0.1 e^- per second			
	quantum efficiency	> 65 % for λ_{peak}			> 50 %
		> 60 % for $\lambda = 1.2$ μm			> 20 %

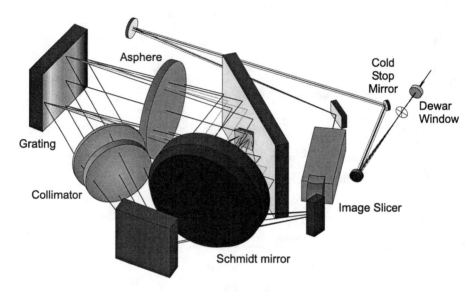

Figure 6. Perspective view of the entire instrument. The entire optics fits into a box of 65 cm × 95 cm × 25 cm.

an assumed median seeing of 1 arcsec the ideal pixel size would be 0.5 arcsec to Nyquist sample the point spread function. For an 8-m telescope this yields a camera f-ratio of $f/0.7$, which is impossible to achieve with the limited choice of glasses in the near infrared. As a compromise, a pixel size of 0.25 arcsec at an 8-m telescope was chosen. At a 4-m telescope, the pixel size would be 0.5 arcsec. The camera is a folded Schmidt design with an aspheric plate and a field and distortion correcting doublet close to the detector.

A reflection grating is used as the dispersing element in the spectrometer. Four reflection gratings on a wheel allow rapid change of wavelength band and resolution. At the moment, three wavelength bands are foreseen: H-band operation with a resolving power of $\sim$ 4200, K-band operation with a resolving power of $\sim$ 4500, and a combined H- and K-band mode with a resolving power of $\sim$ 2000. As the optics is achromatic from 1.1 μm to 2.5 μm, it is possible to include J-band operation or higher resolving powers in parts of the three bands.

The requirement of efficient use of the detector with one spectrum per detector column places a strong constraint on the quality of the spectrometer optics, in particular, it is required that each of the 1024 spectra is dispersed exactly along one column of the detector, with spill-over limited to 5 % over the entire detector. Such spill-over may occur due to image distortion in the collimator and camera, or due to anamorphic behavior of the reflection grating used for dispersion. To minimize the anamorphic effects and to keep them symmetric, it is necessary to operate the grating in Ebert configuration. Anamorphic distortion from the grating is corrected with a doublet close to the detector, which also flattens the image.

The quality of the spectrometer optics limits the spill-over from one spectrum to the neighboring spectrum to less than 5 %, The positioning of the spectra on the detector is achieved by adjusting the magnification of the spectrometer and by adjusting the reflection grating. The length of the fiber slit image on the detector is matched to the detector size by adjusting the focal length of the collimator. Rotating the grating about its surface normal adjusts the dispersion direction to be parallel to the detector columns. Tipping the grating about an axis along the grating surface and perpendicular to the grating grooves moves the spectra along the slit length on the detector so that the spectra can be centered on the detector columns. This alignment needs to be done once for each grating.

References

Eckart, A., & Genzel, R. 1997, BAAS, 191, 9707
Böker, T., & Förster-Schreiber, N. M. 1997, AJ, 114, 1883
Weitzel, L., Krabbe, A., Kroker, H., Thatte, N., Tacconi-Garman, L. E., Cameron, M., & Genzel. R. 1996, A&AS, 119, 531

AOIFU: AOB OSIS Infrared Fiber Unit

J. Guérin

Observatoire de Paris-Meudon, 5 Place Jules Janssen, 92195 Meudon, France

Abstract. AOIFU will provide the CFHT (Canada–France–Hawaii Telescope) with a unique scientific capability for obtaining 2-D visible and near-IR spectra of objects in the I, J, and H bands with a 200 to 2000 spectral resolution and with 0.1″ and 0.4″ spatial resolution. The goal is to link the AOB (Adaptative Optics Bonnette) to the OSIS near-IR spectrograph by a focal enlarger and a microlens-array couple with a fiber bundle.

1. Introduction

Following the MOS/ARGUS optical integral-field spectrograph at the CFHT (Vanderriest 1994) and the IRSIS fiber-fed spectrograph (Dallier et al. 1993), we propose a new integral-field link between the AOB (Arsenault 1994) and the OSIS spectrograph.

AOIFU consists of an enlarger stage fixed at the focus of the AOB and a microlens array coupled with a 216-fiber bundle 23 m long. It feeds the OSIS spectrograph located on the dome floor of the telescope.

The use of the AOB will allow a spatial resolution of 0.1″ to be obtained with a 1.8″×1.2″ field of view. For the 0.4″ sampling, the AOB optics is not used and AOIFU is directly fed by the $f/8$ Cassegrain beam. Objects are observed in the I, J, and H bands with a 200 to 2000 spectral resolution. The wavelength coverage goes from 0.5 to 1.8 μm.

Due to the high angular resolution and the ability to observe important near-IR diagnostic lines, AOIFU will permit several astrophysical topics to be studied, such as:

- Planets and satellites (Titan, Neptune, Uranus, etc.)
- Bipolar outflows and stellar evolution
- Black holes in nearby galaxies
- Extragalactic emission-line nebulae
- The kinematics of high-redshift galaxies
- Gravitationally lensed systems (with sub-arcsecond separation)
- Gravitational lensing (in intermediate-redshift clusters)

This project is a collaboration between the Dominion Astrophysical Observatory with D. Crampton, the Observatoire de Paris-Meudon with P. Felenbok and J. Baudrand, D. Bauduin, F. Gex, J .Guérin, I. Guinouard, L. Jocou, and M. Luet, and the CFHT with G.Barrick.

2. Description of the AOIFU

Figure 1 shows the AOIFU assembly: a focal enlarger transfers the images coming from the AOB onto a microlens array coupled with a coherent fiber bundle to feed the OSIS spectrograph through a pseudo-slit made by the fibers aligned in two rows.

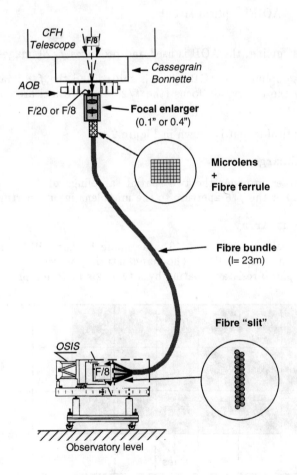

Figure 1.   The AOIFU assembly.

Depending of the spatial resolution requirement:

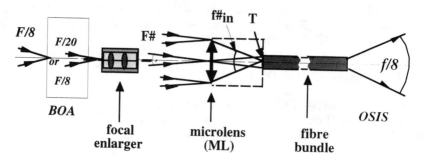

Figure 2. AOIFU optical layout

For the 0.1″ sampling, the AOB is used and produces $f/20$ images

For the 0.4″ sampling, the AOB optics is shifted and the $f/8$ images are produced by the Cassegrain focus (the $f/20$ and $f/8$ foci are located at the same place).

The AOIFU optical layout is shown in Figure 2.

2.1. Focal Enlarger

The focal enlarger provides the two spatial samplings of 0.1″ and 0.4″, and adapts the $f/20$ or the $f/8$ aperture to the microlens input aperture.

2.2. Microlens Array

The square microlens array (Fig. 3) was made by the LPI (Laboratoire de Physique Instrumentale—OPM). The manufacturing procedure is based on the engraving of a photo-resistant coating by a two-axis rastering process.

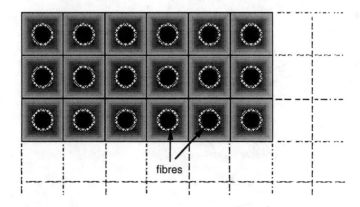

Figure 3. The microlens array.

Characteristics: focal length = 3 mm; aperture = 240 µm.

The thickness of the glass substrate (Bk7 glass) is adapted to the focal length to obtain images on the entrance face of the fibers Ffig. 4).

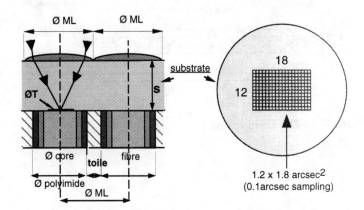

Figure 4.   The microlens array–fiber assembly.

2.3. Fiber Bundle

The fiber bundle has 216 fibers from CeramOptec "super-dry", H-doped type for enhanced blue performance. The core, cladding and polyimide diameters are, respectively, 100, 140, and 160 µm. The 23-m long bundle is sheathed in a flexible conduit.

2.4. Ferrule

Fiber Input Ferrule   Att the entrance of the bundle, a ferrule supports the bundle of 216 fibers arranged to form a 2/3 ratio rectangular format (12×18); the field of view is $1.2'' \times 1.8''$ for the $0.1''$ resolution sampling.

The microlens array is glued onto the fiber entrance face.

Fiber Output Ferrule (OSIS slit)   The slit comprises two fiber alignments (rows) which allow the placement the greatest number of fibers with regard of the 30-mm OSIS slit height (Fig. 5).

2.5. Detectors

The OSIS spectrograph can be alternately equipped with two detectors, depending of the wavelength range: a CCD ($2k \times 2k$, 15-µm pixels) for the visible, or a NICMOS camera ($1k \times 1k$, 18.5-µm pixels) for the near IR. The pitch between two adjacent fibers is 135 µm, which gives a cross-talk between channels lower than 1 %.

2.6. Fiber Arrangement

Figure 6 shows the coherent distribution of the fibers between the input and the output.

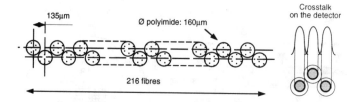

Figure 5. The OSIS slit and cross-talk.

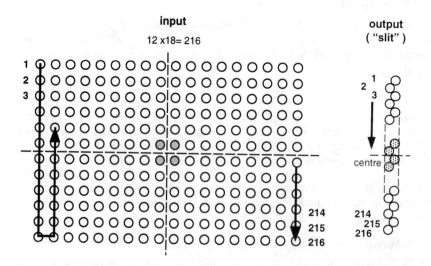

Figure 6. The input and output fiber arrangement.

3. Global Throughput

The expected throughput of AOIFU (at $\lambda = 0.9$ μm) takes into account the following parameters:

- microlens (LPI type)   0.89   (fill factor × transmission efficiency)
- fiber transmission   0.95
- fiber FRD   0.72
- "Fresnel losses"   0.92
- focal enlarger   0.92
= total AOIFU   0.47

4. Simulation of the AOB Image Projected onto the Microlens Array

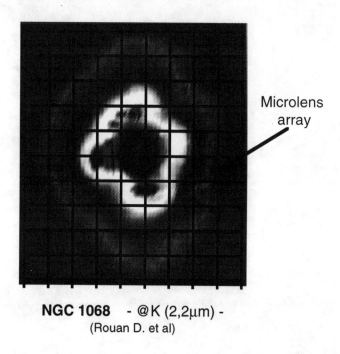

Figure 7.   Simulation: the AOB image projected onto the microlens array.

Figure 7 shows a part of an image (NGC 1068) from the AOB projected onto the microlens array. The area of the figure represents $1'' \times 1.2''$, and the spatial resolution is $0.11''$.

References

Arsenault, R. et al. 1994, Proc. SPIE, 2201, 883
Dallier, R., Cuby, J. G., Czarny, J., & Baudrand, J. 1993, in Astronomy with Infrared Arrays—the Next Generation, ed. I. S. McLean (Reidel: Kluwer)
Vanderriest, C. 1994, Proc. SPIE, 2198, 1376

Fiber Optics in Astronomy III
ASP Conference Series, Vol. 152, 1998
S. Arribas, E. Mediavilla, and F. Watson, eds.

SMIRFS-II: Multi-Object and Integral-Field Unit Spectroscopy at the UKIRT

Roger Haynes, and Jeremy Allington-Smith

Astronomical Instrumentation Group, Physics Department, University of Durham, South Road DH1 3LE, United Kingdom

David Lee

Anglo-Australian Observatory, PO BOX 296, Epping, NSW 2121, Australia

Abstract. SMIRFS-II is a prototype near-infrared fiber system developed for use at the UKIRT in conjunction with the CGS4 cooled-grating spectrograph. There are two fiber bundles for multi-object spectroscopy, fused silica for the J and H bands and zirconium fluoride for the K band. Each contains 14 fibers which can be manually positioned in a circular field of 12 arcmin^2. Small lenses are used at the input and output of each fiber to couple the $f/36$ telescope beam to CGS4. The system was commissioned in 1995 June and again, with upgrades, in 1996 December.

An integral-field bundle, containing 72 fibers, for performing 2-D spectroscopy in the J and H bands is also available. The bundle covers a field of 6 × 4 arcsec with spatial sampling of 0.6 arcsec. Microlenses are used at the input and output to provide nearly 100 % fill factor, enhanced coupling of the $f/36$ UKIRT beam with the fibers and improve the optical coupling with the spectrograph. The integral-field unit was successfully commissioned in 1997 June.

1. Introduction

The SMIRFS-II instrument has been developed as a prototype fiber-based system designed to provide a near-infrared multiple-object and integral-field spectroscopy capability with the CGS4 cooled-grating infrared spectrograph at the United Kingdom InfraRed Telescope (UKIRT). A number of different fiber bundles are now available to couple the telescope focal plane to CGS4. There are two multi-object bundles, each containing 14 fibers, that can be implemented within a 12 arcmin^2 field: a silica fiber bundle for the J and H bands and a zirconium fluoride bundle for the K band. The third bundle provides a contiguous 6 × 4-arcmin field for 2-D spectroscopy.

2. SMIRFS-II Multi-Object Mode Design

A schematic of the SMIRFS-II in its multi-object mode is shown in Figure 1. The field-plate unit (FPU) attaches to the West Port of the instrument support unit (ISU) and its main function is to support the pre-drilled fiber plug plate at the telescope nominal focus. A plate-tensioning device is included, which bows the field plate to ensure that all the fibers in the field point accurately (within ±0.2°) at the secondary mirror, i.e. the exit pupil of the telescope. The fibers are re-formatted into a slit which attaches to the slit projection unit (SPU), which re-images the slit onto the focal plane inside CGS4 via two mirrors. The mirror mounts allow for tip, tilt and translation adjustment to ensure optimal alignment with the CGS4 optics. The SPU requires the removal of the CGS4 calibration unit, so flats and arcs are carried out separately. The guide fiber unit is fed by three guide bundles assisting field acquisition and correction of any field rotation.

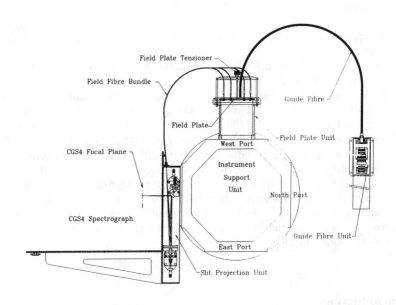

Figure 1. A plan view of SMIRFS-II in multi-object mode, showing the layout around the instrument support unit of the UKIRT.

There are two sets of fibers available each containing 14 field fibers: an ultra-low OH silica fiber bundle for use in the J and H bands, and a zirconium fluoride fiber bundle for the K band. A small calcium fluoride lens is used at the input and output of each fiber to couple the $f/36$ telescope beam into the fiber at $f/5$ for more efficient propagation (significantly reduced FRD) and then convert back to $f/36$ at the output for coupling to the spectrograph. Pupil imaging was used for the lens coupling to maintain the spectrograph beam requirements independent of FRD produced by the fibers. Figure 2 is a schematic of an input ferrule. The optical layout is similar at the fiber output slit. The lenses limit the aperture viewed by the fiber to 4.5 arcsec. However, a field stop can be used to reduce the amount of sky background in good seeing conditions, reducing the aperture to 2 arcsec. The fibers are uncooled, as is the SPU, so the system was designed to take advantage of the Lyot stop and cooled slit within CGS4. These baffle out some of the sky background and most of the thermal background from the SMIRFS-II unit.

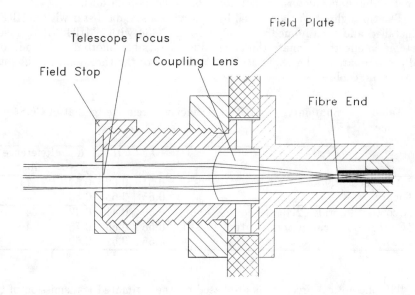

Figure 2. Schematic of a fiber ferrule showing the lens and fiber assembly.

3. Commissioning the Multi-Object Mode

The first commissioning run took place in 1995 June when CGS4 was configured with the short focal length camera (150mm) and the 75-1 mm^{-1} grating. This run was plagued by very poor conditions. The preliminary results from this run are given in Haynes et al. (1995) and Haynes (1995). As a result of the run a number of modifications were made, the most significant of these being the introduction of a reflective mask near the fiber output slit of the zirconium fluoride fiber bundle in an attempt to reduce the thermal background in the K band. The mask contains holes that co-align with the optical path from each of the fibers in the output slit. This reduces the emissivity of the inter-fiber gaps by ensuring that any reflected light originates from within the CGS4 cryostat. Each hole acts as a fiber field stop limiting the effective fiber aperture to approximately 2.5 arcsec on the sky. However the aperture is slightly blurred by FRD effects.

The second commissioning run took place in 1996 December (Allington-Smith et al. 1997) with the same CGS4 configuration as in 1995 June. The weather was partially clear, but much of the time was not suitable for photometric studies which would have allowed a reliable determination of the system throughput. For the remainder the photometric variation was probably around 10 %. This limited the accuracy of all the results presented in this section. A number of K and M giant stars were observed in open and globular clusters to demonstrate the scientific potential of the system and establish a new method of metal abundance determination for cool stellar populations.

The throughput was measured by observing a standard star with SMIRFS-II installed and then immediately after using CGS4 with SMIRFS-II removed. Systematic effects dominate the errors due to variable photometric conditions and are estimated to be about 10 %. A summary of the throughput results can be found in Table 1.

Table 1. Summary of SMIRFS-II efficiency relative to that of CGS4 alone

	J	H	K	Reference fiber
Fused silica (4.9-arcsec CGS4 slit)	0.51	0.62	0.57	7
Theory	0.65	0.65	0.53	
Zirconium fluoride (2.45-arcsec CGS4 slit)	0.39	0.41	0.42	4
ZF (1995 June run with no output mask)	0.56	0.65	0.75	4
Theory	0.55	0.62	0.69	

The theoretical predictions are based on the estimated transmission of the fibers, reflection losses and pupil alignment errors. The silica throughput compares well with the predictions, especially in the H and K bands. However, the zirconium fluoride fiber throughput is significantly lower than predicted. This may be due to vignetting at the output mask caused by misalignment of the star at the fiber input. The results obtained in the 1995 June run are also shown since no output mask was used in that case. The fiber–fiber throughput variations are dependent on the fiber ferrule alignment in the field plate and change with each setup, but are typically 10 % rms.

Because SMIRFS-II is not cooled an additional thermal background is expected in the K band from the following: the telescope and sky; the material between the fibers at the slit; the fiber and lenses (the emissivity approximates to, [1 − transmission] including reflection losses); CGS4 itself (the contribution is expected to be small because it is cooled).

It is important to minimize the thermal background because although it can be removed by background subtraction, e.g beam switching, the overall signal/noise is reduced by the photon noise from the background signal. No quantitative comparison was possible for SMIRFS with and without the fiber slit mask, due to poor conditions on the 1995 June run; however, a large improvement was noticeable in the inter-fiber background after the mask was installed. Previously when observing blank sky the inter-fiber background was often higher than the signal from the fiber, however after the mask was installed the inter-fiber background was less than the sky signal.

4. Science Observations

During the second commissioning run, giant stars in globular and open clusters were observed in the K band. Each object was assigned a fiber and a few were left to monitor the sky. Beam switching was used with a telescope nod of typically a few arcmin. In the example shown (NGC 1904), the field was observed with 2×2 sampling for 12 exposures of 5 s (4 min total exposure time) in each position in the sequence on–off–off–on target positions. This procedure was repeated eight times to give a total on-target exposure time of 64 min. The data were reduced to wavelength scale by removing a skew caused by the CGS4 slit orientation and fitting to an argon arc spectrum. Figure 3 shows the beam-switched image resulting from co-adding all the data in on-target/off-target subtracted pairs. Figure 4 shows extractions of all 14 spectra ordered in K-band flux with an offset of 100 counts between. A list of each object with its broad-band optical magnitude and color is given in Table 2 in the same order as they appear in the figure. Note that two of the fibers were assigned to sky, and that these have effectively zero counts. The spectra generally show Na I (2.204 μm) and weak Ca I (2.258μm) absorption plus CO bands at redder wavelengths (see, for example, Terndrup et al. 1991).

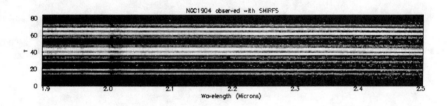

Figure 3. Background-subtracted image of NGC 1904 field after geometric rectification and wavelength calibration.

5. Conclusions: Multi-Object Mode

Though in multi-object mode SMIRFS-II has a throughput typically half that of CGS4, it can provide a significant multiplex advantage in observing efficiency. However, it is a prototype system and would benefit from a number of improvements:

1) Anti-reflection coating of the lenses and direct coupling of the lenses to the fibers, thus avoiding the air gap, improving efficiency and reducing the thermal background from the fiber.

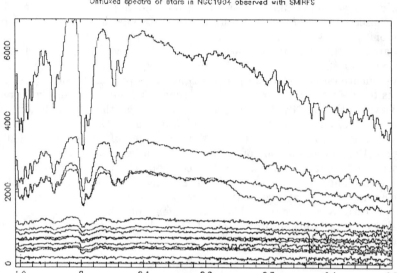

Figure 4. Extractions of all objects in the NGC 1904 image from the faintest (*bottom*) to the brightest (*top*). The spectra are offset from each other by 100 data numbers.

Table 2. Stars observed in NGC 1904 (brightest to faintest)

Fiber	star	V	B − V
8	A41	12.94	1.41
12	A51	13.33	1.39
7	A45	13.84	1.18
13	A53	13.02	1.71
3	A50	13.49	1.76
14	A55	14.93	1.04
5	A42	15.28	0.97
2	A52	14.68	1.06
9	A54	15.78	0.86
6	A47	15.27	0.67
11	A44	16.12	0.82
4	A49	16.07	0.83
10	SKY		
1	SKY		

2) Modification to fiber ferrules and field plate to eliminate pointing errors between the fibers and the secondary mirror, reducing the fluctuation in fiber throughput with each change of field.

3) Improving the target acquisition. Currently, this must be done by *peaking up* the counts from the fibers (i.e. move the telescope to maximize the counts down the fibers), rather than using the cross-head TV, as this does not allow for atmospheric dispersion.

4) The optimal size of holes in the fiber output mask and required CGS4 slit width need to be investigated further to maximize the signal/noise. However, this will depend on the precision of acquisition and seeing conditions at the time.

SMIRFS-II was designed as a prototype technology demonstrator, yet it performs well and points the way forward for new infrared multi-fiber systems for the next-generation IR spectrographs. At present, the number of fibers is limited only by the CGS4 slit length.

6. SMIRFS-II: Integral-Field Mode Design

The SMIRFS-II integral field unit (IFU) fiber bundle provides a 2-D spectroscopic capability at the UKIRT, re-using many of the components of the SMIRFS multi-object system (Haynes et al. 1998). The IFU provides coverage of a 6 × 4-arcsec^2 with 72 hexagonal close packed microlenses, each subtending 0.62 arcsec on the sky. The input field is reformatted into a pseudo-slit which is re-imaged onto the cold slit inside CGS4 via the SPU. The IFU is designed for the J and H bands using low-OH silica fibers, though some useful performance is expected in the K band. Figure 5 shows the image reformatting and a schematic layout of the IFU.

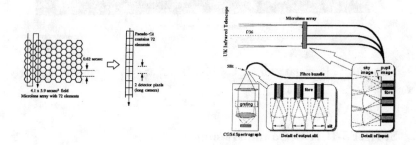

Figure 5. *Left:* Input field format showing how the elements are reformatted at the slit. *Right:* A schematic of the SMIRFS-II IFU optical layout.

The IFU input is held at the nominal telescope focus using the SMIRFS-II field plate unit at the West Port of ISU. A hexagonal microlens array is bonded to the fiber bundle at the input. The fibers are fixed at 412 μm centers in an array of micro-tubes. The input microlens array serves two purposes. It converts the $f/36$ telescope beam to a faster focal ratio for efficient coupling within the fibers and removes the dead space associated with bare fiber systems giving effectively a 100 % filling factor. The fibers are reformatted into a slit at the bundle output, where square microlens arrays are used to convert the beam back to $f/40$ for efficient coupling with the CGS4 spectrograph. Both sets of microlenses, produced by Adaptive Optics Associates, are manufactured from epoxy on a glass substrate. The substrate thickness is such that the microlens focus occurs at the back surface of the glass where the optical fibers have been attached using index-matched UV curing epoxy, reducing air/glass reflection losses. The net effect is the re-imaging the telescope pupil on to the fiber face. The design was constrained by the $f/36$ beam requirement of the CGS4 spectrograph restricting the theoretical throughput to approximately 50 %. Also the 72 pixels available along the CGS4 slit imposed an IFU element limit of 72 when the short (150-mm) camera was fitted.

Proper sampling at the IFU input is achieved when the image quality has a FWHM of 1.24 arcsec or more: i.e. two IFU elements (0.62 arcsec). Ideally, the under-sampling at the detector should have negligible adverse effects, but it can cause some problems because of flexure between CGS4 and the IFU. This should be resolved by the adoption of the long camera (300 mm) in CGS4 on a semi-permanent basis. This doubles the spatial resolution, having 144 pixels along the slit, thus sampling each IFU element with 2 pixels in the spatial direction.

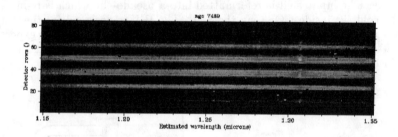

Figure 6. Raw data for NGC 7469. [Fe II] and Paβ emission is clearly visible.

7. SMIRFS-II IFU Commissioning

The IFU was commissioned in 1997 June (Allington-Smith et al. 1998) using the CGS4 short camera. Again we were most unfortunate in that the weather allowed only a very limited time on the sky. However, we were able to determine that the IFU performs close to theoretical predictions. The main results of the run are presented below:

1) The transmission of the IFU unit was 47 % in the J band when averaged over all elements which is close to the predicted 53 %. Most of this shortfall can be accounted for in the flat-field variation discussed below. However, in the H band the value was somewhat lower (33 %). The reason for the low H-band throughput is unclear and will be investigated further on the next run in 1998 February.

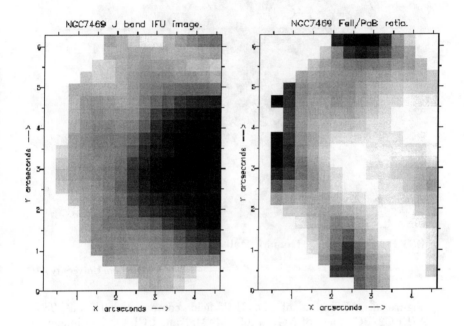

Figure 7. *Left:* Reconstructed J-band image of the Seyfert-2 galaxy NGC 7469. The nucleus is at the right edge. *Right:* spectra from selected elements showing the [Fe II]/Paβ lines. Dark = higher counts (—it left) and higher Fe II/Paβ ratio (*right*).

2) The flat-field shows variations with an rms of 23 %. Much of this is due to four fibers which we damaged during the IFU bundle assembly. Excluding these fibers reduces the rms to 14 %. Most of the remaining fluctuation was in the form of a periodic drop in throughput occurring at the edges of the 12 output blocks that made up the fiber slit. Investigation in the laboratory revealed this to be the result of stress-induced FRD. Subsequent remedial action has significantly improved the situation. We hope to quantify this on the next run.

3) From the point spread functions of the IFU output elements, the effective slit width was determined to have a FWHM of 1.1 pixel. This is very close to the prediction of 1 pixel, which is dominated by CGS4.

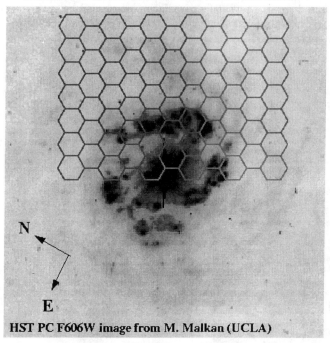

Figure 8. Location of the IFU field compared with an *HST* WFPC2/PC image of NGC 7469 (M. Malkan, UCLA). The image is on a logarithmic scale.

4) It can be seen from Fig. 6 that good sky subtraction can be achieved using beam switching (via a telescope nod). The Fe II and Paβ emission can be clearly seen with little evidence of sky emission-line residuals.

During a brief period of reasonably clear conditions we observed the region just adjacent to the nucleus of a Seyfert-2 Galaxy, NGC 7469. Though the seeing was poor and the atmospheric transmission varied strongly we managed to get some results with a total on object exposure time 400s. The raw sky subtracted data is shown in Figure 6. Figure 7 shows a reconstructed *J*-band image (useful for object acquisition) and a map of the [Fe II]/Paβ ratio. [Fe II] emission is usually interpreted in terms of fast shocks or X-ray excitation while Paβ is an indicator of star formation. Figure 8 shows the IFU field superimposed on a visible *HST* image of NGC 7469 . This suggests that the emission-line ratio is highest near the nucleus, falling off in the starburst ring. This broadly agrees with Genzel et al. (1995) who found [Fe II] 1.644 μm to be more centrally concentrated than Brγ. However, we also see the ratio increase at a larger radius than that observed by Genzel et al. observed.

8. Conclusions: IFU Mode

SMIRFS-II provides the UKIRT with a near-infrared spectroscopic integral-field capability performing close to the predicted performance that is available to the community on a collaborative basis. Though only a modest prototype, the construction of the IFU has been of great benefit in the development of the larger, more finely sampled, integral-field unit we are constructing for the WHT (TEIFU) and the Gemini telescopes (GMOS IFU).

Acknowledgments. We would like to thank Ian Parry for his involvement at the early stages of the project, the staff of UKIRT and JACH for their encouragement and practical help, in particular: Tom Kerr, Gillian Wright, and Tom Geballe. We also thank staff in Durham, especially Ray Sharples, Robert Content, John Webster, and George Dodsworth.

References

Allington-Smith, J., Haynes, R., Lee, D. & Sharples, R., 1997, SMIRFS: Report on second commissioning run, Internal report, University of Durham [http://star-www.dur.ac.uk/ jra/spectroscopy.html]

Allington-Smith, J., Lee, D., Haynes, R. & Content, R., 1998, Commissioning of the SMIRFS-IFU prototype, Internal report, University of Durham [http://star-www.dur.ac.uk/ jra/spectroscopy.html]

Genzel, R. et al. 1995, ApJ, 444, 129

Haynes, R., Sharples, R., & Ennico, K. 1995, Spectrum, 7, 4

Haynes, R. 1995, PhD thesis, University of Durham.

Haynes, R. et al. 1998, UKIRT Newsletter, February 1998

Terndrup, D. M., Frogel, J. A., & Whitford, A. E. 1991, ApJ, 378, 742

COHSI: a Lens Array and Fiber Feed for the Near Infrared

Matthew A. Kenworthy, Ian R. Parry, and Kimberly A. Ennico

Institute of Astronomy, University of Cambridge, Madingley Road, Cambridge CB3 0HA, United Kingdom

Abstract. Two fiber-optic feeds were designed and constructed for the Cambridge OH Suppression Instrument (COHSI). We discuss the techniques used in their manufacture.

1. Introduction

COHSI is designed to suppress the OH airglow in the near infrared 1–1.8 μm allowing observations of faint high-redshift galaxies, gravitational arcs, and other faint extended objects (Piché et al. 1997). Its modular design allows easy transport between telescopes (Fig. 1) it comprises an integral-field unit (IFU), an OH-suppression unit and a cryogenic spectrograph—the last two sit on the floor of the observatory.

Two fiber feeds are used. 110-μm fibers run from the 100-element IFU on the Cassegrain focus and are reformatted into a curved fiber slit in the OH suppressor. A second set of fibers with a core diameter of 200 *micron* runs from the suppressor into the cryogenic spectrograph. The two fiber feeds are physically joined together at the curved double slit so the whole assembly is a single unit (Fig. 5).

2. Protecting the Fibers

Although fibers are strong in tension, they are easily broken with finger-pressure against a sharp corner or when bent into a small radius. With the fibers we use, this radius is about 2 cm.

We use plastic-sheathed steel-wound conduit to protect the fibers from one end to another—they are flexible but can withstand the full weight of someone standing on them, whilst their coil-wound construction makes them flexible. They also have a minimum bend radius of 15 cm, making them well suited for use with optic fibers. Because of their ability to bend and change length we put a box into the fiber path. This "strain relief" box contains a coiled loop of fiber optic that can expand or contract according to the extension of the conduit, thereby preventing tension being inadvertently put onto the fibers.

COHSI: a Lens Array and Fiber Feed for the Near Infrared   301

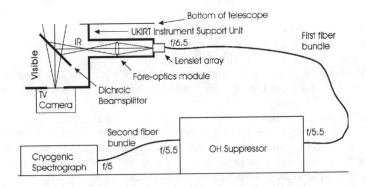

Figure 1.   Layout of the various COHSI modules.

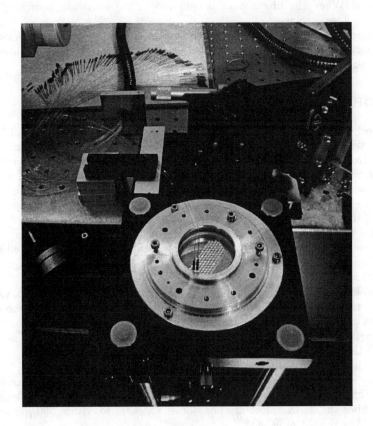

Figure 2.   Fiber ferrules being placed on the back of the lens array.

3. The Integral-Field Unit (IFU)

The IFU comprises 100 hexagonal achromatic doublets cemented onto an optical flat substrate. Individual lenslets were glued onto the flat with an optical glue that cures upon exposure to UV light (Fig. 2).

This substrate is placed into mechanical contact with two other optically flat pieces of glass, the three together forming an achromatic flat whose thickness is equal to the focal length of the doublets.

Concentric steel ferrules are glued over the outer polyamide layer of the 110-μm core optic fiber, providing a greater surface area for gluing of the fiber and aiding the polishing of the fibers. After rough grinding with silicon oxide powder, the fibers are polished with 12-μm aluminum oxide, then with 6-μm, 3-μm and 1-μm oil/diamond suspension on silk cloths.

In order to position the fibers the field lens for the fore-optics is placed in front of the lens array. A back-illuminated fiber placed correctly on the back of the array will form an image of the fiber on a graticule in the focal plane of the field lens. This image is viewed with a microscope eyepiece connected to a CCD camera. When the fiber image is centered in the graticule, optical glue then fixes the fiber in place.

To ensure further mechanical strength the fiber ferrules are encased in re-enterable polyurethane, which supports them but does not apply any lateral force that could move the fibers out of position.

4. The OH Suppressor Double-Fiber Slit

Flexible steel conduit holds the 100 fibers and feeds them into the center of the suppressor, whose design is based on a Schmidt camera with an 86-cm diameter primary mirror (Fig. 3). In this design, the ends of the input and the output fibers sit between the primary mirror and the corrector plate on the curved focal surface of the Schmidt. Due to chromatic field aberrations the 200-μm output fibers have to be within 500 μm of the 110-μm input fibers, sitting on the optical axis of the camera.

In order to get the two sets of fibers within the maximum separation, the two slits were assembled separately then glued together before being polished on a spherically curved polishing plate. The aluminum plates used have a radius of curvature of 834.5 mm.

A small aluminum block with a piece of PTFE sheet glued to it forms a vice which holds the fibers in their respective grooves on the fiber slit. The grooves are cut with a CNC machine and are angled at approximately 6 degrees to the spherical surface. The grooves also are fanned out to ensure that they point correctly to the system pupil.

All the fibers are then inserted between the slit plate and Teflon surface, before epoxy glue is run into the inter-fiber spaces. After setting, the aluminum block and Teflon lift off easily to give the finished fiber slit. After completing the second fiber slit, both halves are positioned together to within 500 μm and epoxy glue fixes the slit (Fig. 4).

COHSI: a Lens Array and Fiber Feed for the Near Infrared

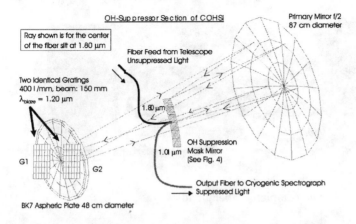

Figure 3. Position of the curved fiber slit in the suppressor.

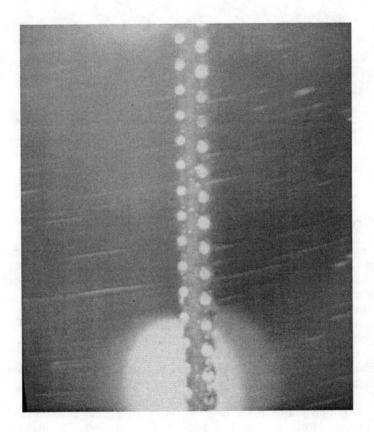

Figure 4. Illuminated fibers in the curved fiber slit.

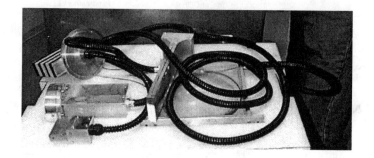

Figure 5.   The completed fiber bundles.

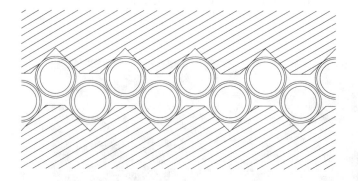

Figure 6.   Schematic of the cryogenic fiber slit.

5. The Cryogenic Fiber Slit

From the output fibers in the suppressor the fiber feed then runs to a strain-relief box attached to the outside of the cryogenic spectrograph. The fibers pass through a hole filled with vacuum-proof epoxy glue and into a 56-cm loop of plastic conduit before forming a staggered output slit that is cooled with the rest of the optics to cryogenic temperatures.

Because of the limited number of pixels available on the PICNIC arrays, we had to pack the fibers as close together as possible. In order to overcome the buffer cladding, the fibers are in a staggered pattern that gives no inter-fiber gap on the array (Fig. 6). This slit is unusual in that the fibers are in contact with the aluminum slit on one side but are supported on the other side by the two adjacent fibers.

To put in the fibers in this slit required thin plastic sheet between the two aluminum groove plates, and careful alignment of the grooves. The fibers were placed with the aid of an optical microscope.

Heating up the epoxy glue makes it cure in a shorter time but makes the glue less viscous. We heated the fiber slit to 70°C and then added the hot glue—gravity and capillary action pulled the glue all through the fiber gaps. The slit surface was then cleaned of excess glue and the fiber slit polished.

6. Performance and Conclusions

We have prepared four sets of fiber ends and they have been successfully mounted into three different types of fiber mounting—the techniques covered have involved curved surfaces with dual fiber slits, fiber alignment with lens arrays, and staggered geometry fiber arrangements for use in cryogenic conditions.

The fiber feeds have all 100 fibers fully transmitting with none broken and no visible chipped fiber surfaces. After preliminary laboratory testing in the suppressor the dual slit feed was shown to be a complete success, the staggered fiber slit performed very well in the cryogenic spectrograph and the alignment of the fibers in the IFU have a positional accuracy of less than 5 $-micron$, well within the tolerance of 10 μm stated in the design. COHSI will first be used in 1998 March.

References

Piché, F., Parry, I. R., Ennico, K., Ellis, R.,S., Pritchard, J., Macka,y C. D., & McMahon, R. G. 1997 Proc. SPIE, 2871, 1332

Part 6: Other Applications

Optical Fibers in Astronomical Interferometry

V. Coudé du Foresto

Observatoire de Paris, DESPA, 5 place Jules Janssen, 92195 Meudon Cedex, France

Abstract. Single-mode fibers can be used in stellar interferometry to recombine independent telescopes coherently, w here they provide substantial advantages over conventional optics. In single-mode waveguides light propagation is forced to follow the structure of the fundamental mode and the spatial turbulence of the atmosphere is filtered out. This enables a gain of almost one order of magnitude in the accuracy of fringe-visibility measurements.

A fiber recombination unit is installed at the IOTA interferometer on Mt. Hopkins in Arizona and is now routinely used for scientific observations. Optical waveguides are also under consideration for other interferometric facilities. Tomorrow, interferometers with large telescopes will combine single-mode integrated optics and adaptive optics to provide maximum precision and sensitivity.

1. Introduction

Multimode fibers are widely used in spectroscopic instrumentation, as is attested by these proceedings. In the last decade, *single-mode* fibers have also found their way into astronomical instrumentation, for stellar interferometry.

Long-baseline interferometry is a powerful technique for obtaining imaging information at a spatial scale which is beyond the diffraction limit of the largest monolithic telescopes. An aperture-synthesis interferometer is made up of several separated pupils: most current optical interferometers operate with two telescopes, but simultaneous operation of three telescopes (thus providing three simultaneous baselines) has been recently demonstrated (Baldwin et al. 1996). At millimeter or radio wavelengths, where the technique is more mature, aperture synthesis can be performed with several dozens of independent pupils.

The beams collected at each pupil are recombined coherently to form a fringe pattern whose contrast expresses the degree of spatial coherence of the wavefront received from the source which, in turn (once the interferometric efficiency of the instrument has been calibrated on a reference source), is directly linked to the Fourier transform of the object spatial intensity distribution (the object complex visibility), at the spatial frequencies defined by the baselines. So the observer ends up with measurements of individual Fourier components of the object's image. They can be used to constrain a model of the source's morphology, or even reconstruct an image of the object with very high angular resolution.

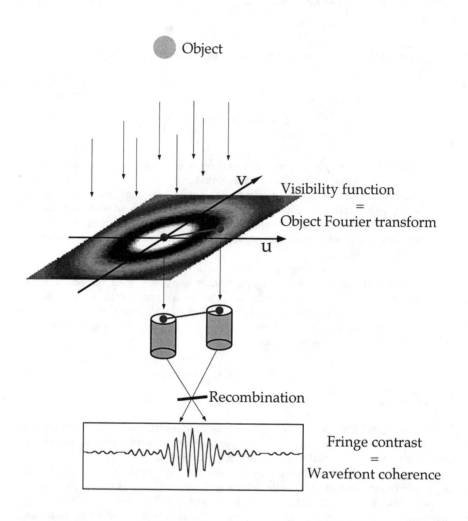

Figure 1. The principle of stellar interferometry.

Coherent recombination implies that the optical path-length difference between the two beams is controlled within a small fraction of the wavelength throughout the bandwidth and for both polarizations. This is meaningful only as far as it is possible to define a single optical path throughout each pupil, despite the wavefront corrugations that are caused by optical aberrations and/or atmospheric turbulence. Therefore a stellar interferometer must perform, at least, the following optical functions:

- Beam collection

- Beam transport towards the recombination station

- Optical delay and optical-path modulation

- Coherent recombination (correlation)

The last three of these functions can be performed by optical fibers. Another very important feature that will be discussed below, and for which single-mode waveguides excel, is the *modal filtering* capability that helps extract a well-defined mode from the corrugated wavefronts.

2. Basic functions

Froehly (1981) was the first to point out the possibility of using single-mode (SM) fibers to transport stellar light coherently. This looked like a very elegant alternative to replace mirror trains on the complicated path that leads from the individual telescope foci to the recombination table.

2.1. Why Single Mode?

In a large multimode fiber each ray has its own optical path, which depends upon its incidence angle at the input of the fiber. When all the rays are scrambled at the output, it becomes impossible to define a single path, and coherence is lost. Therefore multimode fibers are not suitable for interferometry unless one is able to sort the modes and recombine them one by one (as was demonstrated in a laboratory experiment by Shaklan et al. 1991).

A single-mode fiber has a small core (no more than a few wavelengths in diameter) and a large cladding (typically 125 μm). Because the core dimensions are comparable to the wavelength, waveguide theory, not ray optics, is the appropriate tool to understand the fiber properties. Solving Maxwell's equations with the boundary conditions imposed by the core/cladding interface generates, for wavelengths longer than a cutoff λ_c a unique solution for the electromagnetic field, with a well-defined mode profile and propagation constant (Gloge 1971a, b). These are solely determined by the waveguide characteristics (i.e. its shape and the refractive-index profile) and do not depend upon the way light was launched into the fiber. For example, the divergence of the beam at the output is insensitive to the input beam f-number: the focal-ratio degradation is absolute.

Another aspect in which single-mode fibers differ from multimode waveguides is with the localization of the transmitted energy: in multimode fibers it is confined in the core, whereas in single-mode fibers the beam extension into the cladding can be substantial: the mode diameter is somewhat larger than the core diameter.

2.2. Beam Transportation

Silica single-mode fibers (with a transmission from 480 to 1800 nm) have been developed for the telecommunication industry with losses close to the theoretical minimum (0.15 dB/km at $\lambda = 1.5$ μm). For longer wavelengths, fluoride glass (Monerie et al. 1985) SM fibers can be operated in the mid-infrared (1.5 μm $< \lambda < 4$ μm). In fluoride glass, the absorption minimum occurs around 2.6 μm and with the current technology transmissions of better than 90 % for 100 m can be achieved at 2.2 μm (the K band).

Coherent beam propagation over decameter fiber lengths, though, presents some difficulties. There are two important issues to keep in mind when transporting light in an interferometer: avoiding differential dispersion and controlling the polarization. SM fibers suffer chromatic dispersion, which is due on the one hand to the intrinsic dispersion of the glass in which they are made, and on the other hand to the structural waveguide dispersion. Material and waveguide dispersion can be of opposite sign so that the total dispersion in a fiber is usually lower than in the same amount of bulk glass: in some cases it is even conceivable to design the waveguide in such a way that material and structural dispersion cancel each other out over a given bandpass (Coudé du Foresto et al. 1995). In an interferometer, what needs to be minimized is the *differential* dispersion between the two arms. Differential dispersion between segments of the same length would be nullified if the fibers were homogeneous. As it is not always the case, in practice fiber lengths are adjusted until differential dispersion, measured in the laboratory, reaches an acceptable level.

Ideally, in a perfect fiber the fundamental mode is degenerate for the two spatially orthogonal polarizations. Circular fibers in real life behave in a much different manner and exhibit birefringence and mode coupling. Unwanted birefringence occurs in non-perfect fibers because of small random shape and stress anisotropies. Even if it were perfectly manufactured, a circular-core fiber would not be neutral, except when kept straight. Otherwise, bending, twisting and stretching lead to additional birefringence and/or principal-axis rotation. These (and all the polarization irregularities that occurred before light reached the fibers in the interferometer) can be compensated by using a fiber-based polarization controlling device such as the Lefèvre loops (Lefèvre 1980), also known as "Mickey ears".

The best way to control polarization is to use polarization, maintaining highly birefringent fibers. Polarization-preserving fibers have a strong intrinsic birefringence that overwhelms the random birefringence, and well-defined principal axes; they maintain the polarization of the propagating light even when the fiber is bent or twisted. The birefringence is usually induced by a shape anisotropy (elliptical or rectangular core) or a stress anisotropy (Panda, bow-tie fibers) if the circular geometry of the core needs to be maintained. At the output the two polarizations can be observed either separately or simultaneously if an adequate compensator is introduced (Reynaud & Boca 1993). One can refer to Noda et al. (1986) for an overview, with an extensive bibliography, of polarization-preserving fibers.

2.3. Optical Delay

Here a distinction should be made between the variable geometrical delay required to compensate the diurnal motion of the source on the sky (a deterministic, continuous delay with a range comparable to the baseline), and the much smaller (1 mm at most), but also much faster, optical-path modulations that may be needed for fine path-length equalization (to compensate for vibrations in the instrument and the differential piston mode of the atmospheric turbulence), or for a temporal modulation of the fringe pattern.

A controlled change of the optical path can be induced within a fiber by mechanically stretching the waveguide. The traditional solution consists in wrapping the fiber onto a cylinder made of a piezo-electric ceramic that can expand or contract diametrically when a voltage is applied. The stroke is of the order of a few hundred microns (it is limited by the size of the cylinder and its expansion coefficient) and because no moving parts are involved, the response of the system can be extremely fast (> 1 kHz). This is an adequate system for fine path-length equalization and modulation (Reynaud et al. 1992).

Although it is possible to stretch a long piece of fiber by several meters (Simohamed & Reynaud 1997 have demonstrated a 2-m delay line using a fiber wrapped around a pneumatic device), it appears unrealistic to compensate the geometrical delay in a fiber for a wideband interferometer: when only one arm is stretched the differential dispersion that is introduced in this way is too severe.

2.4. Beam Combination

Coherent beam combination can be done in a single-mode directional coupler, which performs the same function as a beamsplitter. Unlike a beamsplitter, however, the coupler is small, rugged, and virtually impossible to misalign.

The basic component is the X-coupler, which features two fiber inputs and two fiber outputs. If one input is used, the coupler distributes the signal between the two outputs. If both inputs are utilized, the coupler acts as a correlator between the two incoming beams. Coupling ratios are usually 50/50 but can be made at different values, depending on applications. Polarization-preserving couplers are also manufactured. Another component using polarization-preserving fibers is a coupler that separates the two orthogonal polarizations. For multiple (≥ 3) beam combination, one can use several stages of X-couplers (Shaklan 1990) or, possibly, single-mode star couplers with $n \times n$ terminals.

Tekippe & Willson (1985) gave a general review of single-mode directional couplers. The two most commonly used manufacturing techniques are polishing and fusion-tapering. In polished couplers, the fibers are polished and held together with the two fiber cores running in parallel very close to each other for a few centimeters, so that mixing of the guided radiation occurs by coupling the evanescent fields in the cladding. In taper-fused couplers, the fibers are twisted together, heated, fused, and stretched. This is the preferred technique for a flattened wavelength response. On the other hand, polished couplers can more easily be tuned to a required coupling ratio at the nominal wavelength (Neumann 1988).

Important parameters for directional couplers in astronomical interferometry are their intrinsic loss (typically 2 % or less for a fused X-coupler), phase dispersion and wavelength response. For example, the best silica wideband X-couplers offer a splitting ratio within the 40/60 boundaries for an operating range $\delta\lambda/\lambda = 0.3$. Currently available fluoride glass couplers are polished and their response is more chromatic (from 25/75 to 75/25 between 2 and 2.4 μm).

2.5. Laboratory Experiments

Several experiments have been performed to demonstrate the validity of the guided-optics approach to stellar interferometry. They showed for example (Connes & Reynaud 1988) that, with a fine servo-control of the fiber length, it was possible to stabilize the optical path in 100 m of silica fibers to 100 nm rms despite an adverse environment (a shaky radiotelescope dish). In a more controlled (laboratory) environment, optical path stability better than $\lambda/200$ with a 0.1 ms response time was demonstrated in the visible, with a full control of differential polarization and dispersion effects (Reynaud & Lagorceix 1996).

Combining silica SM fibers and several stages of couplers, Shaklan (1990) built an five-beam interferometer which produced data good enough to reconstruct images with standard radio-interferometry software. More recently, Delage (1998) demonstrated good phase stability (0.015 rad) on a three-beam interferometer.

3. Coupling to Starlight

One of the key issues in an instrument using SM fibers is the efficiency with which starlight can be injected in the waveguide. The problem was studied theoretically by Shaklan & Roddier (1988), and more recently by Ruilier (1998). When placed at the focus of a telescope, the fundamental guided mode is excited by the electromagnetic field in the stellar image. Energy transfer would be 100 % efficient if the two fields overlapped perfectly, i.e. if they had the same transverse amplitude *and* phase distribution, which can never be the case. The field distribution inside the fiber is dictated by the waveguide structure: to a good approximation, the phase is constant across a transverse plane, and the amplitude presents a quasi-Gaussian profile. The field distribution in the focal plane follows, for an ideal instrument, an Airy pattern with a phase jump in the first ring. When the input f-number is adjusted (at a typical value of $f/3$) so that the width of the Airy pattern optimizes the match with the Gaussian beam, the coupling efficiency is 78 % (excluding Fresnel losses) for a pupil without central obstruction.

When the image is turbulent, the fiber can be coupled to only one speckle at a time, and the coupling efficiency is divided by the number of speckles $(D/r_0)^2$, where D is the telescope diameter and r_0 is the size of a coherence area on the pupil (Fried's parameter). Because of the time evolution of the turbulent image, the injected power presents fluctuations with a typical time scale of τ_0, the coherence time of the atmosphere.

4. Modal Filtering

This behavior points to an additional (and essential) function performed by single-mode waveguides: the "cleaning" of turbulent beams by spatial filtering. Whatever the injection conditions at the input, the output beam is perfectly stable and coherent, since its profile is determined by the waveguide structure. All phase perturbations in the pupil, whether they are static (optical aberrations) or dynamic (induced by turbulence), are removed by the coupling process, and the injected starlight may vary only in intensity. This is very fortunate, as a Michelson interferometer is much less sensitive to intensity variations than to phase corrugations. Moreover, the relationship between the fringe contrast, μ_{12}, and the object visibility, V_{12}, is simple when the system is affected only by unequal intensities P_1 and P_2 into each arm, and can be expressed analytically:

$$\mu_{12} = \frac{2\sqrt{P_1 P_2}}{P_1 + P_2} \times V_{12}. \qquad (1)$$

Because of turbulence, P_1 and P_2 are fluctuating signals. However, it is easy to collect them in real time, for example with the help of an additional directional coupler that deviates part of the light at each telescope. This information can then be used a posteriori to remove the effects of atmospheric turbulence and optical aberrations into object visibility estimates. Because they are free from seeing perturbations, the quality of the measurements improves considerably.

5. Ground Instrumentation

This capability was demonstrated by the FLUOR instrument (d*F*iber *l*inked *U*nit for *O*ptical *R*ecombination), whose prototype version performed the first coherent recombination of stellar light by guided optics and transformed a pair of independent telescopes into an interferometer with a 5.5-m baseline (Coudé du Foresto & Ridgway 1991). FLUOR is now part of the focal instrumentation at the Infrared and Optical Telescope Array (Coudé du Foresto et al. 1998), where it is routinely used for scientific observations with a pair of 45-cm telescopes separated by 7 to 38 m. The precision it achieves on object visibility measurements is 1 % or better, which represents an improvement of one order of magnitude with respect to what can be obtained without spatial filtering. FLUOR uses fluoride glass components and is optimized for the infrared K band ($2\,\mu m < \lambda < 2.4\,\mu m$). Recently, operation has begun in the L band ($3.7\,\mu m < \lambda < 4.1\,\mu m$) as well (Mennesson 1998). The longer the wavelength, the better the coupling efficiency, since r_0 scales as $\lambda^{(6/5)}$: in median seeing conditions and with the 45-cm pupils of IOTA, the D/r_0 ratio is of the order of 1 to 2 in the K band.

The gain in precision for visibility measurements makes it possible to address a new class of astrophysical problems. In stellar physics and with two telescopes, it becomes possible not only to measure photospheric diameters, but also more subtle phenomena such as limb darkening (or variations in limb darkening), surface structures, asymmetries or temporal variations of the diameter (Perrin et al. 1996). A greater accuracy also means a larger dynamic range when observing binary systems, and access to low-mass (stellar or sub-stellar) companions of nearby stars (Coudé du Foresto et al. 1997).

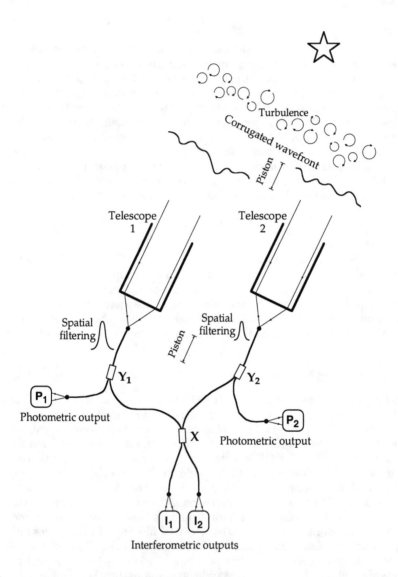

Figure 2. Conceptual design of a stellar fiber interferometer.

Other programs, notably those involving extragalactic objects, are out of reach for the moment because of the sensitivity they require. To improve sensitivity one needs to couple fibers with large telescopes, and maintain a good injection efficiency despite the increased D. This is achieved by phasing the individual pupils with adaptive optics, i.e. increasing the equivalent r_0. The sensitivity is then proportional to SD^2, where S, the Strehl ratio, characterizes the image quality: it is the ratio between the intensity peak in the corrected image and the maximum intensity of a diffraction-limited image of the same source. Using adaptive optics to couple a 3.6-m telescope and a single-mode fiber has already been demonstrated.

Two ground-based arrays with large pupils are currently in the construction phase and will start observing towards the year 2001. The Keck Interferometer on Mauna Kea (Hawaii) has two 10-m telescopes on an 85-m baseline, and the Very Large Telescope Interferometer in Cerro Paranal (Chile) has four 8-m telescopes with a maximum baseline of 130 m. Both are designed to be extended with an auxiliary array of three or more outrigger telescopes of the 1.5–2-m class. A third array, also in construction, is CHARA on Mount Wilson (California), which will eventually feature up to seven 1.5-m telescopes with a maximum baseline of up to 300 m.

All of these large interferometers plan to use single-mode waveguides in their instrumentation. However, the basic functions (beam collection, transport, and delay) will still be performed in a classical manner, using reflective optics. The use of mirrors preserves existing concepts and remains the best way to secure a transmission on the largest possible optical bandpass towards the recombination laboratory, whereas the optical bandpass of a waveguide is usually limited to one octave at most. In the central laboratory, dedicated instruments optimized for each photometric band perform the more complex functions that are required for a good recombination of the beams: spatial filtering, polarization control, optical path modulation, correlation, etc.

6. The Future

Development is still in progress for the production of SM waveguides for the thermal infrared (5–20 μm). Spatial filtering is a key feature of some projects: thus the Darwin mission (Léger et al. 1996), whose goal is the detection and the spectroscopy of telluric exoplanets, is a nulling interferometer: the on-axis interference is destructive, thus cancelling out the light from the central star, while the transmission is optimized off-axis to image the planet. This works with the required dynamic range only if residual phase errors on the wavefront are smaller than $\lambda/25000$ at 10 μm, a specification impossible to meet without cleaning the beams with a spatial filter (Ollivier & Mariotti 1997).

The future will probably see fibers giving way (at least partially) to integrated optics with SM waveguides engraved on a two-dimensional substrate. The optical circuit, that will replace a traditional bench of bulk optics by a plate no larger than 1 cm^2, can easily be placed inside a cryostat and butt-joined to a detector. The technologies involved in manufacturing integrated-optics components are very similar to those employed in microelectronics (photolithography); chemical etching of glasses can also be used. They cover wavelengths ranging

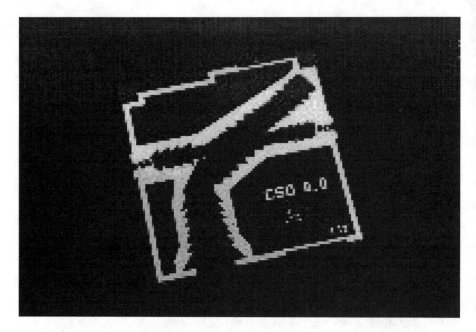

Figure 3. Example of an integrated optics circuit: a Michelson interferometer in 1 cm^2 (LETI, Grenoble).

from the visible to the thermal infrared (10 μm). One can refer to the proceedings of the Astrofib conference, held in Grenoble last year (Kern & Malbet 1996), for an overview of the state of the art in integrated optics and its application to astronomical interferometry.

7. Conclusion

Sixteen years after Froehly's original proposal, single-mode fibers have earned their place in stellar interferometry; their main advantage, however, is not what was originally anticipated (convenient beam transportation), but rather the capability to clean the beams from wavefront corrugations. They represent a first step towards fully integrated optical circuits, which might revolutionize interferometric instrumentation in the future.

References

Baldwin, J. E. et al. 1996, A&A, 306, L13
Connes, P., & Reynaud, F. 1988, in High-Resolution Imaging by Interferometry, ed. F. Merkle (Garching: ESO), 1117
Coudé du Foresto, V., & Ridgway, S. 1991, in High-Resolution Imaging by Interferometry II, eds. J. Beckers, & F. Merkle (Garching: ESO), 731
Coudé du Foresto, V., Perrin, G., & Boccas, M. 1995, A&A, 293, 278

Coudé du Foresto, V.,Mariotti, J.-M., & Perrin, G. 1997, in ASP Conf. Ser. Vol. 119, Planets beyond The Solar System and the Next Generation of Space Missions, ed. D. Soderblom (San Francisco: ASP), 267

Coudé du Foresto, V., Perrin, G., Ruilier, C., Mennesson, B., Traub, W., & Lacasse, M. 1998, Proc. SPIE, 3350, 856

Delage, L. 1998, PhD thesis, Université de Limoge

Froehly, C. 1981, in Scientific Importance of High Angular Resolution at Infrared and Optical Wavelengths, eds. M. H. Ulrich, & K. Kjär (Garching: ESO), 285

Gloge, D. 1971a, Appl. Opt., 10, 2442

Gloge, D. 1971b, Appl. Opt., 10, 2252

Kern, P., & Malbet, F. (eds.) 1996, Proc. Astrofib '96, Integrated Optics for Astronomical Interferometry (Grenoble: Observatoire de Grenoble)

Lefèvre, H. C. 1980, Electron. Lett., 16, 778

Léger, A., Mariotti, J.-M., Mennesson, B.,Ollivier, M., Puget, J.-L., Rouan, D., & Schneider, J. 1996, Icarus, 123, 249

B. Mennesson, 1998, in preparation

Monerie, M., Alard, F., & Mazé, G. 1985, Electronics Lett., 21, 1179

Neumann, E.-G. 1988, Single-Mode Fibers (Berlin: Springer)

Noda, J., Okamoto, K., & Sasaki, Y. 1986, J. Lightwave Technol., LT-4, 1071

Ollivier, M., & Mariotti, J.-M. 1997, Appl. Opt., 36, 5340

Perrin, G., Coudé du Foresto, V., Ridgway, S. T., Mariotti, J.-M., Carleton, N. P., & Traub, W. T. 1996, in Science with the VLT Interferometer, ed. F. Paresce (Garching: ESO)

Reynaud, F., Alleman,J. J., & Connes, P. 1992, Appl. Opt., 31, 3736

Reynaud, F., & Boca, J. 1993, Pure & Appl. Opt., 2, 677

Reynaud, F. & Lagorceix, H. 1996, in Integrated Optics for Astronomical Interferometry, eds. P. Kern, & F. Malbet (Grenoble Observatory), 249

Ruilier, C. 1998, Proc. SPIE, 3350, 319

Shaklan, S., & Roddier, F. 1988, Appl. Opt., 27, 2334

Shaklan, S. 1990, Opt. Eng., 29, 684

Shaklan, S., Reynaud, F., & Froehly, C. 1991, Appl. Opt., 31, 749

Simohamed, L. M., & Reynaud, F. 1997, Pure & Appl. Opt., 6, 37

Tekippe, V. J. & Willson, W. R. 1985, Laser Focus, May 1985, 132

The Impact of Fiber Optics on Photometry: the Design of Two High-Speed Multi-Channel Instruments

H. Barwig, K. H. Mantel, and S. Kiesewetter

Universitäts-Sternwarte München, Scheinerstrasse 1, 81679 München, Federal Republic of Germany

Abstract. The design details are presented of two fiber-coupled photometers for simultaneous multi-color photometry developed at the University of Munich.

1. Introduction

The use of optical fibers as a link between the spectrograph and telescope focus has become more and more common, not only for its utility in avoiding the variation of gravitational forces that attack the stiffness of Cassegrain instruments but also becuase it allows one to provide a constant environment for the spectrograph, thus yielding high spectroscopic stability. Furthermore, for the purpose of multi-object or two-dimensional spectroscopy the application of fiber bundles is essential for an effective optical design.

Applications of optical fibers for aperture photometry are comparatively rare. Unlike the case of spectroscopy, varying the transparency (or vignetting) of the light-path between the telescope and the detector provides the most serious instrumental constraint on the photometric accuracy. In general, geometrical changes of (long) fiber links, e.g. due to the telescope tracking motion, cause changes in the light throughput and in the amount of degradation at the fiber exit. Another problem with using fibers as photometer input channels arises from the uneven flat-field characteristics when using the fiber entrance window as a measuring diaphragm: any dust particle on the fiber surface, which is then identical to the focal plane of the telescope, would cause considerable flux variations even if the seeing disk is moving within the boundary of the fiber core.

Our experiments with fiber-coupled photometers started more than ten years ago when we initialized a research program to analyze eclipsing cataclysmic variables. These stellar systems consist of a late-type secondary that fills its Roche lobe and transfers mass via an accretion disk to a white-dwarf (WD) primary. Due to the geometry of the systems and orbital periods of the order of some hours, sufficiently highly inclined systems give rise to different high-speed eclipse phenomena. For example, the eclipse ingress or egress of the WD of the cataclysmic system OY Carinae with an orbital period of approximately 100 min lasts only 20 s.

In oder to monitor such fast events, the method of classical photometry, i.e. iteratively measuring object, comparison star, and sky background is no longer feasible. If, in additional, five colors are needed, the serial measuring mode yields a time efficiency of less than 7 %, which means, that only a small fraction of the observing time is allocated to a specific filter. It is clear that only simultaneous high-speed monitoring of the program star, a nearby comparison star and the background radiation simultaneously in different filters provides uninterrupted, reliable observations of short-time events.

2. MCCP: a First-Generation Fiber-Fed Photometer

The prototype of a three-channel $UBVRI$ fiber-fed photometer (the so-called "MCCP" = Multi-Channel multi-Color Photometer) meeting these requirements had been developed at the Universitäts-Sternwarte München (Barwig et al. 1987). As shown in the schematic diagram in Fig.1, the instrument comprises four basic units: a) a Fiber input and positioning device, b) a prism spectrograph, c) a detector unit, and d) data acquisition and control.

In the framework of this conference we will mainly focus on a description of the first two units.

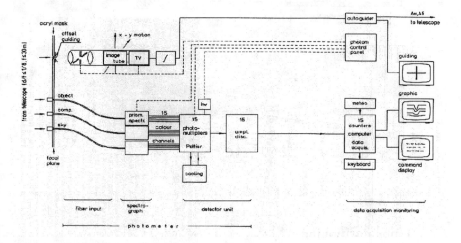

Figure 1.   Schematic diagram of the MCCP.

2.1. Fiber Input Channels

In order to obtain light curves simultaneously in different filters, e.g. similar to the Kron–Cousins $UBVRI$ system, the use of a prism spectrograph for the separation of colors from a large wavelength range turned out to be most convenient. Since the spectral resolution of a (mono) fiber-fed instrument depends on the fiber core diameter, even for a resolution of 5, transfer optic has to be applied to reduce the measuring-diaphragm diameter to the smaller fiber dimensions. As a practicable solution, the input device was chosen was diaphragms of typically ≤ 1 mm (corresponding to $\leq 26''$ on a 1-m $f/8$ telescope) imaged onto quartz mono fibers (QSF 400 W) with 400-μm core diameter by means of two ball lenses with diameters of 3 mm (BK10) and 1 mm (BK7), respectively. The lenses are arranged to match a confocal system that makes the light beam enter the fiber nearly symmetrically with respect to the fiber axis (Fig. 2). This in turn yields a constant light cone ($f/2.8$) at the fiber-exit end, independently of the star's position in the diaphragm. However, due to different reflections on the surface of the ball lenses, the flat-field is distorted towards the edges of the diaphragm. Another disadvantage of this device turned out to be the inhomogeneities (dust particles and/or air bubbles) within the epoxy adhesive which fixed the 1 mm lens to the polished 400-μm fiber core and give rise to high-frequency flat-field distortions. We therefore replaced the two-lens device by a single Fabry lens between the diaphragm and fiber entrance, which projects the pupil of the telescope onto the fiber core resulting in a more homogeneous flat-field without any higher-frequency distortions.

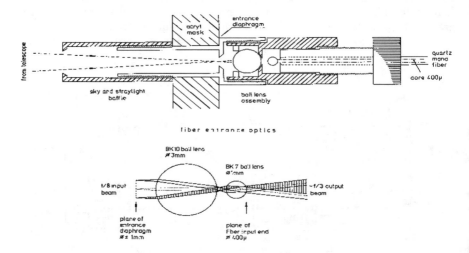

Figure 2. Layout of fiber channel input (two-lens system).

2.2. Fiber Positioning

In the first version of the MCCP the three fibers were positioned in the focal plane of the telescope by means of a mask, which could be prepared on the telescope itself just prior to observation. For mask preparation an acryl plate was fixed in the photometer beside the focal plane of the telescope beneath an automatic drilling machine that was connected to a CCD guiding camera which could be set by remote control to different star positions. Whenever centered on an object, a hole was drilled at the relevant position on the plate. Finally, the mask was inserted in the focal plane and the fibers fixed in the holes.

This economical fiber-positioning device was replaced later by three stepper-motor driven computer controlled X-Y positioning units holding the individual fibers. (It is interesting to note, that this upgrade became necessary when we intended to participate in monitoring the rather spectacular events of the predicted impact of comet Shoemaker-Levy (SL-9) on the hidden side of Jupiter. Our aim was to search for possible light echoes which from the Jovian satellites facing the impact sites. We therefore had to center the fibers on the moons concerned, which in turn were relatively fast moving, thus preventing the use of fixed holes in a mask.)

2.3. Spectrograph Unit

The end faces of the three input fibers form the circular entrance pupils of three small $f/2.2$ prism spectrographs. Each of the resulting spectra (linear dimensions 10×0.4 mm^2) covering the wavelength region 340–900 nm is imaged onto an array of mono-fibers. This array was designed to match the edges of the $UBVRI$ filters approximately. Due to the strong non-linearity of the dispersion of the prisms, five quartz fibers with a core diameter of 1 mm were packed in a row for the U-channel. The other color bands are defined by plastic fibers. Their originally circular faces (1–1.5-mm diameter) were re-formed into an almost rectangular shape. The width of each individual fiber at the spectrograph exit window was adjusted according to its position in the spectrum and the desired filter bandwidth. The required overlap of the filter regions is achieved by the size of the entrance pupil which limits the spectral resolution. To obtain identical color bands for all three spectrographs the respective relative positions of the fiber array and the spectrum were adjusted by means of a monochromator.

The 15 spectrograph output fibers are connected to the detector unit. This consists of a close arrangement of 15 HAMAMATSU miniature photomultipliers encapsuled in a single housing and thermoelectrically cooled down to $-20^circ ircC$. The individual photon counts are processed by a real-time multi-tasking VME-Bus computer system which enables a time resolution of the order of a few milli-seconds. All observing parameters can be defined via different menus. The observer can select which of the 15 channels will be subject to on-line graphics and statistical analysis.

2.4. Data Reduction

The primary aim is the determination of time-resolved intensities of the object (reduced to airmass $X = 0$) relative to a comparison star in the instrumental filter system. Since the MCCP provides data from simultaneous measurements of object, comparison star and sky in five colors, variable atmospheric transparency and wavelength-dependent extinction effects (of first order) can be compensated using the so called standard reduction procedure (Barwig et al. 1987). Any instrumental changes during observation, for example transmission changes of the individual fiber channels, e.g. due to their changing geometry, can only be checked and compensated by performing frequent calibration measurements. For this we need three measuring procedures: measurement of a constant point light source (e.g. a "normal" star during photometric conditions) in the object and comparison channel respectively, measurement of a homogeneous field light source (e.g. sky background) with all three channels, and the dark currents of all the detectors.

2.5. Performance

From our long-time experiences the MCCP can be characterized as follows:

Apart from the Poisson-noise limit the photometric accuracy of the relative intensity for a single uninterrupted run is better than 1 %. On the other hand, the absolute intensities over many observing runs are subject to larger errors due to uncertainties in the calibration to be performed for each run.

One of the most important advantages of the fiber-coupled multi-object photometer is the fact that due to the differential measuring methods useful data can be obtained even if the atmospheric transparency decreases down to less than 10 % (Fig. 3).

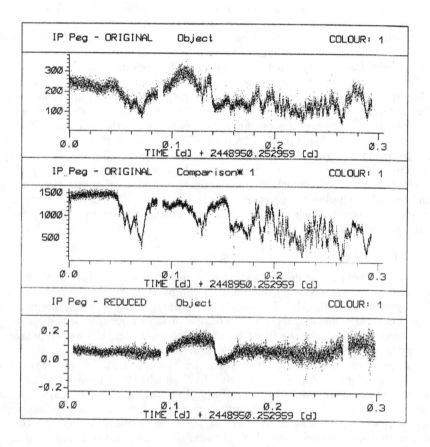

Figure 3. Standard reduction applied to object data obtained during strong transparency variations. The original data for the object and the comparison star are shown in the upper and mid panels, respectively. In the lower panel the reduced eclipse light curve of IP Peg is shown (from Wolf & Mantel 1994).

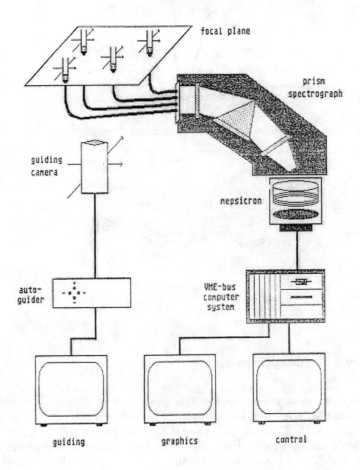

Figure 4. Schematic diagram of MEKASPEK.

3. MEKASPEK: a Second-Generation Spectrophotometer

After all the experiences gained with the MCCP, a more general question arose concerning the photometric accuracy which in principle can be reached using terrestrial broad-band photometry. We learned that this task is hampered by problems associated with the elimination of atmospheric extinction and the color transformation from the instrumental to a standard photometric system. Careful analysis show that these problems can be attributed to an insufficient description of atmospheric extinction effects and to a poor knowledge of the spectral flux distribution of the objects observed (Mantel 1993). Therefore high photometric accuracy requires an optimized instrument to obtain the utmost spectral and temporal information about atmospheric-extinction effects and the program star observed.

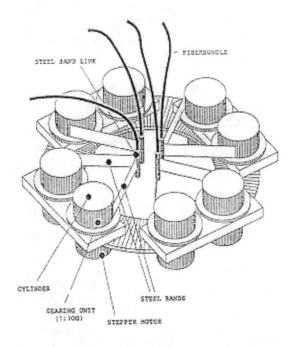

Figure 5. Fiber-positioning unit.

In order to meet these requirements the new four-channel fiber-fed spectrophotometer MEKASPEK was developed at the Universitäts-Sternwarte München (Kiesewetter 1992). Compared to the MCCP, the new design (Fig. 4) uses a fourth fiber channel, which allows one to measure a second comparison star to check the stability of both reference sources. However, most important is the increased spectral resolution ($R = \lambda/\Delta\lambda \approx 50$) of the new instrument, which allows second-order extinction effects to be eliminated and different photometric systems to be synthesized to a sufficiently high accuracy. Furthermore, a time resolution of up to 5 ms, which could be achieved using a photon counting two-dimensional detector (MEPSICRON) and data-compression algorithms, covers the whole frequency domain of variable stellar objects.

3.1. Design Details

For the scope of this conference two details of MEKASPEK will be discussed:

1) In order to reach a sufficient spectral resolution four fiber bundles, each consisting of 130 individual fibers, are used as image slicers. Each input channel consists of an electromagnetically interchangeable measuring diaphragm with diameters up to 2 mm. A Fabry lens mounted behind projects the telescope pupil onto the circular fiber-bundle entrance window (diameter = 0.8 mm). At the double beam–prism spectrograph the individual fibers (core diameter 70μm) of the four bundles are arranged

in two rows forming a long slit of width 250μm. Taking into account the instrumental profile, together with the pixel size of the MEPSICRON detector in the focal plane of the camera lens, a spectral resolution of 50 is achieved within the wavelength region 340–900 nm.

2) Since the four fiber bundles should be simultaneously and independently movable in the focal plane of the telescope (e.g. to follow different planetary satellites) a special fiber positioning unit had been developed. As shown in Fig. 5, each fiber bundle is attached to the link of two steel bands which are wound onto two steel drums. By rotating the drums via stepper motors connected to a reduction gear, the length of the steel bands can be changed and thereby the position of the fiber bundle in the focal plane. This device provides a positional accuracy of 10 μm and a reproducibility of 2μm.

4. Concluding Remarks

The development MEKASPEK was an important step towards the limiting photometric accuracy which can be reached with earth-based telescopes. The application of optical fibers has turned out to be essential to realize the differential measuring principle, which, together with sufficient spectral resolution, provides the highest possible accuracy for terrestrial photometry. Since the completion of MEKASPEK in 1993 numerous scientific observing programs have been performed with this instrument mounted on the 3.5-m telescopes at La Silla and Calar Alto Observatories:
Key projects were:

Eclipse spectrophotometry of cataclysmic variables (CV)

Flux calibration of spectroscopic data for Doppler tomography of CVs

Spectral eclipse mapping of accretion disks in CVs

Simultaneous optical and X-ray high-speed monitoring of flare stars

Phase-resolved spectrophotometry of the Crab pulsar

Light-echo mapping in X-ray binaries

Though still a powerful tool for the investigation of high-speed phenomena in astrophysics, MEKASPEK will presumably be replaced in the not-too-distant future by focal-reducer instruments such as, for example, CAFOSC. They will be equipped with star plates, grisms, and CCD detectors with fast readout facilities and high quantum efficiency covering a broad wavelength region. Adaptive optics may then even render obsolete the application of fibers for light scrambling and image slicing.

References

Barwig, H., Schoembs, R.,& Buckenmeyer, C. 1987, A&A, 175, 327
Kiesewetter, S. 1992, PhD Thesis, Universitäts-Sternwarte München
Mantel, K. H., 1993, PhD Thesis, Universitäts-Sternwarte München
Wolf, S., & Mantel, K. H., 1994, Astron. Posters of the 22nd General Assembly of the IAU, 15-27 August, Den Haag, edt. H.van Woerden, 99.

Overview of Fiber Instruments at ESO

Gerardo Avila

European Southern Observatory, Karl-Schwarzschild-Str. 2, 85748 Garching bei Muenchen, Germany

Abstract. ESO is developing a number of fiber projects for La Silla Observatory and for the well advanced VLT Project at Paranal Observatory. An overview of each of them is given in this paper.

1. Introduction

Two fiber links are being prepared to feed two high-resolution spectrographs for La Silla Observatory. The first one will update the old fiber link between the 3.6-m Telescope and the CES (*C*oudé *E*chelle *S*pectrograph). The second will feed the beam from the 1.52-m Telescope to FEROS (*F*iber-fed *E*xtended *R*ange *O*ptical *S*pectrograph). The latter has been specially designed to be fed by optical fibers.

The present VLT Instrumentation Program foresees 14 instruments to be installed at the Cassegrain and Nasmyth foci. The coherent combination of the four Unit Telescopes is not included. From these 14 instruments, six might use fiber optics to perform mainly multi-object spectroscopy and integral-field spectroscopy. The fibers will work not only in the visible region but also in the near infrared up 2.5 μm. The more relevant instrument regarding the use of fiber optics, will be the Wide Field Fiber Facility. This is a fiber positioner which will be a general-purpose facility to feed dedicated spectrographs. It is envisaged to link, in the first instance, the GIRAFFE, UVES, and, at some future stage, AUSTRALIS. These three instruments will work in multi-object spectroscopy mode. The VIMOS, NIRMOS, GIRAFFE, AUSTRALIS, and SINFONI spectrographs will use integral-field units. FEROS, VIMOS, SINFONI, and AUSTRALIS have been described in detail in these proceedings (see the contributions by Kaufer, Prieto et al., Tecza & Thatte, and Taylor).

2. Fiber Instruments for La Silla Observatory

2.1. 3.6 m–CES Fiber Link

The new instrumentation program for La Silla Observatory includes the refurbishment of the fiber link between the 3.6-m Telescope and the CES. A new Very Long Camera working at $f/12.5$ will allow a resolving power of 220 000 to be reached with a slit width of 128 μm (0.2 arcsec). We will use an image slicer to get this narrow slit. For the old fiber link we used the classical Bowen–Walraven design with excellent results for equivalent slits down to 220 μm. They were

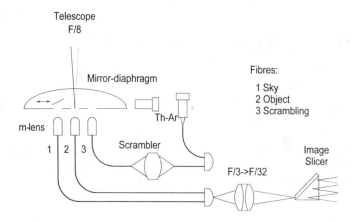

Figure 1. Schematic diagram of the link between the 3.6 meter and the CES.

coupled to fibers with 200 μm core diameter. The new image slicer will work with a narrowed slit and with a smaller fiber (105 μm), it will work at the limit of the performance given by the image quality of the jaws of the slit. It is as well know that this type of image slicer generates the slit inclined with respect to the light beam. Therefore, the image of the slit at the level of the detector is gradually defocused. For relatively wide and short slits, as was our case previously, this defocusing was not very relevant. However, in the new design, the slit will be thinner (128 μm) and longer. The object and sky fibers are close together and will be sliced one after the other by the same image slicer. In this case the defocusing will not be tolerable. To correct for this defocusing error, we had to tilt the CCD by about 13°.

We expect to use the new CeramOptec fiber which transmits in a broader spectral range. Moreover, this fiber will avoid the use of the two classical dry and wet fibers for the red and blue regions, respectively.

The scientific drivers for this configuration are: 1) accurate radial velocities (extra-solar planets), 2) chemical abundances and isotopic ratios, 3) the dynamics of stellar atmospheres, and 4) the interstellar medium.

Figure 1 shows the schematics of the fiber link. To reach the CES from the Cassegrain focus of the 3.6-m Telescope, 35 m of fiber are needed. The fibers will follow the standard Coudé path. A typical mirror-diaphragm will be used to check the positioning of the image of the star in front of the fiber input. The fiber input end is coupled to a rod-lens and is located just below the hole of the mirror-diaphragm. The light from the telescope will be reflected to the *slit viewer* of the Cassegrain station for image acquisition and guiding. When the image of the star falls into the hole of the mirror-diaphragm, the star will disappear from the viewer camera. The corona around the hole will be used to guide the telescope. A blue filter will be used in front of the viewer camera to correct the atmospheric dispersion when observing in the blue range. Indeed, due to atmospheric dispersion, the image of the star will be elongated and the filter will allow the blue portion of the star to be placed in front of the image diaphragm.

A flip mirror in front of the mirror-diaphragm will be used to feed the fibers with a calibration lamp (thorium) and with a halogen lamp for flat-fielding.

A micro-rod-lens will be used to minimize the focal-ratio degradation. It will reduce the $f/8$ beam of the Cassegrain focus to $f/3$ in the fiber. The lens have a focal distance of 0.84 mm and is made of BK7. The spherical surface has a radius of 0.41 mm and a length of 1.26 mm. The diameter of the rod is 0.82 mm.

The scrambled fiber will use a pupil-to-object exchanger relay lens. Further details maybe found in Avila (these proceedings, p. 44).

At the output end of the fibers, the exit $f/3$ beam will be enlarged to $f/32$ to match the CES collimator aperture. The enlarger optics consists of a field lens and a triplet. The glasses used are silica and fluorine to allow a wide spectral range. Table 1 gives the main parameters of the link.

Table 1. Main specifications of the 3.6 m–CES fiber link

Resolving power	80000 to 220000 (with image slicer)
Spectral range	390 to 1000 nm
Operational fibers	Object, sky, and scrambling
Fiber sky aperture	2 arcsec (180-μm hole in mirror-diaphragm)
Fiber diameter	105 μm
Fiber length	35 m
Image slicer	See Table 2
CCD	2688 × 512, 15 μm
Pixel matching	2.7 pixels (equivalent slit)
Fiber link efficiency	68 % (to be confirmed)
Limiting magnitude	12 mag (V) with S/N = 10 and a 2-h exposure (to be confirmed)
Date of operation	1998 May

Three classical Bowen–Walraven image slicers will be used to reach the required resolving powers. Table 2 summarizes the main characteristics of these image slicers. The thickness of the parallel plate is 0.817 mm and it is glued by molecular contact to the base prism at 45°.

Table 2. Image slicers

Resolving power	Slit width	No. of slices	Prism angle
80000	352 μm	3.2	72.25°
110000	256 μm	4.4	77.19°
220000	128 μm	8.8	83.64°

2.2. FEROS

This instrument has been described in detail by Kaufer (these proceedings). The main scientific drivers include: 1) extra-solar planets, 2) astero-seismology, and

3) variability of stellar atmospheres and envelopes. The main specifications are listed in Table 3.

Table 3. FEROS specifications

Resolving power	48000 (with image slicer)
Spectral range	370–860 nm
Operation fibers	Object, sky, simultaneous calibration, and polarization
Fiber sky aperture	2.7 arcsec (300 μm)
$f/\#$ in fiber	$f/4.6$
Image slicer	Modified Bowen–Walraven. 2 slices Slit width = 110 μm
Fiber link efficiency	70 %
CCD	2048 × 4096, 15 μm
Pixel matching	2 pixels (perpendicular to the dispersion)
Limiting magnitude (1.52-m Tel.)	16 mag (V), S/N = 10, 2 h
Date of operation	Early 1999

The way to inject the beam into the fiber is basically the same as for the 3.6 m–CES link: there is a mirror-diaphragm for acquisition and guiding and a rod lens to reduce the $f/15$ telescope beam into a $f/4.6$ in the fiber. The microlens is made of UBK7 with a focal distance of 1.35 mm, spherical surface of radius 0.7 mm, a length of 2 mm, and a rod diameter of 1.1 mm.

The output beam is enlarged from $f/4.6$ to $f/11$ by an optical relay lens. The output of the object and sky fibers are not parallel in front of the field lens, but converge to correct for the pupil distortion at the Echelle level.

3. Fiber instruments for the VLT

3.1. Wide-Field Fiber Facility

The development of this instrument has been rather difficult. It started originally as the FUEGOS project to perform multi-object and integral-field spectroscopy. At present it has been split in two independent instruments, the first being a fiber positioner which will act as a general facility for the VLT to feed diverse dedicated spectrographs. The second instrument will be a medium-resolution spectrograph. The scientific case for the spectrograph including the multi-object and integral-field spectroscopy modes remains more than competitive with respect to similar instruments:

1) Stellar rotation vs. stellar age

2) Stellar system dynamics and structure,

3) Stellar abundances and anomalies in clusters and nearby galaxies

4) The fundamental plane of dwarfs

5) Rotation curves of disks from $z = 0$ to 1.5

6) The dynamics and chemistry of star-forming galaxies

The final specifications of both instruments are not fully defined but the highlights are summarized in Table 4.

Table 4. Specifications of the WF^3

Resolving power	7500 and 15000
Spectral range	370–1000 nm
Operation modes	Multi-object spectroscopy, large integral-field spectroscopy and set of IFU's
Operational fibers	Object, sky, and simultaneous calibration for both operational modes
FOV, MOS fiber sky aperture	21 arcmin and 1.2 arcsec
Large IFU FOV, sampling	11 × 8 arcsec and 0.26 arcsec
IFU's FOV, sampling	5 × 5 arcsec and 0.6 arcsec
Fiber diameter	230 μm MOS and IFU's ; 50 μm large IFU
Fiber length	< 10 m
CCD	2k × 4k, 15 μm
Fiber link efficiency	MOS: 70 % , large IFU and IFU's : 65 %
Date of operation	Early 2001

3.2. Fiber Positioner to UVES Fiber Link

A very valuable and inexpensive configuration will be the feeding of UVES with eight fibers from the Fiber Positioner General Facility. It will allow UVES to work in multi-object spectroscopy mode. The scientific drivers are ambitious:

1. Extra-solar planets

2. Globular-cluster chemistry

3. Young open clusters

4. The dynamics of star clusters

5. Dwarf spherical galaxies

6. QSO absorption lines

7. High-resolution stellar spectroscopy in medium-density fields

A large number of scientific programs originally allocated to FUEGOS at 40000 resolving power will be carried out with this link. Table 5 shows the main specifications.

The strengths of this link are: 1) UVES working in multi-object spectroscopy mode, 2) the Wide Field Fiber Facility and UVES will work in simultaneous observation (a unique VLT capability), and 3) the high-resolution mode (40000) of FUEGOS is partially recovered.

Table 5. Specifications for the Fiber Positioner to UVES link

Resolving power	40000
Spectral range	420–1100 nm
Operational fibers	Object, sky, and simultaneous
Fiber sky aperture	1 arcsec
Fiber diameter	200 μm
Fiber length	$\approx$ 33 m
CCD	2 × 2k × 4k, 15 μm
Fiber link efficiency	62 %
Date of operation	Early 2001

3.3. VIRMOS

VIRMOS is in fact composed by three instruments: VIMOS, NIRMOS, and the MMU. VIMOS (*VI*sual *M*ulti-*O*bject *S*pectrograph) and NIRMOS (*N*ear *I*nfra*R*ed *M*ulti *O*bject *S*pectrograph) will perform MOS, direct imaging and will include integral-field units. The MMU is the *M*ask *M*anufacturing *U*nit, which will make the slitless masks for VIMOS and NIRMOS. The two main scientific goals are ultra-deep surveys and integral-field spectroscopy of protogalaxies. The main parameters regarding the Integral-Field Unit are described in Table 6.

Table 6. Specifications for the Integral-Field Unit of VIMOS

Resolving power	2500
Spectral range	375–1000 nm
Operational fibers	Area spectroscopy, sky
Field of view	1 × 1 arcmin
	24 × 24 arcsec
Sampling	0.75 arcsec
	0.3 arcsec
Fiber diameter	100 μm
Fiber length	$\approx$ 3 m
CCD	2k × 4k, 15 μm
Fiber link efficiency	70 %
Date of operation	Early 2001

The coupling of the telescope beam to the fibers will be done with an array of microlenses. This mask will be made with two crossed arrays of cylindrical lenses. This innovative design is much less costly than the classical array made of spherical lenses arranged in a honeycomb configuration. It is interesting to note that this fiber bundle will use around 15 km of fiber.

3.4. SINFONI

This proposed instrument would be coupled to the NAOS facility (*N*asmyth *A*daptive *O*ptics *S*ystem). The main objective is to perform integral-field spectroscopy of deep fields. The scientific drivers are:

1) Detection/study of quasar fuzz, and distant elliptical primeval galaxies

2) Analysis of distant radio/interacting galaxies

3) Central regions of moderate-z galaxies

4) Regions of stellar formation

5) The Galactic center

6) Surfaces of planets/satellites

The integral-field unit of this instrument will work with fiber optics coupled to an array of microlenses. SINFONI will work in the near infrared up 2.5μm and, therefore, all the units will be cooled down to 80 K. To overcome the mechanical coupling between the array of lenses and the fibers at this temperature, an integrated unit has been designed. Each fiber of 50-μm core diameter will be tapered and fused to the spherical lens. Each lens has a hexagonal section to allow the array to be assembled with the optimum packing ratio. The main parameters of the integral unit are summarized in Table 7.

Table 7. Specifications for the Integral-Field Unit of SINFONI

Adaptive optics	Diffraction limited from 1.5 μm
Resolving power	2000–4500
Spectral range	1.4–2.5 μm
Operational fibers	Area spectroscopy, sky
Field of view (diameter)	1.5 to 9 arcsec
Sampling	0.04 to 0.25 arcsec
Fiber diameter	50 μm
Fiber length	$\approx$ 10 cm
Detector array	1k × 4k, 15 μm
Fiber link efficiency	75 %
Date of operation	End 2000

3.5. Study of an SM Fiber Link for an AO Guide Star

ESO is exploring options to launch the high-power laser beam into the atmosphere and generate artificial stars for the adaptive-optics facility (Bonachini et al. 1998). Three solutions are being investigated: 1) a *standard* mirror train along the coudé path of the telescope, 2) a single-mode fiber, where the laser *is* the fiber itself (the available power would exceed 10 W cw; this option is in the development phase with Litecycles Inc.), and 3) to launch the laser beam into a single mode (SM) fiber to substitute the mirror train of item one.

The third possibility is being studied in collaboration with the Max Planck Institute for Extraterrestrial Physics in Garching. Due to the reduced size of the required artificial star at 50 km from the telescope and the stability of the point spread function, a single-mode fiber must be used. The main problems encountered in achieving the highest coupling efficiency are the procurement of the following elements: 1) diffraction-limited optics to inject the laser beam (coating at 589 nm) 2) sub-micron and high-stability positioning stage to place the fiber input end on the laser waist, and 3) high pure silica fiber to reduce the stimulated Brillouin backscattering (the required length is around 35 m)

Some standard elements may be used to properly launch a laser beam into an SM fiber such as microscope objectives and grin-lenses, but these are not entirely suited to our application. A lens with a focal distance of between 7 and 10 mm is needed. In this range, the most appropriate standard objective would be a 20 × (8 mm focal distance). Such a piece is usually made with several lenses, so the efficiency is not very high (around 70 % at 590 nm) and shows a residual of spherical aberration. Moreover, at high power densities the coating is damaged and, due to the differential glass absorption, the focal distance is very sensitive to the temperature. Graded-index lenses (GRINs) of 0.23 pitch may be a good solution but the choice is very limited and we could not find these lenses with the required focal distance. We have tested all these components with fibers 3 m long and the results are given in Table 8. We are still exploring two further options, the first is the procurement of aspheric lenses specially designed to couple laser beams into fibers. Two of them (8- and 10-mm focal distance) will be tested just after the deadline to submit this paper. The second solution will be the manufacturing of a tapered fiber coupled to a commercial bi-convex lens. The tapered SM fiber would be made by the same manufacturer who is developing the SINFONI fibers. A prototype without the exact specifications has been produced showing the feasibility of tapering an SM fiber and finishing the end with a spherical surface.

Table 8. Coupling efficiency of a 3 W cw laser beam into an SM fiber

Fiber	PCX	10×	Lens 20×	CVI	Grin
7.5μm	40 %	50 %	82 %	65 %	
6	-	70	60	-	-
4.5	-	-	40	-	
Problems	Spherical ab.	Coating damaged	Aberration		bad f

The second problem is the positional accuracy and stability of the fiber in front of the laser waist produced by the lens. Most commercial, standard fiber-positioning stages are unsuitable for our purposes. For our tests, we used one with a differential micrometer, which allowed the fiber to be positioned with a sensitivity better than 2μm. The best option will be to use the sub-micron stages controlled with piezo-electric devices.

4. Conclusion

We have briefly described the instruments under construction and proposed using fiber optics at ESO. Six of them will be integrated into the different telescopes of the VLT. This number represents 40 % of all the instruments of the VLT. This figure indicates that fibers are now a valuable tool for astronomical instrumentation.

References

Bonachini, D., Avila,G., & Hackenberg, W. 1998, Proc. SPIE (in preparation)

A Two-Beam Two-Slice Image Slicer for Fiber-Linked Spectrographs

A. Kaufer

Landessternwarte Heidelberg, Königstuhl 12, D-69117 Heidelberg, Germany

Abstract. A two-beam two-slice image slicer was designed for the fiber-linked high-resolution echelle spectrograph FEROS. It is used at the intermediate focus produced by the fiber-exit focal enlarger which converts the $f/4.5$ beams of the fibers into $f/11$ beams accepted by the spectrograph. The image slicer is based on the classical Bowen–Walraven design but merges two individual slicers into one device optimized to slice the output beams of two fibers (object + sky) simultaneously into two slices. This approach minimizes the defocusing of the sliced images. Air grooves are used to create the internal total reflections; the width of the grooves controls the number of sliced images to avoid light contamination outside the images produced by the two slicers. This is found to be particularly important for image slicers used in echelle spectrographs, which are sensitive to inter-order stray light. In this contribution the very general design of this image slicer is presented in detail. The first measurements obtained with a prototype manufactured by H. Kaufmann Precision Optics already demonstrate the functionality and high efficiency of the device.

1. Introduction

The modified Bowen–Walraven image slicer which is subject of this contribution was developed for the *F*iber-fed *E*xtended *R*ange *O*ptical *S*pectrograph (FEROS) currently under construction for the ESO 1.52-m Telescope at La Silla, Chile. FEROS is a state-of-the-art fiber-fed bench-mounted prism cross-dispersed echelle spectrograph working in quasi-Littrow mode and white-pupil configuration. With two fibers for the object and the nearby sky, the complete optical spectrum from 360 to 860 nm is recorded in one exposure with a resolving power $R = 48000$. For a more detailed description of the FEROS instrument see Kaufer et al. (1997).

The high resolving power of the FEROS instrument depends critically on the image slicer (IS) which is used to slice simultaneously the images produced of the two blank exit surfaces of the object and sky fibers by a focal-enlarger lens system (F/N system).

The fibers have a core diameter of 100 μm and are fed by microlenses on the telescope's side (cf. Fig. 1). Two apertures with 0.29-mm diameter ($= 2.7''$ in the $f/15$ focal plane of the ESO 1.52-m Telescope) are imaged onto 90 % of the fibers' entrance-surface diameter resulting in an effective $f/4.6$ feed which is

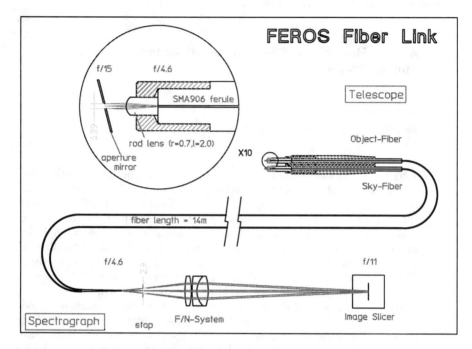

Figure 1. The FEROS fiber link (for details see text).

well suited to the minimization of focal-ratio degradation (FRD) effects on the fiber link. The microlens is a rod lens with a radius of curvature of 0.7 mm and a length of 2 mm and is directly glued with the flat surface to the fiber-entrance surface. The microlens and the fiber are mechanically mounted in a modified SMA 906 connector.

The polished fiber exits are left blank and are re-imaged by the F/N system which converts the $f/4.6$ fiber beams to $f/11$ beams accepted by the spectrograph. The F/N system produces images of the fibers enlarged to 240 μm at the intermediate focus. Therefore, the image slicer described below is located at this intermediate focus and defines the "entrance slit" of the instrument.

The fiber coupling efficiency for an input/output focal ratio of $f/4.6$ is expected to be better than 85 % for a fiber with a small FRD. The total efficiency of the complete fiber link including the microlens, 14 m of fiber, the F/N system, and the AR-coated IS is estimated at 60 %.

2. Opto–Mechanical Layout of the Image Slicer

The FEROS IS is basically a Bowen–Walraven image slicer (Walraven & Walraven 1972) which was modified to slice the two beams emerging from the object and sky fibers simultaneously with a minimum of defocusing introduced by the optical path differences (OPDs) in the slicer. Therefore, an image slicer which merges two individual slicers into one needed to be placed at the intermediate $f/11$ focus feeding the FEROS spectrograph. This requirement is met by us-

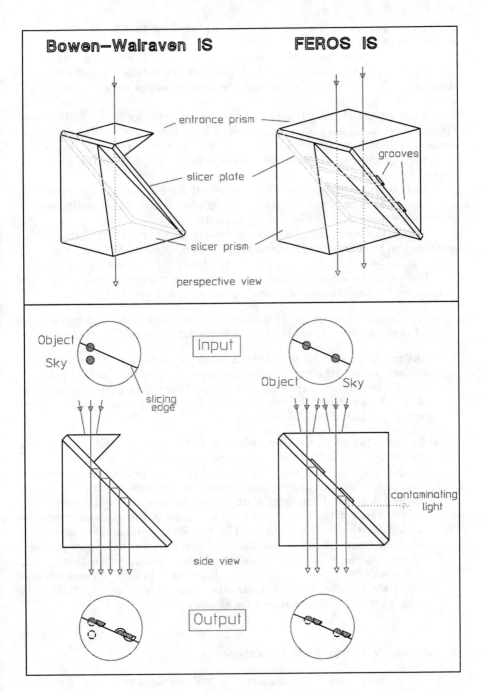

Figure 2. Layout of the classical and the modified Bowen–Walraven IS (for details see text).

ing *two air grooves* on the entrance prism of the slicer that provide the needed internal reflection of the sliced beams.

Figure 2 shows the layout and function of this IS in comparison with a classical Bowen–Walraven IS; Table 1 gives all the details for computing the dimensions for this type of slicer and the specific values used for the FEROS image slicer.

Quartz glass (Homosil) was selected as material for the FEROS slicer. Quartz glass has a very high transmission over the whole optical-wavelength range and good mechanical properties, which allow all surfaces to be polished with high quality—a prerequisite for assembling the parts of the slicer with optical contact only.

An IS completely manufactured from quartz glass in optical contact could be easily broad-band anti-reflection (AR) coated after assembly and testing because of the low expansion coefficient of the material.

Due to the low refractive index of quartz glass, the FEROS slicer prism angle had to be changed from the commonly used 45° to 46.5° to maintain total reflection up to wavelengths of 1 μm for the two $f/11$ beams entering the IS.

The use of two air grooves in the entrance prism results in several additional advantages besides the goal of achieving two slicers in one device:

- The manufacture of the IS is simplified because it then basically consists of two complementary prisms with large flat surfaces.

- the precisely polished flat surfaces allow all surfaces to be connected by optical contact, thereby avoiding any cementing of the parts. Cementing usually reduces the important sharpness of the slicing edge. Further, even a little filling of the air grooves in the entrance prism with cement would result in a loss of total reflection.

- On the other hand, the total reflection in the entrance prism can by controlled by the width of the air grooves, i.e. the maximum number of slices produced by the slicer can be controlled. Light entering a new, unwanted slice (e.g. due to a slight misalignment of the base prism or tolerances in the thickness of the slicer plate) is not directed to the direction of the output beam if the air groove does not provide the glass–air transition needed for the total reflection. The unwanted light is directed perpendicularly out of the slicer and does not contaminate the images produced by the two slicers (cf. Fig. 2, bottom right: "contaminating light"). This is particularly important for image slicers used in echelle spectrographs, which are sensitive to inter-order stray light. Light from a further slice could even affect and contaminate adjacent echelle orders.

3. Results with the First Prototypes

A model of the proposed image slicer was built during the design phase from acrylic with the linear dimensions scaled by a factor of 15. The working principles of the slicer were checked on the optical bench with the acrylic slicer parts mounted together with index-matching oil.

Table 1. Dimensions of the modified Bowen–Walraven IS (all lengths are in mm, and all angles in degrees)

	Formula	FEROS IS
Image-slicer material	quartz glass	Homosil
Refractive index	$n(1\ \mu m)$	1.45
Total-reflection angle	$\alpha_T(1\ \mu m)$	43.6
Entrance-beam focal ratio:	f/d	11
Focal ratio in IS:	$(f/d)' \approx n(f/d)$	16
Max. angle in IS:	$\alpha_T + \arctan(0.5(d/f)')$	45.4
Chosen reflection angle:	α	46.5
fiber image diameter:	D	0.240
→ effective "slit" width:	$w = D/2$	0.120
Transfer distance of sliced image:	t	$1.25D = 0.300$
→ effective "slit" length:	$s = t + D$	0.540
Thickness of slicer plate:	$d = t\frac{\sqrt{2}}{(1+\tan\alpha)}$	0.207
Slice angle:	$\beta = \arccos\left(\frac{D}{2t}\right)$	66.4
Slice angle on slicer prism:	$\beta' = \arctan\left(\frac{\tan\beta}{\cos\alpha}\right)$	73.26
Distance of the two beams:	a_b	0.90
Distance of grooves on entrance prism:	$a_a = \frac{a_b \sin\beta}{\cos\alpha}$	1.2
Width of grooves in slicer plane:	W_a	0.50–0.55
Width of grooves on entrance prism:	$W_a' = W_a \sin\beta$	0.46–0.50 (0.48)
Depth of grooves:	$\delta \geq 10\lambda$	0.025

The first prototypes, of Homosil quartz glass, were manufactured by H. Kaufmann Precision Optics in Crailsheim, Germany. The outer slicer dimensions are $20 \times 20 \times 20$ mm to ease the polishing of the individual surfaces, resulting in lowered tolerances for the prism angles and the thickness of the slicer plate.

Figure 3 shows the CCD images taken with a microscope objective from the unsliced and sliced double-fiber images. The two slices produced from the circular fiber image with 240-μm diameter form an effective 120×540-μm entrance slit for the FEROS spectrograph. The images were taken in the Gunn Z filter to prove the total reflection conditions up to a wavelength of 860 nm. The slicing efficiency was measured as the ratio of the intensities of the individual unsliced and the sliced fiber images neglecting the internal absorption of the quartz glass and the glass–air losses on the entrance and exit surfaces of the slicer. For both fiber images a value of 95 % is found. Therefore, using broad-band AR coatings the total efficiency of this type of slicer should be better than 90 %.

4. Discussion

The modified Bowen–Walraven image slicer introduced here was specifically developed to slice simultaneously the two beams emerging from the object and sky fibers in the intermediate focus available in most fiber-linked spectrographs. At this intermediate focus behind the fiber, the seeing noise and guiding usually introduced by image slicers in the telescope's focal plane are eliminated.

In principle, two fibers can be sliced as one image fed to *one* classical Bowen–Walraven IS. The drawback of this approach is that the defocusing introduced

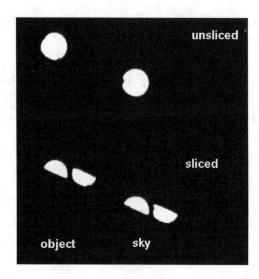

Figure 3.   Images obtained with the first image-slicer prototype.

by the slicer increases for each slice due to the increasing OPD and results in two sliced fiber images with different amounts of defocusing for each individual slice. Therefore, different effective slit widths and resolving powers of the spectrograph are obtained for the two fibers. The modified Bowen–Walraven image slicer basically merges two individual slicers in one device, which gives a minimum and for both fibers an identical amount of defocusing of the slices. For an illustration of the two situations see Fig. 2; note especially the different blur of the output fiber images for the two different slicers.

The use of two air grooves on the entrance prism eases the manufacture of the IS and allows the complete device to be built from quartz glass in optical contact. The resulting mechanically and temperature insensitive slicer can be easily broad-band AR coated.

Measurements with the first prototype show that a highly efficient image slicer with a total efficiency of better than 90 % can be obtained.

Acknowledgments.   The author thanks W. Seifert for long discussions on the new IS design, L. Schäffner for manufacturing the acrylic model in the mechanical workshop, J. Andersen for his suggestion to use an IS for FEROS, and H. Kaufmann for his skills and his dedication to the manufacturing. This work was supported by the DFG (Ap 19/6-1,2, Ka 1421/1-1).

References

Kaufer, A., Wolf, B., Andersen, J., & Pasquini L. 1997, Messenger, 89, 1
Walraven Th., & Walraven, J. H. 1972, Proc. ESO/CERN Conf. on Auxiliary Instrumentation for Large Telescopes, eds. S. Lausten, & A. Reiz. (Geneva: ESO), 175

A Fiber-Linked Four Stokes-Parameter Polarimeter for the SOFIN Spectrometer on the Nordic Optical Telescope

Bertil Pettersson, Eric Stempels, and Nikolai Piskunov

Uppsala Astronomical Observatory, Box 515, S-751 20 Uppsala, Sweden

Abstract. We are building a new, improved version of the Bill Wehlau polarimeter, originally designed and constructed by N. Piskunov for the CFHT. We intend to use the new polarimeter at the NOT, and it should be ready in 1999. The instrument will allow us to measure all four Stokes parameters in three exposures. The polarimeter is designed to be free from instrumental polarization and will feed the light via two fibers to the high-resolution SOFIN spectrometer, thus improving its performance and stability. We include some new solutions related to the orientation of polarizing components and the fiber coupling.

1. Stellar Spectroscopy in Polarized Light

The polarization status of radiation can be characterized by four Stokes parameters: I, the total intensity, Q, the difference between intensities of linearly polarized light at 0° and 90°, U, the difference between intensities of linearly polarized light at 45° and 135°, and V, the difference between intensities of right- and left-circularly polarized light.

Two different approaches can be taken while registering a polarized signal. If the source is bright and the detector supports fast read-out mode one can send the light through a periodic modulator that will extract different polarizations in different parts of the circle. The detector must be synchronized with the modulator. The main advantage of this technique is that all Stokes parameters are registered quasi-simultaneously by exactly the same parts of the detector. The most advanced implementation of such a design is the ZIMPOL instrument for solar polarization measurements (Keller et al. 1992).

In case of a faint target all the Stokes parameters can be obtained in three consecutive exposures if two polarizations can be recorded simultaneously (two circular polarizations in one, and two linear polarizations in each of the others). If a CCD is used as detector, the maximum accuracy would be limited by the quality of flat-fielding unless the image is projected onto exactly the same pixels in every exposure. This can be achieved by using a fiber feed with one end fixed relative to the detector. In fact, in this configuration one can totally eliminate the flat-fielding by splitting each exposure in two and switching the two fibers (Semel et al. 1993). Such a design is typically used for stellar-polarization spectroscopy where small fluxes and slow detector readout prevent the use of a modulator. The polarimeter described here belongs to this type.

The main component of such a polarimeter is the beam-splitter that separates the light, linearly polarized in two perpendicular planes. Therefore, to measure circular polarization it must be converted to linear with appropriate orientation relative to the beam-splitter. This is done with a quarter-wavelength plate (QWP) that adds a phase shift of 90° to the component of electromagnetic field parallel to its axis. Since the circular polarization can be described as a superposition of two linear waves with a phase shift of 90°, the QWP converts a circularly polarized beam to a linearly polarized one. The axis of the QWP must be rotated by 45° relative to the axis of the beam-splitter. Circularly polarized light will come out of such a combination as a single linearly polarized beam. If we now turn the QWP by 90°, the same circular polarization will come out of the beam-splitter as the second beam. When two circular polarizations are present in the input beam, the output will consist of two linearly polarized beams with intensities proportional to the left and right polarized components. Any additional linear polarization can be considered as a superposition of the two opposite circular polarizations and will cancel out in the resulting V value.

The measurement of linear polarization requires a half-wavelength plate. The Q Stokes parameter is recorded by orienting the axis of the half-wavelength plate parallel to the axis of the beam-splitter while for measuring U it must be rotated by 22.5°. As before, any additional circular polarization will be equally distributed between the two output beams and will not influence the result.

2. New Polarimeter for the NOT

The new polarimeter under construction at Uppsala Astronomical Observatory is a Cassegrain-focus instrument that feeds two polarizations to the spectrometer using optical fibers. We aim at using the polarimeter with the 2.6-m NOT on La Palma with the high-resolution SOFIN spectrometer although we will try to make it as portable as possible. SOFIN (Ilyin 1993) is a cross-dispersed echelle instrument with standard resolutions of 180000 and 80000. The optimal slit sizes are 45 μ and 100 μ. The optical design leaves enough space between echelle orders to record the second polarization spectrum.

SOFIN is normally mounted at the Cassegrain focus of the NOT. When used with the polarimeter, SOFIN will remain on the dome floor. The use of a fiber feed will also help to eliminate the flexure problem of SOFIN.

The new polarimeter will consist of the following parts: an entrance pinhole diaphragm, a lens collimator, two QWPs with rotation mechanisms, a Wollaston prism (beam-splitter), a camera, a fiber adapter, and the fibers. Two optional elements are: a Glan–Taylor polarizer with an additional QWP (used for calibrations only) and a fourth QWP that can be used to convert linear polarization to circular after the beam-splitter if the fibers show strong sensitivity to the polarization of the beam. For portability, the collimator can be adjusted to accommodate different focal ratios.

The two super-achromatic QWPs can be combined to form one quarter-wavelength retarder used for measuring circular polarization or one half-wavelength retarder for linear polarization. This is achieved by rotating the QWPs around the optical axis. The rotation of each QWP is performed with a step size of <0.1° and is computer controlled.

The light from the QWPs passes through the Wollaston prism and splits into two orthogonally polarized beams. The camera images the entrance diaphragm on the fibers constructing a $f/5.5$ beam in order to prevent any significant degradation of the focal ratio. On the spectrometer end focal enlargers (one per fiber) bring the focal ratio in agreement with the spectrometer ($f/11$ for SOFIN). Two pairs of fibers, 25 μ and 50 μ, are planned for use with the high- and medium-resolution cameras of SOFIN.

A typical observing sequence includes three pairs of exposures to register the V, Q, and U Stokes parameters. For the first two the axis of the second QWP (the one next to the beam-splitter) is aligned with the axis of the Wollaston prism. We register circular polarization with the first QWP rotated $\pm 45°$ from the second QWP. The operation is repeated with the two QWPs forming a half-wavelength plate with the axes at $0°$ and $90°$ for Q, and at $\pm 22.5°$ for U. The values of Stokes parameters are then found from the relations:

$$\left(\frac{I - V/2}{I + V/2}\right)^2 = \frac{S_1^-}{S_1^+} \cdot \frac{S_2^+}{S_2^-}$$

$$\left(\frac{I - Q/2}{I + Q/2}\right)^2 = \frac{S_1^-}{S_1^+} \cdot \frac{S_2^+}{S_2^-}$$

$$\left(\frac{I - U/2}{I + U/2}\right)^2 = \frac{S_1^-}{S_1^+} \cdot \frac{S_2^+}{S_2^-}$$

where $S_j^\pm$ is the signal per pixel of the CCD registered in the first or second half-exposure (indicated by $+$ or $-$) through fiber j. The ratios on the right-hand side do not depend on the sensitivity of the individual pixels. Hence the degree of polarization, V/I, Q/I, and U/I, can be determined without flat-fielding. A flat-field is still needed to obtain the absolute values of Stokes parameters.

3. Calibration of the Polarimeter

During calibration, the polarimeter will have an additional calibration element that consists of a Glan–Taylor polarizer and a test QWP. It is inserted into the beam just before the first QWP. The light coming through this element becomes circularly polarized. We assume knowledge of the positions of the two standard QWPs to within a few degrees. The problem is to use the polarized light to locate the axes precisely so as to eliminate the cross-talk between the Stokes parameters when the observations are made.

The axes can be precisely located if we can measure the flux in each beam coming from the Wollaston prism. The procedure is based on the fact that the intensities in two fibers is identical for QWP2 = $\pm 45°$. This is independent of the rotation of QWP1 because the output from QWP2 is elliptically or linearly polarized with an axis at $45°$, or circularly polarized, so the two beams from the Wollaston are equal, regardless of the position of QWP1.

1. Set QWP1 to (approximately) $0°$ ($\pm 90°$ will do as well)

2. Set QWP2 to $45°$ ($-45°$ will do just as well). Rotate QWP2 until the difference of the output signals from the two fibers is zero. That happens

when then angle between the axis of the Wollaston prism and QWP2 is precisely 45°.

3. Set QWP2 to precisely 0° (±90° will do just as well). Set QWP1 to (approximately) 0° (±90° will do just as well) and adjust the angle in order to get identical output in two beams. Now the axis of QWP1 is precisely at 0°. Note that the greatest sensitivity to the position of QWP1 occurs when QWP2 is located at 0° or ±90° (i.e. farthest from 45°).

4. Fibers

The fibers should be chosen with care concerning both the spectral transmission properties, damping properties, and the reaction to focal-ratio degradation (FRD). Several manufacturers, produce off-the-shelf fibers with suitable core dimensions, diameters of 50 μ and 100 μ. They also market fibers that have a good spectral coverage with good to excellent transmission in the region 300–1000 nm (high OH content) or 500–2500 nm (low OH content). For observations longwards of ~700 nm the latter fiber should be preferred, at the expense of the UV region, to avoid the strong absorption features produced by the high content of OH necessary for the good UV transmission. This consideration remains necessary until broad-band fibers become available.

The focal-ratio degradation is very dependent on how the fiber ends are mounted. Extreme care must be exercised to prevent excessive stressing here, as this will result in severe FRD. As a rule input f-numbers larger than about 6 will inevitably produce FRD, however staying at $f/5.5$ and transforming the resulting beam to the required $f/11$ should not present a problem. For a discussion of FRD, see, for example, Carrasco & Parry (1993).

The polarization properties of the fibers have to be carefully studied to ensure that different polarization modes suffer the same FRD and intensity losses, as this is important for the beam-switching technique through the fiber pair. Experiences from an earlier version of the polarimeter, built by Piskunov and collaborators have shown that there may be a problem with different amounts of FRD through one and the same fiber, depending on the direction of polarization of the entering light beam. This will cause the illumination of the pixel elements to differ in the different polarization modes and will jeopardize the flat-fielding accomplished through the beam switching. We intend to conduct a series of tests on the fibers we eventually decide to use to clarify this problem.

References

Carrasco, E., & Parry, I. R. 1993, in ASP Conf. Ser. Vol. 37, Fiber Optics in Astronomy II, ed. P. M. Gray (San Francisco: ASP), 392

Ilyin, I. 1993, High resolution échelle spectrograph SOFIN, User's Manual, Helsinki Observatory

Keller, C. U., Aebersold, F., Egger, U., Povel, H. P., Steiner, P., & Stenflo, J. O. 1992, LEST Foundation Technical report No. 53, Univ. Oslo

Semel, M., Donati, J.-F., & Rees, D. F. 1993, A&A, 278, 231

Part 7: Overview

Conference Overview

Paul Felenbok
Observatoire de Paris-Meudon, 5 Place Jules Janssen, 92195 Meudon, France

1. Introduction

Fiber use in astronomical spectroscopy is expanding very rapidly and this conference has given only a modest view of the situation. Many people involved in major programs in this field did not attend this conference, probably thinking that all the problems related to fiber use are now solved, and that no special event should be devoted to this subject. This is an illusion, because it is the nature of researchers always to push back further the limits of observation, and we are now faced with new challenges.

2. New Advances in Fiber Technology and Fiber Testing

Session 1, with five oral contributions on new advances in fiber technology and testing, well illustrates the fact that we have still a lot to learn. Some items, such as FRD, are well mastered, but we are still not at the ultimate sky subtraction level, and my present conviction is that we will very soon be at the slit or mask spectrograph performance, and fibers will be used for detection of very faint objects. Pushing performance is revealing new behavior, especially at high signal-to-noise level, not investigated previously. We have also seen at this conference some new commercial fibers with better performances in the blue and a shallower deep in the red. We learned also that if we enter the mid-IR domain, we have seriously to increase the cladding thickness, up to 20 μm, in order not to allow transmission to fall to unacceptable values. Fiber scrambling properties are used to increase spectrograph stability, mainly with respect to guiding jitter. This capacity is used in the search for extra-solar planets by the radial-velocity variation method. We have seen that radial scrambling performance is input focal-ratio dependent and if we are looking for 99% quality we should not enter the fiber with a more open beam than $f/5$. This is an important result because we are usually pushing in the direction of opening the beam to diminish the FRD. We see that this has a limit in certain types of applications. To improve the scrambling capacity of fibers, double scramblers are built by coupling lenses and fibers and inverting the near and far field. But this design, which works well, consumes a lot of photons and solid scramblers using tapered fibers are now under development.

3. Session 2: Multi-Object Spectroscopy

In Session II on multi-object spectroscopy, six contributions were presented illustrating the advances in this field, but no really new instruments are in view. This is perhaps due to the fact that most 4-m class telescopes are entering production and the 6.5-m HECTOSPEC instrument is now in its testing phase. 2dF is starting production with an impressive goal of recording the redshift of a quarter of a million galaxies to $z = 0.3$, 20 000 QSOs, and a program completion for 2000.

Sloan, will go as deep as $r < 18$, and the goal is to record a million galaxies. The WHT, with Autofib II, is starting to produce science. In the southern hemisphere, we did not have the opportunity to hear about Magellan. The 6dF project on UKS has a goal of collecting 100 000 galaxies in 3 years, but it is still looking for funding. My guess is that at the next *Fiber Optics in Astronomy* conference there will be an overflow of data on galaxy redshifts and the organizers would be wise to allocate sufficient time for the presentation of scientific results.

4. Session 3: Two-Dimensional Fiber Spectroscopy

This session, with eight contributions on two-dimensional fiber spectroscopy, is the densest, and if we compare it with the three contributions at the FOA II conference, we could say that this field is exploding. This effect is mainly due to the fact that in two-dimensional spectroscopy there is no real competition between integral-field and long-slit instruments. This is especially true when high spatial resolution is involved and one does not know where to place a long slit on the object to record the most important features that will be only discovered when the spectrum has been analyzed. There was a competition at CFHT between fiber and lens integral-field units (IFUs), each system having its advantages and drawbacks. With the advent of large telescopes and technological progress in the manufacture of microlenses, this competition is over because all designs will couple fibers and microlenses and enjoy the advantages of both systems. Integral-field units are mounted on most 4-m class telescopes and are now producing science. At CFHT, MOS-Argus has been in full operation for 4 years and is used by visiting astronomers without the help of the building team. INTEGRAL, on the WHT, was used to observe M 31 and the Einstein Cross, and the first results are coming from SPIRAL on the AAT and PMAS at Calar Alto. Evolution in this field is seen to be in two major directions.
- IFU for adaptive-optics spectrographs. On 4-m class telescopes, with a spatial resolution of 0."1, it will be possible to observe only bright objects with emission lines. This kind of observation will become the favorite of 8–10-m class telescopes, especially in the red.
- Multi-IFUs. After multi-object spectroscopy without spatial resolution, people would like to go one step further and analyze extended objects in 2-D. This will be the case for the study of galaxy dynamics in clusters with a full production on large telescopes. On the ground, this will not benefit from adaptive optics because the field corrected will be too small. It would be an ideal instrument for a next-generation space telescope.

5. Session 4: Projects for Large Telescopes

The next session, with six contributions on projects for large telescopes, shows that large telescopes are also in the fiber landscape, but mostly for the integral-field mode. This is the case for GMOS on Gemini, VIMOS on the VLT and at a later stage on Subaru. There are two exceptions: HET with single- and multi-fiber coupling, and FUEGOS at high spectral resolution in Medusa and integral-field mode. I shall recall that a very prestigious member of this family is missing—that of the two Keck telescopes.

6. Session V: Multi-object and 2-D Infrared Fiber Spectroscopy

Session 5, with six contributions on multi-object and two-dimensional infrared fiber spectroscopy, shows that fiber spectroscopy has definitely entered the infrared domain. I was impressed how the IR niche for the Australis project was analyzed. To take as a target objects with a redshift $1<z<2$ is really exciting, and to reach galaxies at $J = 23$ mag with a spectral resolution of $R = 3500$ seems to be at the limit, but still feasible. The goal to reach is a spectral resolution of 10000; an aspect of its intrinsic scientific interest is also a straightforward option of OH suppression by software treatment. The SINFONI approach is also a very innovative challenge, and it looks as though a lot of work has still to be done. As far as the VLT is concerned, this instrument is still at the proposal stage, and at the present time was not approved by the various ESO committees. We also sow the first proposal to use integral-field coupling, in the near IR, with adaptive-optics images. This development will expand with the imminent large telescopes because increasing spatial and spectral resolutions will lead to a great shortage of photons. In this domain, coupling fibers and microlenses is crucial because a thick cladding is needed for a good IR transmission. If I recall that the Dallier paper was the only one in this field presented at FOA II, that shows what a long path has been trodden since then.

7. Session 6: Other Applications

With five contributions, the last session, devoted to other applications of fiber optics, demonstrates that fibers are spreading beyond their traditional domain, if we can really speak of tradition in a so young a field. In astronomical interferometry our colleagues are going ahead with telescope coupling by monomode optical fibers and the first scientific results are imminent. Resolutions of 10 mas (milliarcsec) are achieved with 45-cm telescopes. Multicolor photometry still uses fibers, but we do not notice any expansion of this field. Coupling a spectrograph to a telescope is still a pleasant game and we heard about two new achievements, one at ESO and the other at Calar Alto. In addition to the oral contributions, eleven posters covered a wide field of fiber technology and scientific results.

8. Conclusions

We may conclude, then, optical fibers are now clearly completely integrated into astronomical instrumentation. This is clearly the case if we look for the place of this technology in general conferences on astronomical instrumentation, such as SPIE meetings. There is still some basic groundwork to accomplish but this concerns only instruments that are to be pushed to the utmost limits of performance. Maybe at the next conference, the main emphasis will be on the scientific results coming from fiber instruments and dealing with sky subtraction, S/N limitations, IR instruments, IFU in multi-object mode, and so on. Optical fibers are now completely integrated into astronomical instrumentation.